The
original edition
of this book con-
tained oversized maps.
PDF files of these maps
can be downloaded from
www.elibron.com/maps

Elibron Classics series.

ISBN 1-4212-5691-6 (paperback)
ISBN 1-4212-5690-8 (hardcover)

This Elibron Classics Replica Edition is an unabridged facsimile of the edition published in 1912 by Isaac Pitman & Sons, Ltd., London.

W. Basil Worsfold

THE UNION OF SOUTH AFRICA

WITH CHAPTERS ON RHODESIA
AND THE NATIVE TERRITORIES
OF THE HIGH COMMISSION

Elibron Classics
www.elibron.com

HOUSES OF PARLIAMENT, CAPETOWN

UNIFORM WITH THIS VOLUME

THE "ALL RED" SERIES

Each volume is in demy 8vo, cloth gilt, with full-page plate illustrations, map, etc.

THE COMMONWEALTH OF AUSTRALIA. By the HON. BERNHARD RINGROSE WISE (formerly Attorney-General of New South Wales).

THE DOMINION OF NEW ZEALAND. By SIR ARTHUR P. DOUGLAS, Bt., formerly Under-Secretary for Defence, New Zealand, and previously a Lieutenant, R.N.

THE DOMINION OF CANADA. By W. L. GRIFFITH, *Secretary to the Office of the High Commissioner for Canada.*

THE BRITISH WEST INDIES. Their History, Resources, and Progress. By ALGERNON E. ASPINALL, *Secretary of the West India Committee.*

Other Volumes in preparation.

THE ALL RED SERIES

THE UNION OF SOUTH AFRICA

THE UNION OF SOUTH AFRICA

WITH CHAPTERS ON RHODESIA AND THE NATIVE TERRITORIES OF THE HIGH COMMISSION

BY

W. BASIL WORSFOLD

Sometime Editor of "The Johannesburg Star"

LONDON: SIR ISAAC PITMAN & SONS, LTD.
No. 1 AMEN CORNER, E.C. 1912

PRINTED BY SIR ISAAC PITMAN
& SONS, LTD., LONDON, BATH
AND NEW YORK - - 1912

NOTE

For the gift or loan of official papers, used in the preparation of this book, the author desires to express his grateful thanks to the High Commissioner for the Union of South Africa (The Hon. Sir Richard Solomon, G.C.M.G.) and J. R. Stopford, Esq., of the High Commissioner's Office; the British South Africa Company; the Rhodes Trustees; the Central Mining and Investment Corporation, Ltd.; and the Town Clerk of Johannesburg (J. Taylor, Esq.).

TABLE OF CONTENTS

PART I

THE LAND AND THE PEOPLE

PART II

THE GOVERNMENT

PART III

RHODESIA AND THE NATIVE TERRITORIES OF THE HIGH COMMISSION

PART IV

INDUSTRIAL DEVELOPMENT

PART V

POLITICAL AND SOCIAL CONDITIONS

LIST OF ILLUSTRATIONS

THE UNION OF SOUTH AFRICA

PART I

THE LAND AND THE PEOPLE

CHAPTER I

PHYSICAL CHARACTERISTICS

THE late war and its beneficent sequel, the regaining of the lost solidarity of the European communities, have made the words "South Africa" familiar to the English-speaking world, but it is none the less necessary to distinguish between the two obvious meanings of the name. As a geographical term, then, South Africa is generally understood to cover that part of the continent of Africa which lies southward of the line formed by the river Zambezi and the northern boundary of German South-West Africa. Politically, the term is usually identified with British South Africa; that is to say, with Africa south of the Zambezi, excluding the German and Portugese territories lying respectively on the west and east coasts. South Africa in this sense—the sense in which it will be used in this book—has an area of rather less than a million square miles, and a mixed European and coloured population of 7,500,000; and it to-day comprises (1) The Union of South Africa, (2) the Native Protectorates or Territories not yet placed under the Union Government, and (3) Southern Rhodesia. Beyond the Zambezi are Northern Rhodesia and the Central African

Protectorate, or British Nyassaland, still British territory, but geographically and politically distinct from South Africa, and separated only by some 500 miles of foreign territory from another British system formed by Uganda, British East Africa, the Soudan, and Egypt. Thanks to the genius of Rhodes, the railway has been carried northward from Capetown throughout British South and Central Africa, and to-day it penetrates beyond the boundary of Northern Rhodesia for 100 miles into the Congo Free State—soon to be connected by rail, or steamships on the Great Lakes, with the southward extension of the Soudanese railways, and link up Capetown with Cairo.

The first impression given by the atlas, when we find South Africa on the map of the world, must be modified in one important particular. Africa, as we see, lies right across the Equator, and it spreads a far larger proportion of its total area between the tropics than does any other continent. Even a third of South Africa lies within the tropic of Capricorn. But Africa, thus distinctively the tropical continent, has a high average elevation. It is equal in this respect to Asia, which holds the greatest masses, and the loftiest summits, of the mountains of the world; and it is twice as high as Australia, one-third higher than Europe, and nearly one-third higher than America, North and South. And since elevation is a factor of great importance in climate—it has been calculated that for every 100 feet we ascend the temperature drops one degree[1]—it is obvious that the high average elevation of Africa must largely modify the climatic conditions which its geographical position alone would give to it as a whole; while it renders quite untrustworthy any estimate of the relative commercial and political value of its several regions, which is based

[1] The barometer also falls 1 in. for every 1,000 ft. above sea-level.

merely upon a comparison of their respective distances from the equator. This applies especially to the southernmost portion of the Continent. Lying, as it does, between latitudes 16 and 35 south of the Equator—the latitudes which, north of the Equator, enclose distinctively tropical places, ranging from Cuba, Khartum, Bombay, and Hong-Kong on the south, to New Orleans, the Bermudas, Tunis, Cairo, Jerusalem and Shanghai on the north—visited by the rain-bearing monsoon winds, and washed, on its eastern coast, by the warm currents of the Indian Ocean, South Africa should be a country of tropical and sub-tropical lands, with a hot, moist air, a cloudy sky, an abundant rainfall, and a luxuriant growth of tree and plant. In point of fact it is this for only a fraction of its area—the Eastern littoral between Delagoa Bay and East London, the north-east of the Transvaal, and the eastern borderland of Rhodesia. Elsewhere the conditions natural to a country thus relatively near to the Equator are largely modified or altogether absent. Not only is the sun's heat diminished by the high average elevation of the inland regions, and tempered by the cool winds and ocean currents that, uninterrupted by any land, reach the low-lying western districts from the Antarctic zone, but the great mountain masses, lying in close proximity to the east coast, intercept with their steep eastern escarpments the monsoon winds, causing them to precipitate most of their moisture before they pass over the great Central plateau; and thus three-fourths of South Africa is left with a brilliant sky, a spasmodic rather than insufficient rainfall, and a sparse and scanty vegetation.

Turning now from the map of the world to that of South Africa itself, we notice, first, the disposition of the great mountain masses, which, as we have seen, determines more than any other single feature its physical characteristics. From the Limpopo southward, they lie to the east, and run at distances varying from 200 to 50 miles, in a succession of ranges roughly parallel to the east and southern coasts. At the south-west corner of the Continent, where is the actual Cape of Good Hope, lesser ranges rise northward with a westerly trend that keeps them also relatively near to the opposite, or western, coast line. Of the eastern ranges the loftiest and the longest is the Drakenberg, which forms the eastern escarpment of the Transvaal and Free State provinces, and culminates in Basutoland with its almost Alpine grandeurs of snowy peak and rushing torrent. Westward of the great ranges the land slopes gently to the central table-lands, and then declines to sandy plains; eastward it falls abruptly to the coast in terraces, the sides of which are formed by lesser ranges. The rain-bearing winds come from the east and the south, and hence the marked difference in the character of the country east and west of the watershed. Westward the land is dry and vegetation is sparse; but the land about the ranges themselves, and the terrace lands eastward to the coast, are clad luxuriously with tree and herbage.

From this relative proximity of the great mountain masses to the east and south coasts, taken together with the high average elevation of South Africa inland, there emerge two significant facts. First, the climate improves as we advance inland from the coast; and second, the rivers of South Africa are, with rare exceptions, singularly useless for the purposes either of navigation or irrigation. The great central plateau which stretches from the valley of the Zambezi on the north to the coastal ranges of the Cape province in the south, has an average elevation of no less than 4,000 feet; and while from the Zambezi to the Limpopo it slopes downwards from west to east, at the valley of the Limpopo it changes the direction of its tilt, and from this point southwards it slopes from east to west. From the Limpopo to the Cape the ranges

THE ZWARTBERG PASS IN THE CAPE COLONY

that form the watershed run near the east and south-east coasts. East and south of the dividing ranges, therefore, the rivers descend precipitously over the broken land that separates the watershed from the sea. Westward the Orange, and its tributaries the Vaal and Modder, flow with longer courses and less abrupt descents; but while they fling great volumes of water down to the sea during the few rainy months, for the rest of the year they sleep in shallow reaches, or trickle in narrow streams through the sand and pebbles of their mile-wide beds. Northward, where the central plateau changes its tilt, the Limpopo, the Sabi, the Pungwe, and the Zambezi flow from west to east with longer courses and easy descents, but the low-lying districts of Mozambique through which they flow are infested with insects injurious to man and beast, and with the exception of the Zambezi, these rivers also are useless for navigation. The Karoos of the Cape, the table-lands of the Free State, the high veld of the Transvaal and the uplands of Rhodesia vary in altitude from 1,000 to 7,000 feet above sea level, and this high elevation both tempers the heat of the sun's rays and increases their life-giving properties. As the area covered by the coastal belts and the mountain ranges is relatively small, a dry and stimulating air, bright sunshine, little rain, constant and boisterous winds, and a moderate temperature, seldom much warmer than an English summer or much colder than an English spring, are the qualities that give the climate of South Africa its distinctive character. The forests of the Knysna, the coast lands of Natal warmed by the Mozambique current, the sub-tropical valleys of the mountainous regions of the north-east Transvaal, have each their seasons of dry weather and moderate temperature, when the rich luxuriance of their vegetation may be enjoyed, but it is the veld, the wind-swept, grey-green sea of heaving uplands, with its wide horizon and its glittering lights, that pictures South Africa to the eyes of her adopted sons.

Nearly twenty years ago I wrote of the veld. Since then I have visited, with pencil in hand, the high veld of the Transvaal, the Matopos, and the uplands of Mashonaland; but I do not think I can improve upon this record of my earliest impressions of South Africa, received now nearly thirty years ago:

> Once beyond the barrier ranges, the undulating plains spread on every side desolate and illimitable. The surface of the earth is broken only by rounded and flat-topped masses, hills with the contour of mountains, weird distortions which serve only to confuse the vision. There is neither tree nor shrub, homestead nor boundary, to arrest the eye. At most, a line of mimosa bushes marks the barren tract of the periodic water course, and the brown earth at our feet is studded here and there by stunted bushes. Such is the veld, and such is the characteristic landscape of two-thirds of settled South Africa. Yet, even so, the human spectator experiences no sense of depression; for over his head the great sun is shining in his might.[1]

A few figures and comparisons will give precision to this account of the rainfall and temperature of South Africa. The average annual rainfall of the British Isles is approximately 30 inches; in France, omitting abnormal stations, the rainfall varies from 24 inches in Paris to 52 inches in the Department of the Lower Pyrenees. Over the Union of South Africa (*i.e.*, the four provinces of the Cape, Transvaal, Free State and Natal) there falls an average of approximately 23 inches of rain for the year. In the Cape Peninsula the heaviest rains come in the winter; in the south-east coast districts rain falls equally throughout the year; but with these exceptions the South African rainfall occurs mainly in the summer months, October—March, which, as South Africa is south of the line, correspond roughly to April—September in Europe. We have also to remember the marked difference between the rainfall east and west of the great

[1] *South Africa: A Study, etc.*, 1895.

ranges, and how from these ranges westward the rainfall gradually lessens as we approach the opposite coast. Thus the average rainfall for the year is from 40 to 50 inches in Natal, 24 inches at Pretoria and Bloemfontein, from 8 to 9 in Bechuanaland, and from 3 to 4 in Namaqualand. Or again, crossing the Cape province from east to west, we find an average rainfall of 27 inches in its eastern, 17 in its midland, and 9 in its western districts. In this last respect South Africa resembles Australia, where the mountain masses also lie to the south-east, and the rain clouds are brought by the trade winds when they set westwards. In New South Wales, for example, Sydney on the east coast has an annual rainfall of 50 inches, while that of Milparinka, in the west of this state, is only nine.

The South African rainfall, therefore, is not only something like one-third less in volume than that of England or France, but it is also much less evenly distributed. This, however, is not all that is to be said about it. Another peculiarity remains to be noticed which is both good and bad. Not only does considerably less rain fall in South Africa than in western Europe, but this lesser volume of rain falls in a much shorter time than that in which the same volume would fall in England or France. Thus, for example, while over the Transvaal there falls on an average in the summer from 15 inches of rain in the west to 25 inches in the east, the actual duration of the rainfall for the whole year, as observed for five years, was only 213 hours, and the number of rainy days (*i.e.*, days on which 0.01 inches or more of rain were recorded on the plate) at Pretoria and Johannesburg were respectively 80 and 84.[1] This almost constant presence of the sun in a country where the rainfall is nevertheless considerable in volume, is to be attributed to the fact that

[1] As stated by Mr. R. T. A. Innes, Director of the Transvaal Observatory, in *The Times* of November 5th, 1910.

2—(2139)

heavy downpours of rain, often accompanied by thunderstorms, are the order of the day—downpours in which great volumes of water are discharged rapidly from clouds that themselves quickly disperse to leave the sky in its accustomed purity. The fact is that while the climate of South Africa is not tropical, the rainfall is. It is tropical, that is to say, not in amount (except on the eastern coast), but in character. What South Africa, west of the great ranges, suffers from is not so much a deficiency of rainfall, as the spasmodic and irregular manner in which this rainfall comes, and in particular the wasteful rapidity with which, in the absence of any artificial means of water-storage and irrigation on an adequate scale, the rivers carry the rains down from the central plateau into the sea. Something is being done already, as we shall see in a subsequent chapter,[1] to save a part of the rainfall; but it remains to create in South Africa a great public and national system of water storage and irrigation comparable to those with which India, Egypt, and other tropical countries have been provided. Such a system would add many millions of acres to the cultivable area of the Union.

Passing now from rainfall to temperature, we find that while the mean annual temperature of England and Wales is 50° F., and that of the various climatic districts of France varies from 48·2° F. in the Vosges to 57·5° F. on the Mediterranean, in all parts of South Africa it ranges somewhere between 60° F. and 70° F. The temperature of South Africa, therefore, is considerably higher than that of England or France, but the difference is due mainly not to an excess of summer heat, but to an absence of winter cold. With, however, this relatively low variation within the twelve months of the year is combined a relatively high variation within the twenty-four hours of the day. It must also be remembered that

[1] Part IV, Chap. IV, pp. 361-4.

the dryness of the air, which, with the few exceptions already noted is characteristic of South Africa as a whole, renders a high thermometrical heat less enervating here than even a considerably lower degree of heat would be, if, and when, it occurred in a country where the atmosphere is charged with moisture. These characteristics are readily illustrated by the observations recorded. Capetown, for example, has a mean annual temperature of 62° F.—the mean summer temperature of England, and almost identical with the mean annual temperature of Sydney (61·7° F.) in Australia—the means of its extremes of heat and cold are respectively 91° F. and 40° F., and its highest and lowest recorded temperatures average respectively 101° F. and 34° F. over a period of years. Although it is 34° south of the Equator—that is to say, nearer to the equator than the most southern points of Europe—it has the same temperature as Nice or Naples. With an annual mean of 62° F. as against the 56° F. of Madrid, it is at once less hot and less cold than the Spanish capital, where the extremes of heat recorded average 103° F., and the extremes of cold 20° F. And it has the advantage in equability even over Melbourne, which has a maximum of 105° F. and a minimum of 30° F. Pretoria, with a range between 97° F. and 24° F., and an annual mean of 64° F., and Johannesburg, with average extremes of 94° F. and 23° F. and a mean of 59° F., are almost as equable. Bloemfontein and Kimberly are less equable. The average of recorded extremes of the Free State capital are 109° F. and 16° F., but its annual mean is the same as that of Capetown—62° F. Kimberley's average of highest recorded temperature is 108° F. and of lowest 20° F.; its mean is 65° F.—3° F. above the mean of Capetown and 6° above that of Johannesburg—yet the singular dryness of its atmosphere so tempers the heat that there is no place in South Africa where men can work with more vigour of hand or brain. Natal has a mean annual temperature hardly higher than that of the Cape Province, since its summer heat is reduced by the frequent thunderstorms, accompanied by heavy rains, which occur during this season. Durban, on the coast, has a mean temperature for the year of 72° F., with an average of recorded extremes of 111° F., and 41° F.; Maritzburg, the capital of the Province, lying forty miles inland at an elevation of 2,000 feet above sea level, has a range of temperature which varies on the average from 32° F. in winter to 95° F. in summer. In Rhodesia the air is less dry than that of the southern regions of the great central plateau, and the rainfall is more evenly distributed throughout the year. Salisbury, the administrative capital, is no further from the Equator (17°) than Khartum; but it is nearly 5,000 feet above sea level, and the greatest extreme of heat recorded by Major Forbes in the two years 1901-2 was 93° F. in October, 1901; while in June, 1902, the thermometer sank to 34° F. Bulawayo, the largest town, is 3,800 feet above sea level, and has a temperature which ranges between extremes of cold which average 41° F., and extremes of heat which average 106° F. Thus, even Rhodesia, near as it is to the Equator, owing to its high average elevation has a temperate climate, in which, to use the words of one who knows it well,[1] "European children grow up with rosy cheeks and apples are not flavourless."

The greatness of the diurnal range of variation is a marked characteristic of the climate of South Africa throughout the great central plateau, and it is a beneficial factor, since from the point of view of health it compensates for the relative absence of seasonal variations. With the exception of the low-lying lands of the east and south-east coasts the nights are cool everywhere, however hot the days may be; and on the higher table-lands and in mountainous districts bright sunny days, with the

[1] Mr. F. C. Selous.

temperature of an English summer, are succeeded by nights in which the temperature falls several degrees below freezing point. If the precautions necessary to guard against the ill-effects of sudden changes of temperature are not neglected, this alternation of cool nights and warm days is wholly good. Not only are the functional organs stimulated to activity, thereby enabling the system to throw off liver and other complaints to which Europeans are subject in warm climates, but the strain of the long periods of bright sunshine and cloudless skies—delightful in themselves, but making heavy demands upon the brain—is largely mitigated by the nightly refreshment of cool hours in which sleep can readily be enjoyed.

Before this chapter ends, it will be convenient to gather into focus the two or three great outstanding facts which are sufficient by themselves to give an idea of South Africa's position in the world of nations, and to show how she is distinguished from the sister dominions of the British Empire. It remains, however, first to complete our review of the physical characteristics of the country, by asking what account geology has to give of the manner in which the material fabric thus reviewed was built up.

Long ages ago—so the geologist tells us—the rocky foundations of South Africa, consisting largely of granite, were overlaid by sedimentary accumulations, left behind by the retiring waters, which became horizontal quartzose strata. This was the first stage. In it water was the chief agent, and for its fulfilment an immeasurable period of time was required. In the next stage fire was the prime worker, and what was done was done quickly. Both the foundation rocks and the layers of quartzose strata were pierced and broken up by the intrusion of igneous rocks, forced upward by the expansion of the molten interior of the earth. In the third stage water again assumed the leading part. Over the now broken and distorted surface of the land, hills and mountains of sandstone, with veins of quartz, pounded, or worn away, into sand and fragments, were laid down in horizontal layers. The sandstone was reduced to its original particles, but the fragments of the hard quartz were ground and worn into pebbles, varying in shape and measuring from 3 inches to ½ inch in diameter. Between two layers of sandstone of enormous depth lie the quartz pebbles, cemented together by their own detritus and crushed into a conglomerate, and forming a narrow layer only 15 feet at the thickest. The deposits thus laid down sandwich fashion—with sandstone top and bottom, and the thin layer of conglomerate in the middle—are four or five miles, or even more, in vertical depth. It was the period when there were great inland seas, whose waters lashed and pounded for countless æons against the mountain masses which enclosed them, and laid down softly and smoothly the spoils they won from the land, to form a carpet deep down beneath their waves. Then fire again took up the task. This, the fourth stage, was the era in which the deposits of diamonds and of gold were made, or raised to the surface, or both made and raised to the surface at the same time. The diamonds were brought up in the pipes and craters of the volcanoes which burst into activity under the waters of the inland seas. The gold-bearing solutions, or the actual grains of gold washed down among the quartz pebbles, were brought with the streams of molten rock that were driven upwards by the same Titanic agency through the crust of the earth and emptied into the beds of these same inland seas. After this there came a mighty convulsion, which, heaving from beneath, poured the waters of the inland seas over and through the barrier ranges into the ocean, leaving the craters of the diamond-bearing, and now long extinct, volcanoes as dry land, and forcing up the edges of the broken sections of the horizontal strata, so that the edges of the gold bearing conglomerates also came to the surface of the earth. Then the scarred face of the land was smoothed and healed by time.

When, later on, we come to discuss the subject of mineral resources, it will be necessary to look a little closer at the actual geological history of the diamond and gold mines. For the moment it is sufficient to note on how Titanic a scale both volcanic action and denudation have been employed in the making of South Africa.

We have now to gather up the separate impressions left by these accounts of climate and rainfall, rivers, mountains, and the action of fire and water, by the aid of those two or three great central facts which "place" South Africa. First, then, this, the youngest of the Oversea British Dominions, differs from her compeers in that she provides no export of wheat and a very small export of raw materials. All other new Anglo-Saxon countries, including the United States, find their readiest form of wealth to lie in the development of the soil, and all are large exporters of food and raw materials to England and other closely populated European countries. In 1909, to take the latest returns furnished by the *Statistical Abstract* of the trade of the Empire, the Union of South Africa sent from her ports only 3·7 million pounds worth of wool, as against Australia's 26·7, and New Zealand's 6·3, and no wheat at all, as against the 13·1[1] million pounds worth of Canada's, and the 6.6 of Australia's respective exports under this head. And this comparison, it must be added, omits to take account of minor items of food, such as Australia's large export of butter, New Zealand's frozen mutton, and Canada's four million pounds worth of cheese. But there is another side to the picture. South Africa, with a European population of

[1] The figures are for 1910 in the case of Canada.

little more than a million and a quarter, has a total export trade which is about a fourth only less than those of Australia or Canada ; although the European population of the former is nearly four times, and that of the latter is nearly six times, as numerous. The balance is redressed—or, more correctly, the scale is weighed down heavily—in favour of South Africa by her altogether exceptional exports of gold and diamonds. Thus, in 1909 again, while neither Canada, Australia, nor New Zealand won any diamonds, South Africa exported them to the value of 6·3 million pounds, and in the same year her export of gold amounted in value to no less than 33·1 million pounds, as against 3·0 from Australia, 2·0 from New Zealand, and 1·2 from Canada (1910).

In a word, South Africa contributes little to the old world's supply of food and raw materials, but she provides to-day no less than one-third of the world's total output of gold. And of this vast contribution nine-tenths comes from a single province of the Union—the Transvaal, and ninety-five per cent. of this Transvaal gold is won from a single district, the Witwatersrand. She is a country of undeveloped, rather than insufficient, agricultural resources, but of altogether exceptional mineral wealth.

No less distinctive is the mingling of races and nationalities in South Africa. In Canada there are two nationalities, British and French ; but the British are twice as numerous as the French and the two combined form a European nation of seven millions, with a quite negligible population of red-skinned Indians. In Australia there is a solid Anglo-Saxon population of 4,500,000, with a handful of natives and Asiatics. In New Zealand, again solid Anglo-Saxon, there are 1,000,000 Europeans with a very small, and stationary,[1] population of Maoris.

[1] The recent increase in the returns of the Maori population appears to be illusive—being due to fuller enumeration, not actual increment.

Now note the constituents of the 6,000,000 which make up the population of the Union of South Africa—or of the 7,500,000 of South Africa as we defined it. The vast majority of this total are dark-skinned Bantu, who outnumbering the Europeans in the proportion of four to one in the Union, and five to one in South Africa as a whole, form a native population, which, instead of dwindling away in the presence of European civilisation, increases by natural increment alone more rapidly than the white man does by natural increment combined with immigration. The European minority, some 1,300,000,[1] is not a solid British community like the Commonwealth of Australia, or the Dominion of New Zealand, but is made up of two distinct nationalities, Dutch, or more correctly Franco-Dutch, and British. In this respect it resembles Canada, but there is a very important difference between the mixed European populations of the two countries. Whereas in Canada the British are twice as numerous as the French, in South Africa the Dutch are in a majority.

South Africa, therefore, has both its "native question" and its nationality difficulty. The former has no counterpart in any other oversea dominion; the latter arises in a form infinitely more embarrassing than it does in Canada, where also the European community is composed of two nationalities. The visitor, or the settler, therefore, must be prepared to find on his arrival in South Africa a country in which whole regions are inhabited by dark-skinned and primitive tribes, and one in which there are European towns and villages where a Dutch patois, and not English, is the language spoken by the people. Even in the large towns, where the white population is mainly

[1] The Census of 1911 gave the number of the white population of the Union of South Africa as 1,278,025, and that of the Bantu, etc., as 4,680,474. Southern Rhodesia has 23,606 Europeans and 750,000 natives, etc.

British, he will read Dutch names on the street corners, and study public notices printed in Dutch side by side with English. Some will be repelled by these signs that the African native has by no means renounced his claim to the land of his forefathers, and that here, for once, the British colonist is second in the field. But to most men the presence of these Dutch neighbours, and the dusky faces of the natives, will add a zest of novelty to the buoyant anticipations with which the British Dominions, and certainly South Africa not least among them, never fail to inspire all those who are resolved to give them of their best.

The existence of these diverse elements in the population brings to mind the fact that South Africa has a remote past upon which history is not entirely silent. The annals of Canada furnish episodes of the highest importance in the drama of the long struggle between France and England for trade supremacy in the eighteenth century. New Zealand has her Maori wars, and neither these nor the more peaceful incidents of Australia's development are by any means barren of interest. The brief and fragmentary records of South Africa cannot pretend to compare in fullness or dignity with the great volume of English history; but they begin nevertheless 1,000 years before the birth of Christ, and thus South Africa, alone of the British Dominions beyond the Seas, has points of contact with the great nations of the ancient world—with Israel, Phœnicia, and the Egypt of the Pharaohs.

South Africa weaves many spells. To the settler she offers broad lands and the speedy acquisition of wealth; to the sportsman, the variety and abundance of the wild creatures that still haunt her forests and uplands; to the invalid, the life-giving air of her central tablelands; to the philanthropist, missionary effort upon a great scale and with great results; to the man of science

By permission of *W. Basil Worsfold*

THE VICTORIA FALLS

(FROM THE SOUTH END OF THE RAILWAY BRIDGE LOOKING TOWARDS THE BOILING POT)

and the financier, the problems and prizes of gold and diamond winning; to the politician, the adolescence of a new nation; to the traveller, the glamour of the shining veld, the forest of perpetual rain, the rainbow brilliance and thunderous echoes of the Victoria Falls, and the battle-fields of the great war. But in the Phœnician temple-fortresses of Rhodesia, and in the seventeenth century homesteads of the early Dutch settlers at the Cape of Good Hope, she offers a source of interest which in a "new country" is as unexpected as it is alluring.

CHAPTER II

THE NATIVE RACES

THE great outstanding fact that five-sixths at least of the population of South Africa are coloured, or dark-skinned people, has already been brought to the notice of the reader. It is a circumstance that has a double significance. It means that a burden of responsibility is placed upon the comparatively small number of Europeans in South Africa from which the European inhabitants of the other dominions are free; but it also means that the Europeans in South Africa possess an industrial resource enjoyed by no other similar community. The physical strain of a century of almost continuous native wars has now been succeeded by the moral strain of a realised obligation to raise these backward races in the scale of civilisation. And here interest and duty combine; since the industrial utility and contentment of the Native population as a whole depend upon the successful fulfilment of this task.

Some mention of the methods employed to fit the natives for a partnership in industry with the Europeans will be made in due course. For the moment it is enough to notice that, owing to the presence of the native races, the Europeans in South Africa are placed in somewhat the same position as that held by the citizens of ancient Greece and Rome. Just as the communities of freemen in the age of the Græco-Roman civilisation handed over to their slaves all the rough work of life, so to-day the white man in South Africa looks to the "Kafir boys" to relieve him of the necessity of mere manual labour. Whether this condition of affairs is to the social and political advantage of the Europeans is a matter upon which two opinions are held; but the fact remains that for the time being existing economic conditions and the necessity for maintaining the prestige of the European community as a whole, unite to make it seem undesirable for the white man to enter into competition with the Kafir in the unskilled labour market. In these circumstances the base of the industrial structure will continue to be formed by the coloured man: and, so long as this is the case, the individual European, merely because he is a European, must maintain his position as a member of a superior race. If he fails to do this, he will degrade not merely himself, but the European community as a whole, in the eyes of the natives, and thereby add materially to the difficulty of the already formidable task of native administration.

The presence of the native population is, therefore, the cardinal fact upon which the whole fabric of the social and political life of South Africa turns; and it is one which gives to the European inhabitants of this country a unique position among the English-speaking oversea communities. It is here, too, that any review of the past history and the present conditions of South Africa, however brief, must find its most appropriate starting-point.

When we speak of "the natives" of South Africa, we think naturally of the virile and prolific Bantu, who form the vast majority of its non-European population, and are familiar to Englishmen under the names of one or other of the great South African branches of the race—Zulus, Matebele, Bechuanas, Basutos and so on. The earliest inhabitants of South Africa, however, were not a dark-skinned people, but the yellow-skinned Bushmen and Hottentots. Of these the former were a race of pigmies, to whom the first Dutch settlers gave the name "Bushmen," because they found them lurking in the shelter of the underwoods at the base of Table

Mountain. This primitive race knew no form of government other than the parental, had no settled habitations, and owned no property except the few rude implements with which they satisfied their elementary needs. They neither pastured cattle nor tilled the soil, but subsisted on roots, berries and honey, with the flesh of such wild creatures as they could kill with bone-tipped, poisoned arrows, sped from their tiny bows. They were incapable of civilisation, and their condition had not altered when Pringle[1] came to South Africa among the Albany Settlers in 1820, nearly 200 years after the Dutch Settlers had first found them. In "The Song of the Wild Bushman" he thus describes their life in the desert-regions of the Cape Colony of his day.

Let the proud White Man boast his flocks,
  And fields of foodful grain;
My home is 'mid the mountain rocks,
  The Desert my domain.
I plant no herbs not pleasant fruits,
  I toil not for my cheer;
The Desert yields me juicy roots,
  And herds of bounding deer.

* * * *

My yoke is the quivering assagai,[2]
  My rein the tough bow-string;
My bridle curb is a slender barb—
  Yet it quells the forest King.
The crested adder honoureth me,
  And yields at my command
His poison bag, like the honey bee,
  When I seize him on the sand.
Yea, even the wasting locust-swarm,
  Which mighty nations dread,
To me nor terror brings nor harm—
  For I make of them my bread.[3]

[1] Afterwards Secretary to the Society for the Abolition of Slavery.

[2] Assagai = *hasta*, through the Portuguese. It was the javelin or stabbing spear of the South African natives, first seen (and so named) by the Portuguese at the end of the fifteenth century.

[3] *Ephemerides*, by Thomas Pringle, 1828.

THE HOTTENTOTS

The Hottentots were some degrees higher in the scale of humanity. They were a nomadic people, each clan recognising the authority of a chief; and they both cultivated the soil and pastured herds of cattle. As, however, the institution of private property was unknown to them, the fruits of their tillage and their cattle were the property of the clan, or tribe, as a whole. In appearance they were not far removed from the Bushmen, but their greater mental development enabled them to adapt themselves to the ways of civilised life. The place from which they came, and the time and manner of their coming, are hidden in the dimness of the remote past; the one fact known with certainty is that the Portuguese navigators, when they reached the Cape of Good Hope in the fifteenth century, found the Hottentots established there. The tractability of the Hottentot, as compared with the wildness of the Bushman, is well indicated in Pringle's sonnet:

> Mild, melancholy, and sedate, he stands,
> Tending another's flock upon the fields,
> His fathers' once, where now the White Man builds
> His home, and issues forth his proud commands.
> His dark eye flashes not; his listless hands
> Lean on the shepherd's staff; no more he wields
> The Libyan bow—but to th' oppressor yields
> Submissively his freedom and his lands.[1]

Although the Dutch East India Company imported both eastern labour (Malays) and slaves from Central Africa for the service of their settlers at the Cape, the Hottentots remained practically the sole "native" population of the Cape Colony until the early part of the nineteenth century. Even at the beginning of the period of British occupation, when the westward expansion of the Colony had brought the Europeans into contact with the Kafirs, they remained the characteristic people of the country. Thus Lord Charles

[1] Sonnets: "The Hottentot."

Somerset, who was appointed governor of the Cape in 1814, was said by a contemporary wag to have been sent out "to fleece the Hottentots on £10,000 a year"; and the native regiment of the Colony at this time was a Hottentot corps—a circumstance which gave rise to the complaint on the part of the Dutch farmers, that the English had "armed the Hottentots against them." At the present time very few pure types of either Bushmen or Hottentots survive, but the characteristic qualities of these yellow races are to be recognised in certain families among the considerable population of mingled native and native and European blood, known collectively as "coloured persons" in distinction to the "natives" of the Bantu race. According to the census of 1911 there are in the Union of South Africa some 500,000 of these "coloured persons," of whom the great majority (some 400,000) are naturally to be found in the Cape Province.

In the fifteenth century, the epoch which saw the first beginnings of European settlement—South Africa was inhabited, first, by wandering tribes of these primitive yellow races, scattered along its west and south coasts and on the banks of the Orange River, and, second, by branches of the characteristic dark-skinned native African race, known to ethnologists under the name of Bantu.[1] At this time, however, these latter were not to be found on the central plateau further south than a line drawn from Walfish Bay to the head waters of the Vaal, although

[1] The word Bantu (*aba-ntu*) is the plural of *um-ntu*, a word which means a human being in the dialect of the Tembu, Pondo, Zulu, and other East Coast tribes. It was adopted by the late Dr. Bleek as a generic term for the dark-skinned African natives of Central and South Africa. "In the division of mankind thus named are included all those Africans who use a language which is inflected principally by means of prefixes, and which, in the construction of sentences, follows certain rules depending upon harmony of sound."—Theal, *The Beginning of South African History*, 1902.

in the fertile and well-watered country between the great eastern ranges and the sea they had already pushed their way to what is now the south-west border of Natal. Nor did the Dutch settlers of the Cape Colony come into serious conflict even with the Bantu of the east coast until the closing years of the eighteenth century. To the Englishman, however, the Bantu, known in common speech under the name of "Kafirs," which was given to them by the Mohammedans—the word means "Unbeliever" and is the counterpart of the term "Infidel" as applied to Europeans—are familiar enough as the "natives" of South Africa: since, during the nineteenth century, British settlers and soldiers have been employed almost continuously in repelling the attacks of such warlike tribes as the Xosas, the Zulus, and the Matabele, and generally in establishing order and settled government among these dense masses of dark-skinned people as a whole.

We have no certain information as to the original cradle of the Bantu race; but we know that the South African tribes of this family must have made their way gradually southward from what is now the main seat of the race in Central Africa; and it is believed that it was in the course of their migration that the warlike east coast tribes, to whom the term "Kafir" most properly applies, acquired the considerable infusion of eastern blood, which is apparent in their mental and physical characteristics. On this assumption, the Kafirs of South Africa would owe their superiority over the Central African Bantu to the circumstance that their progenitors intermarried with one or more of the eastern peoples, the Sabæans or Phœnicians, Arabs, and Indians, who from time to time established settlements upon the east coast of Africa, and not to an original Semitic strain carried with them from some remote Asiatic home. However this may be, the South African Bantu, although naturally

3—(2139)

a lustful and cruel people, possess some good and even noble qualities, from whatever source acquired; and Pringle's sonnet to the Kafir exhibits him in marked contrast to the Hottentot and Bushman.

> Lo! where he crouches by the cleugh's dark side,
> Eyeing the farmer's lowing herds afar;
> Impatient watching till the Evening Star
> Lead forth the Twilight dim, that he may glide
> Like panther to the prey. With freeborn pride
> He scorns the herdsman, nor regards the scar
> Of recent wound—but burnishes for war
> His assagai and targe of buffalo-hide.

The characteristics which reveal the position of the Kafirs in the scale of humanity form an exceedingly interesting subject of study, but in a work of this kind they can only be indicated very briefly. They are:

(1) A simple but completely developed tribal system of government, with a body of customary law, tribunals of justice, and a system of land tenure.

(2) An adequate and musical language, the use of which is determined by rules of grammar. Oral narratives, containing the traditions of the tribe, and folk-lore represent their development of the poetic faculty, since they have no literature, nor any acquaintance with written signs for the conveying of ideas.[1]

(3) A religious sense founded upon the belief in the survival of the soul after death, and in the power of the spirits of the dead to work good or evil to the living; and a deep-seated belief in witchcraft and other forms of supernatural agencies. Ancestor-worship is, of course, common to all primitive people, and in China, as Taöism, it is one of the three established cults (those of Brahma and Confucius being the other two).

(4) A knowledge of agriculture with the possession of domestic animals; horned cattle being their distinctive

[1] This, of course, refers to the condition of the Bantu in their natural state. Under Missionary and State education, they now use Kafir text-books, and publish native newspapers.

form of wealth, and the common medium in which tribute to the chief and the bride-price are fixed and paid. In addition to this they have some acquaintance with the arts of the smith, the potter and the weaver. The typical " house " is a bee-hive shaped hut, thatched with reeds or grass laid upon a framework of boughs, and having a single low and small opening for ingress and egress. A collection of such huts, called a " Kraal," forms a native village, or town, which for purpose of defence may be surrounded by a stockade. The women wear wrappers or fringes of leather about the middle ; young children and men find no need of clothing in warm weather, but on the table-lands, and generally when requiring protection from cold, the men wrap around them the kaross, a square of skins, sewn together, and of the size of an ordinary rug, or blanket. It is with the kaross thrown over his shoulders that the Kafir is seen crossing the mountains and the high plateaux, or sitting round the fire talking and jesting with his fellows in the open air.

(5) The possession of primitive but effective weapons of war, with an aptitude for military discipline, which enabled powerful chiefs to organise the whole of their adult males into a single fighting machine. Thus Ketchwayo possessed at the time of the Zulu war a " military organisation which enabled him to form out of his comparatively small population an army, at the very lowest estimate of 25,000 perfectly trained and perfectly obedient soldiers, able to march three times as fast as we could, to dispense with commissariat of every kind and transport of every kind, and to fall upon [Natal] or any part of the neighbouring colony [the Transvaal] in such numbers, and with such determination that nothing but a fortified post could resist them, making no prisoners and sparing neither age nor sex."[1] First among these weapons is the assagai, a short, light javelin,

[1] Despatch of Sir Bartle Frere, March 1st, 1879.

with a long, flat, iron head, sharpened on both edges, and fixed to a light wooden shaft. It was used both as a missile—the end of the shaft being furnished with cow's hair, to give precision of aim in the manner of the "grey goose feathers" used by the English bowmen of the age of Agincourt—and as a weapon for stabbing at close quarters. When an impi, or Kafir regiment, advanced in its characteristic formation of a half-circle, the assagais were hurled from a distance, but when the "horns" of the impi had closed, and the enemy was surrounded, they were used as stabbing spears. The complement of the assagai is the long, oval-shaped shield of hide, studded with metal bosses, and enriched with tassels of cow-hair plumes. In addition to these weapons of serious warfare, the Kafir possesses a wooden club, with a short handle and a massive round head, termed a knob-kerrie, which, like the assagai, can either be thrown from a distance, or wielded in the hand, as occasion demands. The knob-kerrie is a more inseparable part of his everyday equipment than the kaross. Not only does he use it to "knock over" any stray game that he may come across, or to settle a dispute, however trivial, with a member of his family or a neighbour, but he carries it about with him in his hand on all occasions, much as an eighteenth-century gentleman wore his sword, whether likely to be wanted, or not.

The manner in which these weapons were used against the white man is pictured to us by Pringle's "Makanna's Gathering."

> Then come, ye Chieftains bold,
>   With war-plumes waving high;
> Come every warrior, young and old,
>   With club and assagai.
>
> * * * *
>
> Fling your broad shields away—
>   Bootless against such foes;
> But hand-to-hand we'll fight to-day
>   And with their bayonets close.

Grasp each man short his stabbing spear—
And, when to battle's edge we come,
Rush on their ranks in full career,
And to their hearts strike home!

Wake! Amakósa, wake!
And muster for the war;
The wizard-wolves from Keisi's brake,
The vultures from afar,
Are gathering at Uhlanga's call,
And follow fast our westward way—
For well they know, ere evening-fall,
They shall have glorious prey! [1]

By the side of this poetic picture we may set two vivid prose descriptions furnished by Mrs. Harriet Ward, who was in the Colony at the time of the "War of the Axe" (1846–48).

The Kafir, at the first onset, is, perhaps, less ferocious than cunning, and more intent upon improving his own interests by theft than in taking life from the mere spirit of cruelty; but once roused, he is like the wild beast after the taste of blood, and loses all the best attributes of humanity. The movement of a body of these savages through the land may be likened to a "rushing and mighty wind." On, on, they sweep! like a blast; filling the air with a strange *whirr* reminding one, on a grand scale, of a flight of locusts.

An officer [in the war of 1835] had his attention suddenly arrested by a cloud of dust; then a silent mass appeared, and lo! a multitude of beings, more resembling demons than men, rushed past. There were no noises, no sound of footsteps, nothing but the shiver of the assagais, which gleamed as they dashed onwards.

The second passage gives an account of the attack on Fort Peddie, an incident in the "War of the Axe." It was furnished by an eye-witness, and published in a local South African paper at the time. Fort Peddie is described as "a mere earthen embankment" held by a handful of British soldiers. It is noticeable that at this

[1] In the Kafir war of 1817–19 Makanna led the Xosa chiefs and their warriors into the Cape Colony, and, after minor successes, attacked Graham's Town, the head-quarters of the British troops on the eastern border of the Colony.

date the Kafirs on the eastern border of the Cape Colony had in part acquired the use of European weapons.

> Were it not that life and death were concerned in it, I should have pronounced it [the attack] a most beautiful sight. The Kafir commanders sent their *aides-de-camp* from one party to another, just as you would see it done on a field day with European troops. The main bodies were continually increasing with horse and foot men, and soon after eleven the array was truly terrific. The largest body was to the westward; finding their schemes of drawing the troops out did not succeed, small parties advanced in skirmishing order, and then the two divisions of Pato and the Gaikas moved towards each other, as if intending a combined attack on some given point. Colonel Lindsay was superintending the working of the gun himself, and, as soon as a body of Gaikas moved within range, a shot was sent into the midst of them. . . . The Kafirs now extended themselves in a line six miles in length. These, advancing at the same time, so filled the valley, that it seemed a mass of moving Kafirs; rockets and shells were poured rapidly on them, and presently a tremendous fire of musketry poured over our heads. The enemy, however, did not come near enough for the infantry to play upon them, and only a few shots were fired from the infantry barracks. . . . The actual fighting was between the Fingoes [1] and the Kafirs; the troops could not have gone out without exposing the forts to danger, as there were masses ready to pour in from all quarters. [2]

The emigrant Boers had a method of their own of resisting the rush of the Kafirs. This was to form up their huge travelling wagons in a square, and shoot down their savage assailants from behind the barrier thus provided. As the phrase to "laager the wagons" has become historical,—the neglect of this precaution cost the British nation the loss of almost the entire 24th Regiment at Isandhlwana—it will be interesting to recall Cloete's account of this useful manœuvre. "These laagers, or camps," he writes, [3] "were formed by their wagons being brought up into a square, the poles and

[1] A tribe of loyal Kafirs.

[2] *Five Years in Kafirland.*

[3] *Five Lectures on the Emigration of the Dutch Farmers, etc.*, Capetown, 1856.

wagon 'gear' of one wagon being firmly secured under the 'perch' of the next wagon; and, when time admitted, branches of the thorny mimosas were also wattled in under each wagon, so that no entrance could be effected into the enclosure without forcibly tearing up all these impediments."

Thus the laager of the Voertrekkers on Vecht Kop in 1836 resisted the onset of the Matebele Zulus under Moselekatze; and in 1838 the only Boer camps on the Blaauwkrantz River in Natal that escaped wholesale destruction at the hands of Dingaan, the treacherous Zulu King, were the few in which the occupants had been warned of the despatch of his ten regiments of trained warriors in time to laager their wagons. Later in the century the British Army adopted the hollow square formation for the same purpose as the Boers went into laager; and by this means small bodies of troops were enabled to resist successfully the furious, enveloping attacks delivered by the Kafirs, the Dervishes, and other barbarous or half-civilised enemies. It was by advancing and fighting in this formation that Lord Chelmsford retrieved the disaster of Isandhlwana at Ulundi. His force, which consisted of not more than 4,000 European and 1,000 native troops, was drawn up in a single hollow rectangle, with the infantry supported by guns on the outside, and the cavalry and stores enclosed within. To complete these pictures of Kafir warfare I may, perhaps, be allowed to reproduce the description of this battle which I gave in my *Story of South Africa.*

Against this square, Ketchwayo's warriors rushed with splendid discipline and courage. As the Zulus approached, leaping and shrieking, it seemed as though the black wave must flow over the slender lines of red and grey. But as it drew nearer and the crack of the rifles rang out, the wave grew less dense; then, when it had come quite close, it bent before the hissing discharge of the gatlings, and fell broken at the feet of the British line. But, even so, before this withering fire, here and there a Zulu warrior would reach the line, and animated by the desperate courage of his race, grasp the bayonets, and thus, by fixing their points in his breast, try to open a path for his comrades.

When at length the Zulus had broken and taken to flight, the lines opened and the cavalry rode out to complete the work.

It is a remarkable testimony to the indomitable spirit of the South African Bantu that a century of conflict with disciplined troops, armed with the increasingly lethal weapons provided by science, left them still ready to try conclusions with the white man in the twentieth century. The last native war, the Rebellion in Natal, which took place only five years ago (1906–7), bears sinister witness to the fact.

Dr. Theal divides the Bantu of Africa south of the Zambezi into three groups. In the first he places the tribes of the east coast from the Sabi river southwards to the Cape Province, with the offshoots of these tribes that made their way inland to the central plateau. The group contains the "first-rate fighting men" that made South Africa known to the British tax-payer before its diamonds and gold gave it a happier reputation—the Xosas, the Zulus, and the Matabele, together with the Tembu, the Pondos, the Swazis, and other warlike and virile clans. The second group covers "the tribes that at the beginning of the century occupied the great interior plain and came down to the coast between the Zambezi and Sabi rivers." In it there occur the names of natives prominently associated with missionary enterprise, such as the Batlapins, the Barolongs, the Bamangwatos, and the Makalanga, and the Basutos—perhaps the most interesting of all the native African peoples. The third group consists of all the Bantu tribes between the Kalahari desert and the Atlantic. The Hereros—known to us from their revolt against the Germans—and the Ovampos are the chief among them; and the people of this group are distinguished as having no admixture of Asiatic blood, and being in general "blacker in colour, coarser in appearance, and duller in intellect" than those of the two preceding groups.[1]

The first and second of these groups, which comprise the Bantu of our South Africa, correspond with the method of division adopted by the late Mr. John Mackenzie, the famous missionary. According to this authority the Bantu tribes of South Africa are to be distinguished as "military" and "industrial." The military tribes—those of Dr. Theal's first group—are characterised by their absorption in the business and practice of war, and their tribal organisation and customs were, accordingly, such as would be expected where war and the preparation for war form the chief element in the life both of the individual and the community. Thus the power of the paramount chief, or king, was almost absolute; since, although he consulted his council of lesser chiefs, he was not bound to follow their advice. The "town," or chief seat of the tribe, was planned for defence; the king's hut and the cattle pen being placed in the centre of an outer circle of huts, and the whole often surrounded by a rude fence, or stockade. Among these military tribes it was the custom to cultivate only just so much land as was required to satisfy their temporary needs, and to depend upon cattle for their permanent supply of food: while for the maintenance of their herds, they looked rather to constant and successful forays upon the weaker and more settled tribes than to the results of their own care and industry. Among the industrial Bantu—the tribes of Dr. Theal's second group—the authority of the paramount chief was limited both by the council of lesser chiefs, and, in matters of great importance, by the will of the whole body of tribesmen, as declared in the "pitso," or general assembly. The dwellings of which their "town" consisted were disposed merely as chance or convenience suggested, and were

[1] Theal, *The Beginning of South African History*, p. 31.

unguarded by any stockade, although they were often larger and more commodious than the single-roomed bee-hive shaped hut of the military tribes. Not only were the fertile lands around the " town " well and permanently cultivated, but among these industrial tribes crafts, such as the working of metals, pottery, and even the weaving of a coarse cloth of cotton, were practised.

If we except the Basutos,[1] who have held their own in their mountain fortresses against all assailants, not excluding the British regulars, the industrial Bantu, although more advanced in the arts of peace, were quite unable to resist the attacks of the military tribes. Their sole means of preservation was flight, and their hope of security lay in settling in distant regions where the land was too poor to arouse the cupidity of their enemies. The power and ruthlessness of the great chiefs of the military Bantu may be realised from the record of one of their number—Tshaka, the grandfather of Ketchwayo. It is estimated that this one chief caused the death of no less than 1,000,000 human beings, and devastated thousands of square miles of country, between the years 1812 and 1828. The passage of his army was marked by the bones of human beings and animals, picked clean by the scavenger birds and beasts, which formed the sole remains of the tribes which he had attacked ; for, with the exception of a few of the handsomest boys and girls whom he incorporated into his people, and such cattle as he chose to carry off, he left no living thing alive.

The general distribution of the military and industrial Bantu is, therefore, the outcome of their respective characteristics. The former, as we have seen, are to be found in possession of the well-watered and fertile country between the great mountain ranges and the Indian Ocean,

[1] They owed their national existence—they are a composite people—mainly to the sagacity and courage of a single chief, Moshesh.

and in the sub-tropical region, to the north-east of the Transvaal; the latter, with the exception of the Basutos, have been for the most part driven westward to the arid and almost rainless plains of the central plateau and the deserts beyond it.

But weak as were the industrial Bantu in comparison with the more virile tribes of the east coast, they nevertheless succeeded in reaching a more advanced stage of civilisation, and in the place of warrior despots such as Tshaka and Ketchwayo, they have produced sagacious rulers like Moshesh and Khama. The former built up a respectable native state largely out of fugitives from other tribes, many of whom were deserters from the ranks of Moselekatse, the chief of the Matabele Zulus, and himself a runaway general of Tshaka's army. In 1852, after an engagement at Mount Berea in which the Basutos had more than held their own against the British troops despatched against them by Sir George Cathcart, the Governor of the Cape Colony, Moshesh gave proof of his exceptional political wisdom. Instead of being deceived by his temporary military success, he determined to utilise his good fortune to obtain favourable terms of peace, and with the assistance of his missionary adviser, Mr. Casalis, he sent the following diplomatic communication to the British lines:

"THABA BOSIGO,
"Midnight, *December* 20*th*.

"Your Excellency,—This day you have fought against my people, and taken much cattle. As the object to which you have come is to have a compensation for Boers,[1] I beg you will be satisfied with what you have taken. I entreat peace from you. You have chastised—

[1] *i.e.*, the Boers of what was then the Orange River sovereignty, and afterwards the Free State. They were at this date under British authority, and it was the duty of the British Government to protect them from the depredations of their native neighbours, the Basutos.

let it be enough; and let me be no longer considered an enemy to the Queen. I will try all I can to keep my people in order for the future.

"Your humble Servant,
"MOSHESH."

And sixteen years later, when the Basutos had ceded the rich corn-lands known to-day as "The Conquered Territory" to the Boers of the Free State, and were threatened with dispersion, if not extinction, he sought the protection of the Imperial Government in diction which has rarely, if ever, been surpassed in poetic beauty. "Let me and my people rest and live under the large folds of the flag of England before I am no more." So Moshesh prayed; and Basutoland from that date up to the present —with the exception of nine years during which it was annexed to the Cape Colony—has remained under the direct authority of the High Commissioner for South Africa.

A charming picture of Khama, who perhaps stands first among the enlightened and capable rulers of the industrial Bantu, is given by Mr. Theodore Bent. He is drawn as he was seen by the author of *The Ruined Cities of Mashonaland* in his new capital of Palapye[1] in the Bechuanaland Protectorate. "Khama pervades every thing in his town," writes Mr. Bent. "He is always on horseback, visiting the fields, the stores, and the outlying kraals. He has a word for everyone; he calls every woman 'my daughter,' and every man 'my son' he pats the little children on the head. He is a veritable father of his people . . . . In manner the chief is essentially a gentleman, courteous and dignified."[2]

[1] Before the occupation of Bechuanaland (1884), Khama's tribe, the Ba-mangwato, were shut up in the valley of the Shoshong river, which he could defend against the attacks of the Matabele; but with the establishment of British authority, he was able to remove to the more fertile district, in the midst of which he founded Palapye.

[2] *Ruined Cities of Mashonaland*, p. 26.

As before indicated, accounts of the chief civilising agencies now applied to the South African natives will be given in subsequent chapters. They include (1) Missionary enterprise, (2) Industrial and domestic employment under European mastership, and (3) Education and Training in crafts furnished by the State. It will be seen, therefore, that these are subjects which can be discussed, each in its proper context, most appropriately as part of the life of the European community. The reader, however, will find it convenient to learn at once what has been the attitude of Great Britain towards the South African native population as a whole during the past century, and how this attitude has varied from time to time, either under compulsion of events, or merely to suit the political exigencies of the Home Government for the time being in office.

In the Native policy of Great Britain, then, five stages of development may be distinguished, which reveal on the whole a gradually increasing assumption of responsibility. They are in chronological order as follows:

I. From the date of the permanent occupation of the Cape (1806) to the Great Trek (1835-8).

II. From the Great Trek to the recognition of the independence of the Boers under the Conventions of 1852 and 1854.

III. From the Conventions to the Discovery of Diamonds (1870).

IV. From the discovery of diamonds to the Bechuanaland Settlement (1884–5).

V. From this date to the Union (1910).

During the first period the aim of British Statesmen was to interfere as little as possible with the natives outside the Cape Colony, but to encourage missionary enterprise as the agency by which the gradual advance in civilisation of the Bantu tribes could be secured. It was the time when public opinion in Great Britain and Ireland was affected powerfully by the great humanitarian and philanthropic movements of the early nineteenth century, of which the abolition of slavery within the Empire in 1833 was a signal result. The consistent pursuit of this policy was, however, rendered impossible by the necessity of protecting the lives and property of the Dutch and British inhabitants of the eastern districts of the Cape Colony against the periodic incursions of the warlike Kafir tribes beyond the frontier. In the fulfilment of their duty the British Government was reluctantly compelled to enter upon more than one Kafir war, to extend the frontier of the Cape Colony eastwards, and to enter into definite political relations with the paramount chief of the powerful Amakosa tribes whose territory neighboured with its own. During the whole of this period there lurked in the mind of the Home Englishman the disquieting, but utterly erroneous idea that the colonists and not the natives were the aggressors; that, in fact, these costly Kafir wars were provoked by the European settlers to serve as excuses for gaining more land, or to provide opportunities for commercial profit in supplying the soldiers with food and transport. The period ended in disaster. Hintza, the paramount chief, proved unable, or unwilling, to restrain the Kosas, 15,000 of whom, on Christmas Day, 1834, invaded the colony. The measures taken by the Governor, Sir Benjamin Durban, after six months' hard fighting to secure a defensible frontier were reversed, and compensation was refused by the Home Government to the colonists, Dutch and English, whose houses had been burnt or pillaged, and whose cattle, sheep, and horses, had been carried off by the Kafirs. In the course of the next two years some 10,000 Dutch farmers, in disgust at this failure to protect them from the Kafirs, and distressed by the emancipation of their slaves, left the colony to find new homes beyond the Orange River.

II. In the second period the attempt to secure order among the natives by means of alliances with powerful chiefs was pursued on a larger scale. This "larger scale" was rendered necessary by the fact that the "Emigrant farmers," or Boers, had made settlements in what are now the Transvaal, Free State, and Natal provinces, and in the course of doing so had driven Moselezatze and his Matebele Zulus beyond the Limpopo in 1838, and destroyed the power of Dingaan, the treacherous king of the Zulus, by their victories over him in 1838 and 1840. The difficulties of the British Government were enormously increased by the Great Trek. The existence of the small but warlike and ambitious communities founded by the Boers, on the very edge of the great masses of the Bantu population, constituted a double peril. On the one hand, there was the danger of the Boers arbitrarily dispossessing the natives of tracts of land which rightfully belonged to them, and without which they would be unable to maintain themselves and find pasture for their cattle; on the other, there was the possibility that one or other of the Boer "Republics," in spite of the courage and skill in shooting which the Voertrekkers displayed, would be at any moment exterminated by a sudden attack of the Bantu in overwhelming numbers. While desiring, therefore, to keep their responsibilities within the narrowest limits, the British Government were nevertheless compelled to provide the Boers with a protection against the natives sufficient to leave them no excuse for seeking to become the subjects of another European power, and at the same time to satisfy philanthropic opinion in England by shielding the natives from general injustice, and in particular from needless or wrongful loss of territory, at the hands of the Boers. In order to accomplish these objects, Natal, the only

Boer settlement which, having a seaboard, permitted of direct communication with Europe, was occupied, and under the inspiration of the missionaries the policy of entering into alliances with native chiefs was largely extended. In addition to Sandile, the legal heir of Gaika and paramount chief of the Kosas, treaties were made with Adam Kok, the head of the Griquas, with Moshesh, the chief of the Basutos, and with Faku, chief of the Pondos ; and these chiefs in return for the support of the British Government undertook to maintain order within their respective territories. In this way a barrier of native states was to be placed between the northern and eastern borders of the Cape Colony and the great mass of uncivilised natives, disturbed as they were by the movements of the Emigrant Boers. This policy proved unsuccessful. The chiefs were too weak to protect themselves against the Boers, and too unscrupulous to fulfil their obligations to the British Government. In the end both the Boers and the chiefs had to be reduced to order by force of arms, the frontiers of the Cape Colony were advanced northwards and eastwards, and the country between the Orange and Vaal rivers was declared British territory under the name of the Orange River sovereignty. The most disastrous effect produced by this policy was the infliction of the terrors and losses of another Kafir invasion upon the eastern settlers of the Cape Colony. The " War of the Axe " (1846-8), due to the faithlessness of Sandile, was scarcely concluded before the restless Kosa tribes again revolted ; and for another two years, commencing at the end of the year 1850, the British and Dutch farmers and the British troops fought side by side. At length, after a desperate struggle, European control was again established over the wild and treacherous mass of Bantu population that lay immediately eastward of the Colonial border. But the war was costly to England —both in men and money. In a single disaster, the wreck of the transport *Birkenhead*,[1] some 400 soldiers were drowned; and heavy compensation was paid to the border farmers for the losses inflicted upon them by the Kafirs—losses due to the reversal of the measures taken by Sir Benjamin Durban in 1835 for the protection of the frontier. The question whether the burden of these constant Kafir wars was not altogether out of proportion to the value of South Africa was raised, and a "cold fit" followed, which led to a change of policy. The British Government now took upon itself the permanent administration of the natives on the eastern border of the Cape Colony and in Natal, while at the same time they recognised the independence of the Boers, both in the Transvaal and Orange River sovereignty, under the terms of the Conventions of Sand River (1852) and Bloemfontein (1854), and virtually decided to leave the Boers and the natives outside the three British possessions—the Cape Colony, Kaffraria, and Natal—to settle matters for themselves.

III. From the date of the Conventions to the discovery of diamonds (1870) the non-intervention policy was maintained with fair consistency. It was strongly condemned by Sir George Grey, who was Governor of the Cape from 1854 to 1862, mainly on the ground that it led to a definite scism in the small European community of South Africa—or the "dismemberment" of South Africa, as he called it. Nevertheless it gave a welcome breathing space to the mother country so far as her South African responsibilities were concerned, and it allowed the local government of the Cape Colony to make a serious beginning of the task of civilising the Bantu people in

[1] She was wrecked off Danger Point, eastwards of Capetown, on February 26th, 1852, as she was bringing reinforcements to the frontier. The soldiers stood at arms on the deck of the sinking ship to allow the women and children to be got off in the boats, and remained motionless until the word to leap overboard and save themselves was given.

Kaffraria. Just as a belief in the poverty of the resources of South Africa, and a fear that the country would never be able to attract and support a considerable British population, had lain at the bottom of the non-intervention policy, so a surprising demonstration of the falsity of this belief led to the abandonment of the policy. The discovery of diamonds brought the establishment of British authority over Griqualand West—that is, outside the limits of the three colonies. The British Government believed that they had got a good title to this new territory from a native chief, Waterboer; but the Free State Boers claimed the Diamond Fields as a part of their republic, and in the event it was shown that they and not Waterboer were the rightful owners. In any case this step—apart from the dispute with the Free State—would have impelled the British Government to take an active part once more in regulating the conflicts which had arisen between the Boers and the natives. The most serious of these arose out of the circumstance that the western border of the Transvaal had never been properly delimited. If the Boers had been allowed to displace the peaceful natives lying on their western border, Bechuanaland, the high road to Central Africa which Livingstone had so far kept open,[1] would have been closed. Nevertheless it is extremely probable that the Boers, by dint of mere pertinacity, would have established themselves permanently in Bechuanaland, had not the discovery of diamonds drawn the British flag northward, and created a powerful British community close to the territory which the Boers coveted.

IV. But the Boers of the Transvaal were in conflict not only with the industrial tribes on their western border. On the east they had very different neighbours.

[1] "The Boers resolved to shut up the interior, and I determined to open the country." Livingstone (after the plunder of his house by a Boer commando in 1852).

There, both within and without their Republic, were great masses of military Bantu, and among them were the Zulus, who in the years of non-intervention had grown up into a formidable power under Ketchwayo. In view of the seriousness of the situation arising out of the weakness of the Transvaal and the increasing power of the military Bantu, the British Government set about to undo the work of the Conventions, in order that the handful of Europeans might present a united front against the menace of the dark-skinned masses. In 1877 the Transvaal was annexed, and at the same time active measures were taken for bringing about a union of the Boer Republics with the British Colonies in a federal system. The Federation proposals of Lord Carnarvon failed; but in the endeavour to put them into effect, Sir Bartle Frere performed a service of the utmost importance alike for South Africa and for England. In 1878 he checked the threatened disturbance of the solid mass of Bantu population lying between the Cape Colony and Natal, and saved the settlers of the eastern province of the former from a long protracted Kafir war, by his prompt suppression of the insurgent Gaikas and Galekas; and in the year following he rescued the Colony of Natal from invasion by Ketchwayo, and by the destruction of the Zulu power frustrated a concerted movement of revolt among the military Bantu from the Limpopo to Kaffraria. The movement was nothing less than the manifestation of a "common purpose" and a "general understanding" among the natives of South Africa to shake off the domination of the Europeans; and of the movement itself Ketchwayo was "the head-centre," and his 25,000 perfectly trained and disciplined warriors the "main strength." By the resolute action of Sir Bartle Frere the supremacy of the white man was established in South Africa. Since this date (1877–80) there have been local native wars or rebellions from time to time—the last was in 1906—but there has been no general movement of revolt on the part of the dark-skinned population as a whole against the Europeans.

The Zulu war, necessary as it was to South Africa, was both costly and inopportune for England. It was followed by the failure of Lord Carnarvon's federation proposals, and the revolt of the Transvaal Boers. These unwelcome events produced another "cold fit." The Transvaal was given back to the Boers in 1881, and the British Government declared its intention of not again intervening by force of arms in the affairs of South Africa, unless they were assured of the active support of the majority of their own subjects in that province of the Empire. The ambition of the Transvaal under President Kruger, and in particular the renewed attempts of the Boers to extend their western borders to the detriment of the commercial interests of the Cape Colony, alienated the sympathies of the Colonial Dutch, and led to a direct appeal for Imperial intervention, in response to which the British Government occupied Bechuanaland by an armed force under Sir Charles Warren in 1884-5. The attempted expansion of the Transvaal westwards had been accompanied by the establishment of Germany in South-west Africa; and these events, which in combination threatened to cut off the Cape Colony from Central Africa, compelled the British Government to declare that the whole of the central region intervening between the Diamond Fields and the Zambezi, was within their "sphere of influence." In taking this step to secure the trade route to the interior a very important advance in native policy had also been made. This was nothing less than to place formally all the natives of South Africa who were not enclosed within the borders of the two Republics or the British Colonies, under the direct control of the Imperial Government. The significance of this new departure will be seen from Article IV of the new commission, which was then (1884) issued by Queen Victoria to the Governor of the Cape and High Commissioner for South Africa.

And we do hereby require and empower you . . . to take all such measures, and to do all such things in relation to the native tribes in South Africa with which it is expedient that we should have relations, and which are not included within the territory of either of the Republics, or of any foreign power, as are lawful and appear to you to be advisable for maintaining our possessions in peace and safety, and for promoting the peace, order, and good government of the tribes aforesaid, and for preserving friendly relations with them.

V. From the time of the Bechuanaland Settlement up to the date of the Union the development of the European communities has been accompanied by the establishment, or extension, of the simple but fairly effective system of European administration which to-day embraces the entire native population of South Africa. As the British Colonies grew in wealth and population they gradually relieved the Imperial Government of its responsibility for large areas peopled wholly, or mainly, by Bantu. Thus in 1887 Natal annexed Zululand. In 1895 the Cape Colony incorporated into itself the southern part of Bechuanaland, and by the annexation of Pondoland assumed the administration of the whole of the dense native population between its eastern borders and Natal. The occupation of Mashonaland by the Chartered Company in 1890, and the subsequent foundation of what is now Southern Rhodesia, brought something like 500,000 Bantu under the control of European magistrates, without adding to the immediate responsibilities of the Imperial Government. Thus gradually between the year 1884 and the outbreak of the great Boer war (1899–1902) a network of European Magistrates and Commissioners was spread over all the native areas. To-day the Union Government has 4,680,474 "other than Europeans," of whom the great majority are natives of the Bantu race, under its control. Outside of the Union, the Imperial

Government remains responsible for more than 500,000 dark-skinned people in Basutoland, Swaziland, and the Bechuanaland Protectorate; and Southern Rhodesia has a native population of 750,000. Thus we have a total of 6,000,000 "other than European" people for South Africa as defined; but it must be remembered that in Northern Rhodesia—*i.e.*, the territory of the Chartered Company northward of the Zambezi—there is yet another 1,000,000 of Bantu population.

The character of the European control thus finally established over the natives of South Africa after a century and a half of constantly recurring wars and rebellions, naturally varies with the circumstances of the various tribes or races. The position of the Basutos, and of the Bechuanas, where the chief rules under the guidance of an Imperial Commissioner is almost analogous to that of a "protected state" in India. In the province of Natal, in the remote districts lying to the north and north-east of the Transvaal, in Pondoland, and in the "reserves"[1] generally, considerable authority has been left to the chiefs, and native (customary) law is administered by them and by the European Magistrates, who are, of course, supported by native and European police. On the other hand, the majority of the semi-civilised natives living within the four provinces of the Union, with the exception that they are subject to special laws and regulations applicable only to them and not to Europeans, are controlled by the ordinary legal and administrative authorities of the state. In this class are included the hundreds of thousands of African natives, who are regularly employed in manual labour of various kinds under European masters. And lastly, apart from both the "red," or tribal natives, and the semi-civilised manual labourers, there is a considerable population of educated

[1] *i.e.*, areas kept for the sole occupation of natives. Europeans are excluded from them, and the land cannot be alienated.

and civilised Bantu and "coloured" people. In the Cape Province these civilised natives are admitted to the Parliamentary franchise; but elsewhere within the Union the educated native, with rare exceptions, although placed on an equality with the European in legal and administrative matters, is still deprived of any voice in the government of the country.

This brief outline of the character and conditions of the native population, and of the policy pursued by the British Government in dealing with them, suggests two reflections.

The control of the native races of South Africa, and the consequent possibility of raising them in the scale of civilisation, have been won by the sword. In this, and in their unhappy conflicts with the colonists of the second European nationality, the British of South Africa have had a larger and more varied experience of war than any other similar community within the Empire. In the necessary and long-protracted task of establishing European authority over the native races the colonists of both nationalities, Dutch and British, have played a part, but by far the heaviest share of the burden both in lives and in treasure has been borne by the latter. The British of South Africa are, therefore, a people who have passed through their baptism of fire.

The second reflection is this. The native population is at present the determining factor in the social and political life of South Africa. The foregoing review will have revealed how largely the action of the British Government in this province of the Empire has been inspired and conditioned by its relations with the native races. The difficulty of these relations has been enormously increased by the presence in the country of Europeans of another nationality, whose sentiments in respect of native affairs were markedly less liberal than those of the people of the British Isles. The native question will still bulk largely in the life of the young

South African nation which has just, as it were, attained its majority, although its problems will be of a different order. The task of the Union Government will be made the easier by the fact that it inherits the honourable traditions of the British Government. On this aspect of the native policy of Great Britain I may perhaps be allowed to recall some words written fifteen years ago.

> To the honour of England it stands written on the page of history that, from the first assumption of the government of the Cape of Good Hope, she has resolutely set herself the task of meting out justice between the conflicting claims of the Colonists and the Natives ; that by assuming this attitude she rendered her government unacceptable to the mass of the original European inhabitants ; but that, in the face of the difficulties and the bitter opposition thus created, she again and again compelled the most stubborn of these European offenders to do justice to the coloured races whose champion and protector she was.[1]

[1] *Contemporary Review*, 1896.

CHAPTER III

THE SEMITIC OCCUPATIONS

It is to Egypt, the one seat of civilisation which existed in Africa prior to the Græco-Roman era, that we naturally turn for information upon the earliest history of other parts of the continent. But what Egypt has to tell us of South Africa is both vague and meagre. Ethiopia, the Abyssinia of to-day, was known to the Egyptians 2,000 years before the birth of Christ; but the first record of any acquaintance with the continent further south is that which is afforded by the monuments of the temple of Queen Hatasou, at Deir-el-Bahari. In these sculptures, referred variously to a date ranging from 1700 to 1200 B.C., we have an account of the conquest of the Land of Punt, and the spoils of the conquest as pictured on the hieroglyphics were ebony, ostrich feathers, the tusks of elephants, and ingots of gold. If we assume that these articles were the produce of the conquered districts, and further that the land of Punt was identical with the Mashonaland of to-day, and one of the sources of supply of King Solomon's Ophir, then it follows that the Egyptians from the age of Queen Hatasou onwards were acquainted with the region south of the Zambezi where lie the oldest gold-workings of South Africa.

To the knowledge of South-east Africa obtained by the conquest of Punt, and by whatever commercial and political intercourse may have arisen out of it, the Egyptians appear to have added nothing until the rise of the Phœnicians as a maritime power. These people, seated at Tyre and Sidon on the east coast of the Mediterranean, and drawing supplies of cedar and oak for their shipbuilding from Lebanon, became a great world power; and, in addition to the business arising from their colonies and trading stations in Africa, Spain and Britain, maintained a carrying trade which brought the great empires of Babylon and Assyria in Asia into commercial relations with Egypt and Greece. Utica, the first colony of the Phœnicians, on the north coast of Africa, was founded in 1100 B.C.; Carthage, the rival of Rome, about 800 B.C.; and 300 years later Hanno, the Carthaginian admiral, established the authority of his Republic as far down the West Coast of Africa as Sierra Leone. On the east coast of Africa the Phœnicians originated, or supervised and developed, the extraction of gold in the mineralised regions between the Zambezi and the Limpopo. The Phœnicians proper, while apparently sometimes transporting the gold and other produce of South-east Africa in their own ships to the Red Sea ports, and thence to Tyre and Sidon, appear for the most part to have maintained communication with this particular "Tarshish," or wealthy region, through the commercial system of that branch of the Phœnician family which had settled in Yemen, or Southern Arabia.[1] It was by these latter, the Sabaeo-Arabians,[2] that the actual occupation of the gold-bearing country was made, and the general work of mining carried on, through the agency of African, or imported Asiatic labour, or of both. In addition to the development of South-east Africa, the Sabæans also collected the produce of India and the Far East, and both in the age of Soloman (*circa* 1000 B.C.) and later in the era of Græco-Roman civilisation, they made their seats in Southern Arabia—the Ophir of Solomon and the Arabia Felix of the Romans—the emporium from which the gold

[1] The Phœnicians are supposed to have migrated from Chaldæa to the Persian Gulf, and thence to Arabia and Syria.

[2] The kingdom of *Saba* (Hebrew, *Sheba*) of the Bible.

and precious things of the East were distributed to the great centres of civilisation on the coasts of the Mediterranean. Phœnicia herself was the first to benefit; and her principal city, Tyre, in the words of Zechariah, "heaped up silver as the dust, and fine gold as the mire of the streets." In the Græco-Roman age, after the fall of Tyre and Sidon, Alexandria and Rome became the centres to which the riches of the East were attracted. The ignorance of the civilised world of 2,000 years ago as to the precise localities from which it drew its supplies of the precious metals is largely to be attributed to the jealousy with which the Sabæans maintained their position as the commercial intermediaries between the Mediterranean powers and the East—both India and South-east Africa. India was known through the conquests of Alexander; but it was Arabia and not Africa with which the educated Roman of the time of Horace was familiar as the source of the world's gold supply. And so the Roman poet speaks of Arabia when he tells the millionaire of that day—

> Intactis opulentior
> Thesauris Arabum et divitis Indiæ[1]—

that "if he had all the wealth of Arabia and the sumptuous East to himself," it could not save his soul from fear, nor his neck from the halter of Death. With the commencement of the modern era, when the Portuguese navigators had explored the coast of Africa and opened an ocean pathway to India, a glimmering of the truth came to Europe. A century later we find Milton in *Paradise Lost* reflecting the common surmise of his day, that the gold of the old world came actually from South-east Africa, and not from Arabia. When Michael in consoling Adam for the loss of Paradise shows "what shall come in future days" to him and to his offspring, and makes him see from the

[1] *Odes, III*, 24.

highest hill in Paradise "all earth's kingdoms and their glory," we read :

> . . . Nor could his eye not ken
> The empire of Negus to his utmost port
> Ercoco, and the less maritime kings
> Mombasa, and Quiloa, and Melind,
> And Sofala,[1] *thought Ophir*, to the realm
> Of Congo . . . . .

The Phœnician occupation of the gold-bearing districts of Rhodesia gives us what seems to be our first complete page of South African history. It is the old and familiar story of Solomon and the Queen of Sheba. A thousand years before the Christian era began, King Solomon, who was in alliance with Hiram, King of Tyre, built a fleet at Ezion-Geber, his port on the Red Sea, to bring a supply of gold from an "Ophir" lying to the south. The records of this enterprise, and its sequel, are, of course, to be found in the books of Kings and Chronicles ; but it will be well for the reader to have some idea of the nature of the evidence which connects the Phœnicians with this part of South Africa, before he considers the narrative itself.

The most valuable and reliable contribution to our knowledge upon this interesting subject is naturally to be found in the data provided by the archæological researches, which have been carried on since the foundation of the colony in 1890, and are still in progress. At the same time a renewed study of the accounts of South-east Africa written by the Portuguese commanders and missionaries of the sixteenth and seventeenth centuries, and an examination of the traditions of the Bantu population of the country, have yielded a body of evidence which at once supplements and interprets the testimony of the ancient remains. Obviously it is only the bare headings of this threefold evidence that can be indicated

[1] Sofala is a little south of Beira, the port of Rhodesia. The Portuguese town was on the site of the ancient port from which the Sabæans shipped the gold from the mines in the interior.

within the space at our disposal; but even such a summary is impressive.[1]

Archæological. In the gold-bearing area of Rhodesia, 700 by 600 miles in extent, there are found a number of rock mines which collectively form "the most extensive gold mines sunk to depth on rock during some prehistoric times yet known to the world," and some 500 ancient ruins. The former are tunnelled and excavated to water level, *i.e.*, a depth of 150 to 200 feet, and sunk on hard refractory rock, and mining experts have estimated that at least £75,000,000 worth of gold must have been extracted from them. The methods of mining pursued in all these prehistoric workings are the same—a fact which points to a single source of exploitation—and they are identical with those of ancient miners in Arabia and India. None of the gold was used in the country, but it was all exported; and the line of communication between the mining area and the coast was the Sabi river, and not the Zambezi.

> Most of the large and deep mines of the Selukwe, Gwanda, Belingwe, Sabi, and Manica districts are undoubtedly ancient, while those in the Mazoe district are of both prehistoric and mediæval times.

Among the ruins are "many scores of colossal buildings of dressed stone blocks." Mr. Hall's summary of the features of the original Zimbabwe (Phœnician temple-fortress) type is this:

> Finest construction; granite blocks dressed by chisels and hammers; elaborate drainage system; evidences of Nature worship: birds on beams, conical tower, ornate phalli, cylinder or *linga*, sun images, monoliths; foreign decorative designs;

---

[1] A full account of the Rhodesian remains so far explored, and of the very interesting questions to which they give rise, is to be found in the works of Mr. R. N. Hall: *The Ancient Ruins of Rhodesia* (Hall & Neal, published 1902), and *Prehistoric Rhodesia* (Hall, published 1909). The statement in the text is based upon the facts and conclusions contained in the latter volume. Where quotation marks are used, the words are those of Mr. Hall.

oldest class of relics, such as *astragali* ingot mould ; great wealth in chaste gold ornaments ; no woodwork left.[1]

The late Theodore Bent, who had conducted archæological researches in Asia Minor and Persia before he made his examination of the Rhodesian mines in 1891, sums up the results of his explorations in the following sentence :

> Here, near the east coast of Africa, far nearer to Arabia than India and China, and other places which they (the Sabæo-Phœnicians) were accustomed to visit, not only is there evidence of the extensive production of gold, but also evidence of a cult known to Arabia and Phœnicia alike : temples built on accurate mathematical principles, containing kindred objects of art, methods of producing gold known only to have been employed in the ancient world, and evidence of a vast population devoted to the mining of gold.[2]

In addition to these prehistoric remains, there are stone buildings of an inferior order, in which the characteristic Phœnician features are not found. These, and other remains of a still ruder type, are believed to have been constructed at much later periods by Bantu in imitation of the prehistoric buildings, and with, or without, Asiatic supervision. Thus Mr. Hall holds that " beyond all possible shadow of doubt " there were " three periods of both mining and building in Rhodesia." The first was the rock-mining, or prehistoric period. In the second period, which was partly prehistoric and partly historic, the river-sand washing for gold was carried on. The ruins of this period are poor, and without gold or phallic emblems. The Bantu tribes had begun to enter the country from Central or Northern Africa, and the process of " Kafirisation " had set in. The third period is marked by the abandonment of washing for gold, and the ruins are those of " native made barricades on hills and stone walls of cattle-kraals."

The ethnological evidence is mainly of a negative character. Native African tradition points to the

[1] *Prehistoric Rhodesia*, p. 473, *et seq.*
[2] *The Ruined Cities of Mashonaland*, pp. 193-4.

buildings of the Zimbabwe type as being of extreme antiquity; since it contains no trace of any knowledge of the people who built the Zimbabwes, and no suggestion of any Bantu race having had a hand in the work. In this it is supported by the historical evidence of the earliest Portuguese accounts of the country. When the Portuguese navigators arrived in 1505, they were informed by the Moors, who had been in occupation of the Sofala coast since the eleventh century of the Christian era, that the Great Zimbabwe was "very ancient," and that they believed that the "barbarians," *i.e.*, the Bantu, had nothing to do with its construction.

On the other hand, the light-skinned, industrial negroids, believed to have settled on the central plateau of Southern Rhodesia 1,000 years ago, and known as the Ma-Karanga, display "physical and linguistic features" that indicate a strain of semitic blood acquired in prehistoric times. Mr. Hall says that there are "over forty distinctly semitic customs to be found among this family of the Bantu race, most of which are of pre-Koranic origin." He writes: "Livingstone and all European scientists consider these as dating from some altogether indefinite time prior to the Koran—A.D. 600—and as such not to be explained by any influence as late as that of the Islamic influence of the tenth and eleventh centuries." The explanation of this phenomenon which Mr. Hall suggests is this. These industrial Bantu having arrived South of the Zambezi about the tenth century, or some 100 to 200 years later than the military tribes, inter-married with, and absorbed a decadent remnant of the Sabæan, or Indian, or Persian gold-mining population of the prehistoric era, whom they found in the country.[1]

An outline of the historical evidence (other than the Biblical record) which connects the Sabæans with Rhodesia will be given in due course. One item, however,

[1] *Prehistoric Rhodesia*, p. 84, note, *et seq.*

to which Mr. Hall attaches great importance, may be mentioned here. This is the fact that the area of the mediæval kingdom, marked on the early Portuguese and Dutch maps as "Monomotapæ Imperium," does not correspond with that of the gold-bearing region occupied by the Sabæan, or other prehistoric, miners. The word "Monomotapa" was admittedly the title of a Bantu dynasty, and the "Empire" of this line of great chiefs was a Bantu system, but the gold mining which the subjects of the Monomotapa carried on during the period of the Portuguese occupation of the Zambezi region (1505–1760), was conducted by primitive methods within the intelligence of the Bantu race, and had no connection with the rock mines and stone-built remains of the prehistoric miners. The contention that the prehistoric remains may have been the work of a native African, or Bantu, people, which is based upon the supposed identity of the site of the Great Zimbabwe and the "capital" of the Monomotapa's kingdom, therefore falls to the ground. On this head Mr. Hall writes: "The Portuguese records most explicitly show that:

"The Ma-Karanga did not occupy any stone buildings, 1505–1760, or within any traditionary times.

"The Monomotapan Capital, and also the centre of the gold trade of the Ma-Karanga (1505–1760, and even in traditionary times), was at Masapa in the Mazoe district, over 300 miles from Zimbabwe.

"The 'very ancient' ruins of Zimbabwe were in the district of Toro (= ancient) in the Kingdom of Sabia, where were 'the most ancient mines known in the country' (De Barros), which Kingdom of Sabia the records declare was never visited by the Portuguese, who never saw Zimbabwe, and from which kingdom the Ma-Karanga, it is stated, traded no gold whatever."[1]

This evidence is sufficient at least to show that there

[1] *Prehistoric Rhodesia*, p. 99.

is nothing extravagant in the belief that the Ophir of Solomon was Arabia Felix, and the actual source from which the Sabæans of this era drew their gold supply, the mineralised region south of the Zambezi now forming a part of Southern Rhodesia. The Biblical narrative, which in this setting assumes a new interest and meaning, is almost identical in the Kings and the Chronicles. In the former book it runs:

> And King Solomon made a navy of ships in Eziongeber, which is beside Eloth, on the shore of the Red Sea, in the land of Edom. And Hiram sent in the navy his servants, shipmen that had knowledge of the sea, with the servants of Solomon. And they came to Ophir and fetched from thence gold, four hundred and twenty talents, and brought it to King Solomon.[1]

In both the Kings and Chronicles this account is immediately followed by the narrative of the Queen of Sheba's visit, which in both books is introduced by the words, "And when the Queen of Sheba heard of the fame of Solomon . . . . she came to prove him with hard questions." As the result of the good relations established by the personal intercourse between the heads of the Jewish and Sabæan peoples, it would appear that Solomon was subsequently allowed by the latter to trade direct with the countries in which their supplies of gold and other valuable commodities were actually obtained. For later on in the narrative we are told, in explanation of the extraordinary abundance of gold which marked the reign of Solomon, that "the King had at sea a navy of Tharshish with the navy of Hiram." Or, as it is expressed in the Chronicles, "the King's ships went to Tharshish with the servants of Hiram."[2] The products obtained from "Tharshish" were "gold and silver, ivory, and apes and peacocks," and the time occupied in each expedition was three years; "every three years once came the ships of Tharshish." As the word "Tharshish,"

[1] 1 Kings, ix, v. 26–28.
[2] 2 Chron., ix, v. 21.

5—(2139)

like "Ophir," is a generic term used to indicate any source from which rich natural products or merchandise might be obtained, the countries visited by Solomon's merchant fleets in this three years' voyage may well have been both the "Tharshish" of the Sabæan goldfields in South-east Africa, and the "Tharshish" of India and the Far East. In any case, the length of the voyage shows that the maritime commerce of Solomon's Red Sea ports, when fully established, embraced not only the Sabæan emporium in South Arabia, but the actual countries from which this emporium was itself supplied with gold, precious stones, spices, and the rest.

In addition to the colonisation or occupation of the north and east coasts, and the development of the gold mines of South-east Africa, Phœnician seamen accomplished the circumnavigation of the entire Continent from east to west, at a date stated to be 600 B.C. The account of this remarkable feat of ancient seamanship is to be found in Herodotus. After comparing Europe to Asia and Africa, he continues:

> For Libya (Africa) is clearly surrounded by sea, with the exception of the space covered by the isthmus which joins it to Asia. Neko, the Egyptian king, was the first to demonstrate this. Upon the abandonment of work upon the canal from the Nile to the Arabian Gulf, he despatched some Phœnician navigators with instructions to sail back round the Continent to the northern sea through the Pillars of Hercules, and thus reach Egypt. The Phœnicians accordingly started from the Red Sea and sailed over the Southern Sea. And whenever their supplies gave out, they put in to the shore, and proceeded to sow the land with seed. They did this at the several points of the Continent which they reached in the course of their voyage, and waited for the harvest. After they had reaped the corn, they set sail, and thus, after two years had gone by, in the third year they bent their course through the Pillars of Hercules, and came to Egypt. And they said—what I refuse to believe, though others may do so if they like—that in sailing round Libya they had the sun to the right of them. The fact of Libya being surrounded by sea was first established by this expedition.[1]

[1] Book IV, p. 42.

As a point of evidence, it is interesting to notice that to us to whom the regions south of the equator are almost as familiar as those lying to the north of it, the statement which Herodotus refuses to believe as being contrary to all human knowledge—viz.: that the Phœnician mariners, sailing west, had the sun on their right—brings conclusive proof of the reality of this feat of ancient seamanship. Since it is obvious that no one who wished to obtain credence for an invented story would have introduced an incident which he must have known would have seemed impossible to the civilised world of his day.

Later, Herodotus tells us,[1] Sataspes was despatched by Xerxes on the same errand; only in this case the navigator started westwards from Egypt. Having passed safely through the Pillars of Hercules, he rounded a promontory called "Solium," on the west coast of Africa, and then proceeded southwards until his ship "stuck fast," and he was forced to return by the way he had come without accomplishing his task. His account of the furthest point to which he sailed suggests that he reached the region inhabited by Hottentots; since the people on the shore were a "diminutive" but not a "black" race, clothed in garments made from palm trees, who fled to the mountains on the approach of the Carthaginian ship, leaving Sataspes to possess himself of their sheep. This supports the belief, based on other sources of evidence, that the Bantu did not appear in South-Central Africa until approximately the ninth century of the Christian era; and in particular that the "natives" of Rhodesia during the whole period of the Sabæan occupation were either Bushmen or Hottentots, or both.

One other paragraph from Herodotus is worth quotation. Speaking of Æthiopia, which means for him the most easterly, as well as the most southward, of the inhabited regions of Africa, he says:

[1] Book IV, p. 43.

> It produces much gold, elephants' tusks so large that it takes both hands to carry them, all kinds of wild trees, ebony, and the largest, finest, and most long-lived of mankind.[1]

Five hundred years later than the time of Herodotus, in the first century of the Christian era, Egypt and all Northern Africa formed a part of the vast political system of which Imperial Rome was the centre; Roman expeditions penetrated southwards on both sides of the Red Sea; Roman tourists travelled up the Nile, and inscribed their names upon the walls of the Ptolemaic Temple of Isis at Philæ. The civilised world of this day knew the valley of the Nile as far south as Meroe, the fertile region stretching from Khartum southwards and enclosed by the fork of the White and Blue Niles; and it was familiar with the main features of the east coast of Africa as far south as Zanzibar.

The author of the *Periplus*, or *Round-Voyage*, of the Erythrean Sea (*i.e.*, the Red Sea and Indian Ocean), who is now said to have been a certain Basil of Alexandria, living in the reign of Nero (A.D. 54–68), takes us further south. This writer not only gives accounts of the chief ports on the east coast of Africa, and mentions the island of Menuthias, identified with Madagascar, but reveals two facts of cardinal importance. The first is that his knowledge of the east coast extended far enough southwards to enable him to state definitely that the Indian Ocean curved westwards, and united with the sea upon the opposite side of the Continent.[2] The second is that at this epoch the Sabæans were still in occupation of the east coast of Africa. The Sabæan King Kharabit was in possession of the coast from Mombasa to an indefinite extent southwards. His possession was based upon

[1] Book III, p. 114.

[2] At Prasum, a promontory of Æthiopia, "an ocean curves towards the sunset, and, stretching along the southern extremities of Æthiopia, Libya and Africa amalgamates with the Western Sea." (*Translated by Bent.*)

"ancient right," and the Sabæans were in the habit of sending transports, manned by sailors who were familiar with the places on the coast and with the language of the natives. As further evidence of the maritime supremacy of the Sabæans of this day in the eastern seas, he tells us that Musa, Aden, and other ports near the entrance of the Red Sea, were the recognised *entrepôts*, where the merchandise and produce collected by the Sabæan fleets were transhipped into Egyptian and Phœnician vessels to be carried to Egypt and finally to Rome, the centre of the civilised world.

A hundred years later than the date of the *Periplus*, another and more famous student of Alexandria, Claudius Ptolemaeus (*circa* 150 A.D.), drew the maps and wrote the eight books of geographical description, in which he sought to present a view of the entire surface of the inhabited world before the minds of his readers. In picturing Africa he makes the southern part of the Continent extend eastwards in unknown tracts until it unites with further Asia, thus enclosing the waters of the Indian Ocean by land in the manner of his own Mediterranean. It is strange that one of the brilliant group of learned Alexandrines—and one moreover whose aim was to gather together the fruits of Greek geographical researches—should have fallen into a cardinal error from which Herodotus's record of the Phœnician voyage round Africa alone should have saved him. With this exception Ptolemy's account of Africa to the sixteenth parallel of South Latitude, or roughly to the Zambezi, was based upon knowledge. The great tract of territory which he called Agisumba, has been identified with the country between the Great Lakes and the Zambezi, and the word itself is regarded by Father Torrend as indicating the presence of a group of Bantu tribes; the Ma-Zimba, Ma-Sumba, La-Sumba and others.[1] In

[1] *Comparative Grammar of the South African Bantu Languages.*

September, 1869, Livingstone, then fresh from his explorations of these regions, wrote that Ptolemy's references to the Central African Lakes were "substantially correct geography."[1] In other respects the Ptolemaic conception of the world represented a great advance in geographical knowledge, and it was on the basis of his work, made known during the Middle Ages through Arabic translations of the Greek original, that the explorers and students of the modern era began that more scientific and extended study of the earth's surface to which the polar expeditions of to-day will bring completion.

The little that has to be told of Africa during the centuries intervening between the partition of the Roman Empire (A.D. 395) and the commencement of the modern era can be set down in a few sentences. The period in which Egypt remained a part of the Eastern, or Byzantine, half of the Roman system was closed in A.D. 638 by the invasion of the Saracens. On December 22nd, A.D. 640, Alexandria opened its gates to Amr Ibn-el-'As, the victorious general of the Khalif Omar. Near the site of Memphis, Fostät (Cairo), the Mohammedan capital of Egypt, was founded, and in the course of the remaining years of the seventh century the Saracens overran Mediterranean Africa. During the eighth and ninth centuries the Mohammedans of Arabia and Persia established themselves in Abyssinia and spread down the east coast of Africa. The Sultanates of Magadosho, Brava, Melinde, Mombasa, and Kilwa (Quiloa), with settlements in Madagascar and Sofala, marked the permanence of their rule; and Arabian travellers explored the interior, traversing the Continent from east to west, until ultimately the Moslem states on the north and west coasts were united by a system of trade routes with those of the east coast.

[1] Letter in the Grey Collection, Capetown: Quoted by Mr. Hall in *Prehistoric Rhodesia*, p. 357.

The great library of Alexandria had been destroyed, but the Arabs in their period of greatness were skilled in literature and the sciences. The information gained in the course of their political and commercial expansion in Africa was duly recorded in writing. Alexandria remained the chief emporium through which the merchandise of the East passed to the European world; and Cairo and Aden were the centres and depositories of the geographical knowledge of Eastern Africa and the Far East, which enabled Vasco da Gama in 1497 to find an ocean pathway from Europe to India.

CHAPTER IV

EUROPEAN COLONISATION—PORTUGAL IN AFRICA

THE history of modern Africa is a record of the gradual establishment of European authority over almost the entire area of the Continent. The justification for this process, which in its later phases has been aptly called the "partition of Africa," is to be found in the circumstance that the great mass of the inhabitants of the tropical Continent are markedly inferior, both mentally and physically, to the rest of mankind. Nor is the tutelage of the original inhabitants of Africa any new thing. Before the rise of Greece and Rome—the most effective agencies for civilisation known to the ancient world—Africa owed its partial development to the presence of Asiatic races. During the period intervening between the conclusion of the Græco-Roman and the commencement of the modern era, whatever of civilisation it maintained it owed again to the intrusion and settlement of new populations from the East. With the modern era the turn of Europe to take Africa in hand came round once more. Only now the south, and not the north of the Continent was destined to be the chief seat of European influence and colonisation.

The record begins with the enterprise and foresight of one who was closely related by blood to the royal family of England—the Infante Henry of Portugal. The mother of this prince (b. 1394, d. 1460) was Philippa of Lancaster; Henry IV of England was his uncle, and Henry V, the victor of Agincourt and conqueror of France, was his cousin. The Moors, against whom he had fought in northern Africa, were his hereditary foes; and the task he set himself was nothing less than, by the discovery of an ocean route to India, to render his own country in particular, and western Europe in general, independent of the services of both Moors and Turks, as the middlemen through whose hands the produce of the East then reached the merchants of the West. With this object in view, he founded an observatory and a naval college, and in other respects devoted himself to the promotion of geographical research and the art of navigation. In the successive voyages upon which Portuguese seamen were despatched by him the coast of Africa was marked with the *padroes de descoberta* (crosses of discovery) which gradually crept further and further south. In 1448 a fort was erected on the coast of Guinea, and John II of Portugal, by permission of the Pope, assumed the title of "Lord of Guinea." When Henry died his task was still unfinished; but the work which he had begun was none the less destined to be accomplished. In 1471 the Equator was reached; and still the arid coast line stretched, endlessly as it seemed, southward between the Portuguese navigators and the eastward course they sought. Even Bartholomew Diaz would scarcely have rounded the Cape, in 1486, but for an accident. Driven helplessly before the wind for thirteen days southward, he steered east and failed to make the land. Altering his course to the northward, he reached the coast of Africa again a little eastward of Cape Agulhas, the southernmost point of the Continent; and then sailing east crept along the coast as far as the little island in Algoa Bay, which from the "cross of discovery," duly planted by Diaz, bears to this day the name of "Santa Cruz." This was progress indeed; and John II was not slow to recognise that the end was now in sight. To Diaz, who proposed to name the dreadful headland where the storm had seized him, the "Cape of Tempests" (Cabo Tormentoso), he replied: "Not so; but call it rather the Cape of Good Hope, for by this Cape shall we sail to India." And at the same time he sent two agents to Cairo and Aden to gain information concerning the east coast of Africa and the route to India.

Twelve years later the goal was reached. In July, 1497, in the reign of Emanuel the Fortunate, Vasco da Gama set sail from Portugal; by November he had reached the Cape, on Christmas Day he anchored off the coast of Natal,[1] and early next year he brought his ship to port at Calicut, on the west coast of Hindostan. In September, 1499, he was back in the river at Lisbon with his ship laden with the silks and spices, pearls and gold, that he had taken on board at Indian ports. Henceforward the merchants of Lisbon, and of Western Europe, could trade direct, and no longer through the Moors as middle-men, with the wealthy communities of India, Java, Malacca, China and Japan.

It had taken much time and long-continued effort to fulfil the purpose of Henry the Navigator. But in the course of this effort much more had been accomplished than the attainment of the immediate end in view, important as it was. While the Portuguese navigators were endeavouring to reach the Far East by sailing eastwards, Spain, closely followed by England, had accidentally discovered the new world of America by despatching Columbus to reach the Far East by sailing westwards. It was six years after Diaz had rounded the southernmost point of Africa and found a clear course eastwards, that Columbus brought the little fleet of Ferdinand and Isabella to anchor off the Bahamas, and learnt that a huge continent stretching almost from pole to pole lay between him and the shores he sought. In eight years more Da Gama had completed the work of Diaz, and thus the closing decade of the fifteenth century witnessed the two discoveries which were destined to make the nations of western Europe into colonising

[1] Hence the name [*dies*] "Natalis."

powers, and eager rivals for the supremacy of the seas.

In this seaward expansion of western Europe Portugal led the way. In her successful effort to open up maritime communication with the Far East, not only were the coasts of Africa occupied, but Brazil was colonised, Mexico conquered, and the Straits of Magellan navigated. On the east coast of Africa, Sofala was taken from the Arabs in 1505, and soon afterwards the flag of Portugal was flying over Melinde, Mombasa, the island of Zanzibar, and Magadosho, together with Sena and other ports on the Zambezi. By 1520 the Portuguese had wrested the east coast ports between Cape Gardafui and Delagoa Bay from the Arabs, and possessed themselves of the east coast trade. In the middle of the century their settlements in India, Java, and Malacca were united into a single system, over which a Portuguese Governor-General of India, living at Goa, presided; and the East African ports, with a Governor at Mozambique or Sofala, continued to form a part of this Indian dominion until a hundred years later, when they were made a separate administration, the present province of Mozambique.

This display of energy on the part of the little kingdom of Portugal lasted only 165 years; that is from the capture of Ceuta in 1415 to the union of the crowns of Spain and Portugal in 1580. Barely sixty years sufficed to create this "colossal Portuguese domination, extending from the Atlantic Ocean to the shores of China"; the "giant edifice" which has been since then "levelled from one century to another by the breath of events, as if it had been made of sand."[1] The Portuguese were closely followed to the East by the seamen and merchants of Spain, Holland, France, and England. In the development of this new trade certain points on the coast of Africa, and certain convenient islands, were

[1] M. Guillain, in *L'Afrique Orientale*.

occupied by these nations as harbours and victualling stations for the ships they employed ; but by the close of the sixteenth century the merchants of Holland and of England had alike determined to make an organised effort to obtain supremacy in the Eastern markets. The English East India Company received its charter from Queen Elizabeth in 1600, and the Dutch East India Company was founded only two years later. The original objective of both these great trading corporations was Java, not continental India ; and both in turn, in order to secure their interests and possessions in the East Indies, were compelled to establish themselves at the Cape of Good Hope. In the case of the Dutch Company, the unexpected discovery of a temperate climate in the southern extremity of the Continent caused their occupation to develop into an effective and permanent colonisation ; and when, at the end of the eighteenth century, England was constrained by the same strategic considerations to occupy the Cape, it was on the basis of the colonisation thus accomplished by the Dutch East India Company that the fabric of the European South Africa of to-day was gradually, and laboriously, erected.

Portuguese East Africa, even more directly than the Cape, was the offspring of India. But the circumstance which gave permanence to the Dutch occupation of the Cape, the temperate character of the climate, was wanting in East Africa—or, more correctly, the Portuguese believed it to be wanting. Then, as now, at a distance of a few hundred miles from the east coast there were high plateaux or mountainous regions where Europeans could live and thrive ; but except where the water-way of the Zambezi gave easy communication with the coast, they established no inland settlements. In particular they failed to penetrate from Sofala to the escarpment of the central plateau of South Africa, behind which lay the fertile and temperate uplands of what is now Southern Rhodesia. For them, therefore, the back-country of Mozambique meant only the low-lying and pestiferous lands of the coast belt. In these circumstances it is not surprising that the Portuguese made little effort to establish themselves securely in the *hinterland* of their east coast ports. The little that was done in the way of exploration and development was the work of the missionaries; first the Jesuits and then the Dominicans. The gold still won by river-washing and rude methods of quartz crushing known to the Bantu, stimulated them to enter into relations with the Bantu potentate styled the Monomotapa; but neither military expeditions nor treaties enabled them to obtain even a moderate revenue from this source. One solid advantage, however, they gained from their relations with the "Monomotapæ Imperium."[1] By the deed of gift of 1607, and again by a second treaty in 1630, the control of the "gold mines" was formally granted by the Monomotapa to the King of Portugal. It was in virtue of this latter document, produced with the Bantu chief's mark duly affixed to it, that the Portuguese were successful in the MacMahon Arbitration[2] of 1875, thereby winning Delagoa Bay, the natural port of the Transvaal Province, and a factor of great importance in the industrial and political development of South Africa during the last forty years. By the middle of the seventeenth century the Mozambique authorities had practically abandoned their endeavours

[1] I quote the words from a map of 1623, the original of which is in the Vatican.

[2] The dispute between England and Portugal for the coast north of Tongoland up to the Limpopo was submitted to Marshal MacMahon, then President of the French Republic. In his award, issued on July 24th, 1875, it was declared that the Portuguese territory on the East Coast of Africa extended southwards to a line parallel to 26° 30′ of south latitude, drawn from the coast to the Lebombo Mountains, which form the boundary between Swaziland and the Transvaal.

to find and work the gold mines of the Monomotapa, and even ceased to carry on the collection of gold dust from the natives of the interior as a State industry. In the place of this illusory and disappointing enterprise they found a new and more lucrative source of revenue in the capture and exportation of native Africans as slaves; and from the year 1645 onwards for a century East Africa supplied large numbers of these unfortunate people to the Brazils and other Portuguese colonies, and occasionally to the French and Dutch possessions. But the commencement of the slave trade inaugurated, and was in a large measure directly responsible for, the decline of Portuguese influence in East Africa. The slave raids drove away the native population from the neighbourhood of the Portuguese settlements, and all other commercial dealings with the tribes of the interior, including the collection of gold, became more difficult and less profitable. The Indian dominion was menaced by the increasing activity and success of the Dutch and British in Eastern waters. The East African ports escaped permanent capture; but they owed their immunity to the pestiferous climate of the regions in which they lay, and to the circumstance that by the time the Dutch and English East India Companies were fully established, the east coast trade of Portugal had become too insignificant to offer a profitable field to these rich and powerful corporations. Moreover, both the Dutch and British merchantmen had by this time provided themselves with more healthy and more convenient "half-way houses" to India—the former using the Cape, and the latter Saint Helena, as a victualling station and port of call.

But although the Portuguese are guilty of introducing the slave trade, with all its train of attendant evils, into Africa, it must not be forgotten that the original impulse which carried their navigators through their first perilous voyages, was largely religious. The discovery of the maritime route to India was a signal victory for the Cross in the long struggle between Christendom and Islam. The crosses of discovery planted on successive headlands of the Continent of Africa were no meaningless emblems. They marked the extension of the authority of the Vicar of Christ no less than the acquisition of new territory for the King of Portugal: and in the period of their prosperity the Portuguese were active in the fulfilment of their religious obligations both in India and in Africa.

The efforts of their missionaries to spread Christianity among the natives in East Africa were fruitful of geographical discovery, and among the most interesting records of the period are the letters and reports written by the Jesuit and Dominican priests to their ecclesiastical superiors in Portugal and Rome. The Monomotapa, or Kafir, mission was founded in 1560 by a party of Jesuit priests, despatched under the leadership of Gonzalez Silveira, the Provincial designate of the Order, from Goa, the seat of the Indian administration. In the year following Silveira, having reached the chief kraal, or capital, of the Monomotapa, was there treacherously murdered by order of this Bantu potentate. The news of the tragic event in due time reached Europe, and ten years later an expedition under Francisco Barretto was sent out from Portugal to Mozambique, thence to proceed by way of the Zambezi against the Monomotapa. Its objects were threefold: "to spread the knowledge of the Christian religion; to secure a source of revenue which would defray the expenses of the Indian administration; and to punish the murder of Silveira." The Monomotapa made his submission upon the arrival of the Portuguese force at Sena; but before the negotiations were concluded, Barretto himself, and practically the whole of the 700 men under his command, had perished of fever. Among other conditions imposed on the Monomotapa was an undertaking to receive the missionaries; and from this time onwards the work of Christianising the natives was prosecuted with energy and success. The organisation of the missions in Mozambique as a whole was, however, transferred by the Papal authorities from the Jesuits to the Dominican Order; and it is from a priest of this order that fifty years later we get a report on "the state of Christianity in the rest of Africa on the other side of the Cape of Good Hope." At this date (*circa* 1625) there were twenty-two places, at each of which there were some Christian natives, while most of them possessed parochial churches or Monastic establishments. At Sena, the head-quarters of the Portuguese in the Zambezi districts, there "were more than 200 Christian houses; a parochial church with ten priests; the Church of St. Dominic, with three or four friars; and the Church of St. Antony in charge of a priest." Equally interesting is the account of the Kingdom of Monomotapa which is contained in the same report, as forwarded by Monsignor Lorenzo, the Colletore of Portugal, to the Cardinal Præfect of the Propaganda at Rome.

> The kingdom of Monomotapa is very large and full of people, nearly all Pagans, and without knowledge of religion. It is rich in gold mines, ebony, and ivory. And in the opinion of many it is the ancient Ophir, where Solomon sent his ships, which sailed through the Red Sea to the coast of Africa. A very easy navigation and full of ports.
>
> The extent of the kingdom is not known, but it is believed to be bounded on one side by the kingdom of Angola, and on the other by that of Prester John . . . .[1]

But, as already noticed, from the middle of the seventeenth century onwards the power of the Portuguese both in East Africa and in India rapidly

[1] The above quotations are taken from the translations of the actual documents in the archives of the Propaganda at Rome, obtained by Mr. A. Wilmot, and published by him in Appendices to his book (*Monomotapa*, by the Hon. A. Wilmot: London, 1896).

declined. The causes of their failure in East Africa are not far to seek. They never succeeded in finding the real gold mines; that is, the ancient Sabæan workings, concerning which they remained in complete ignorance, although they were much nearer to them than the so-called "mines" of the Monomotapa. Nor did they ever discover the cardinal geographical fact which would have made an effective occupation of the hinterland possible—the fact that inland, beyond the fever-stricken coast belt, there were high plateaux and mountain regions where valuable produce could be grown, and European colonies could be planted. As it was, they were soon driven to content themselves with the bare maintenance of their hold upon the ports and islands of the coastline, leaving the natives—with the exception of periodic slave raids—to manage their own affairs. The Portuguese did, however, make two attempts to introduce an element of permanence and settled industry, into their east coast possessions—attempts, which, as representing the sum of their actual work of colonisation, deserve a moment's consideration.

The first was the grant by the Governor-General of Portuguese India of the exclusive right to trade between Diu and Mozambique to a company of Banyan (Hindu) merchants. This measure, which was well calculated to promote the mutual prosperity of the Portuguese possessions in India and East Africa, was taken in 1686; and in addition to the trading monopoly, the Banyans were allowed to nominate persons to act as judges in the event of any member of the company being involved in any civil or criminal proceedings, while residing in East Africa. The persons chosen for this purpose were Jesuits from the establishment of the Order at Goa. The sequel is noticeable in view of the recent controversy as to the status of the British Indians in the Transvaal. The Banyan traders grew so prosperous that they incurred the hostility of other classes in the community, and the Jesuit fathers, rightly or wrongly, were accused of complicity in their alleged misdoings.

> One Governor-General wrote of them (says Consul O'Neill) that they (the Banyans) were selfish, false, and cunning, given to lying and usury ; that they knew not how to keep a contract ; and that it was a part of their religious creed to deceive and rob a Christian. And of the Jesuits a Secretary of State for Foreign Affairs, writing from Lisbon, stated in a despatch to the Governor-General of the Colony, that " His Majesty the King was perfectly aware, and his royal sense of piety had received a severe shock therefrom, that the missionaries had degenerated into a mere association of smugglers." [1]

The Jesuits accordingly were sent back in 1759 to Goa, where they were punished by imprisonment and the confiscation of their property ; and twenty years later (1777) the Banyans themselves were deprived of their trading and other privileges. In spite, however, of the withdrawal of official support the Indian traders continued their operations in East Africa, and as recently as 1882 the British consul at Mozambique reported that they were " in sole possession " of the trade of the Portuguese coast line. They do not make colonists, he added, because their caste prevents them from remaining permanently outside their own country, nor does their religion allow them to take their women abroad with them. But whatever encouragement to industry the native Africans of the Portuguese hinterland have received, has come from these Indian traders, who, unlike the European merchants, do not confine themselves to wholesale transactions, but make their way by river and caravan into the interior, where they collect the indigenous produce of the country in exchange for articles of European or Indian manufacture.

[1] Quoted from a paper, published in the *Journal of the Royal Geographical Society* for 1882, by Mr. H. E. O'Neill, H.B.M. Consul at Mozambique. I am indebted to this source for the facts upon which this account of these two colonising efforts of Portugal is based.

In the second of these attempts we have the sole direct effort of the Portuguese, during the period here in question, to settle a European population in South-east Africa. Early in the eighteenth century the fertile districts lying north and south of the Zambezi were apportioned into Crown grants (Prazos da Coroa), which were bestowed upon Portuguese women, with succession through female offspring for three generations, on condition that (1) they married a Portuguese of European birth, and (2) actually resided with their husbands upon the lands respectively assigned to them. Subsequently, when a sufficient number of European Portuguese were not forthcoming, women of mixed European and Indian or African blood were accepted. In this modified form the system of Crown grants, though originally well conceived, was worked so ineffectively by the local Portuguese officials, that it produced abuses which not only defeated the main purpose of the measure, but threatened altogether to subvert the authority of Portugal in these regions. The heiresses, or their Portuguese husbands, were not content with one grant, but succeeded in obtaining several. Having established themselves in possession of the great tracts of land thus acquired, they and their husbands, by means of hired soldiers and retinues of slaves, set up the state of mediæval nobles, made themselves masters of the surrounding country, and defied the Portuguese authorities. In these circumstances it is not surprising that legislation was passed in Portugal in 1836, and again in 1854, to abolish the system of Crown grants, and deprive the then existing grantees of their rights and privileges. Owing, however, to the weakness of the Portuguese administration in East Africa, the possessors of these large estates were enabled to defy the authority of the government until quite recent years.

None have judged the comparative failure of the Portuguese in Africa more severely than those contemporary writers among them, who have striven lately, and not altogether unsuccessfully, to recall their nation to a sense of its past greatness. Of the seventeenth century administration Andrade Corvo says:

> We dragged out a sad existence, without progressing in colonisation, without developing commerce or industries, and without the famous gold and silver mines giving the marvellous results which were expected from them.

And again:

> The early Portuguese did no more than substitute themselves for the Moors, as they called them, in the parts that they occupied on the Coast; and their influence extended to the interior very little, unless, indeed, through some ephemeral alliance of no value whatever, or through missionaries, or without any practical or lasting results. The true conquest is still (1885) to be made.[1]

It remains to state very briefly the few facts which serve to link up the Portuguese East Africa of the past with the South Africa of to-day. Shortly after Mozambique was emancipated from the control of the Portuguese Governor-General of India (1752), the Jesuits, as already mentioned, were expelled, and the Dominican missionaries, although they remained longer in the country, gradually declined in numbers and importance, until in 1830 they too finally disappeared. Meanwhile, in 1763, the seven chief towns of the province were created municipalities. In 1838 the Governor-General of Mozambique received fresh powers from the Home Government, including the appointment of the District governors; and twenty years later a "Junta," or representative council of the Province, was established, to which Sena, Sofala, Inhambane, and Lorenzo Marques, each contributed one, Tete two, and Mozambique itself seven members, thus providing collectively a total of thirteen for the whole province.

Under the MacMahon award, to which an allusion has been made above, the possession of Lorenzo Marques,

[1] As translated by Mr. Scott Keltie in *The Partition of Africa.*

with the fine harbour of Delagoa Bay, was confirmed to Portugal in 1875. The occupation of Mashonaland by the pioneer expedition of the British South Africa Company in 1890 led to immediate and serious disputes between the British and Portuguese authorities. But although for a time there was much diplomatic friction between Great Britain and her ancient ally, the entire frontier of Portuguese East Africa was eventually delimited to the satisfaction of both parties under the terms of the Anglo-Portuguese Convention of June 11th, 1891. By this agreement, while her just territorial claims were generously recognised by her powerful neighbour, Portugal, on her part, bound herself to observe certain conditions intended to prevent her from using her legal but ineffective possession of the coast belt to hinder the development of the British interior, or unduly tax its future colonists. The most important of these undertakings were: (1) To recognise the Zambezi as a free water-way; (2) to construct a railway from Beira, the natural port of Mashonaland, to the interior; or, failing this, to allow the construction of such a railway by others; and (3) to limit the duties leviable upon goods in transit from the coast to the British sphere to a maximum of 3 per cent. *ad valorem.*

Since the date of this Agreement the prospects of Portuguese East Africa have steadily improved. The profits of the transit trade arising from the rapid development of Southern Rhodesia and, on a far greater scale, the Transvaal, have rescued the finances of Mozambique from their chronic deficits, and created the modern ports of Beira and Delagoa Bay. Apart from this indirect assistance, British capital and energy have largely and directly contributed, through the agency of associations such as the Mozambique and Nyassa Companies, to the development of the agricultural and mineral resources of the Portuguese territory. At the same time, by providing the Witwatersrand mines with a great part of their African labour, and the Transvaal with its nearest and most economic port, Mozambique performs an indispensable service to the Union of South Africa. Indeed the material interests of the Transvaal and Mozambique have been so closely associated in the recent improvements of the railway and dock accommodation of Lorenzo Marques, that, although politically separate, the province of Mozambique has come to be recognised, and treated, as a necessary part of the British industrial and commercial system. Nor should it be forgotten that within the last twenty years the Portuguese themselves, stirred by such writings as those of Andrade Corvo, have shown a revived interest in their East African possessions. It cannot be said that the Portuguese Government has learnt as yet to avail itself fully of the opportunities presented by the rapid expansion of a great industrial state upon the borders of these once worthless territories; but there are signs that, as the result of this closer association with British statesmen and men of affairs, the Mozambique authorities at all events have acquired a more enlightened political outlook, and are sincerely desirous of reforming their administrative methods.

CHAPTER V

THE DUTCH EAST INDIA COMPANY

WITH the appearance of the English and Dutch in Eastern waters the central current of South African history returns from the east coast to the Cape of Good Hope. A tragic incident is said to explain the otherwise extraordinary neglect of the Portuguese to form any settlement upon the temperate south coast of Africa. In 1509–10 Francis d'Almeida, the first Governor-General of Portuguese India, when returning to Europe after the signal defeat of the Moors which gave Portugal the mastery of the Indian Ocean, put in at Table Bay with four ships. Here he himself and no less than sixty-five of the flower of his men-at-arms were killed by the missiles of the agile Hottentots in the bush at the foot of Table Mountain. The effect of this unexpected and disproportionate disaster was so great that in the future the Portuguese ships rarely came to land between St. Helena and the east coast.

The first Englishman to "round" the Cape was Sir Francis Drake, who, passing it in 1580 on the homeward course of his famous voyage round the world, reported it to be "the most stately thing and fairest cape we saw in the whole circumference of the world." The pioneer of the English East India Company was Sir James Lancaster, who had served under Drake in the defeat of the Spanish Armada. Sailing for the East Indies in 1591, he returned to England after many adventures in 1594, but only to set out for the same goal a few months later at the head of an expedition fitted out by the merchants of London. This time he was away only two years, and the results of the expedition were so favourable that an association of merchants was formed for the purpose of trading with the East Indies. The new company received its charter of incorporation from Queen Elizabeth on the last day of 1600 (o.s.), and a few weeks afterwards Lancaster sailed in command of four ships—the first fleet of the East India Company—for Bantam, in the island of Java, where he presented a letter from the Queen to the King of Bantam, and established the first trading agencies of the company.

The English ships which from this time onwards to the opening of the Suez Canal regularly passed the Cape on their outward and homeward voyages to and from India, used Table Bay as an occasional port of call; and it is noticeable that their crews seem to have encountered no hostility on the part of the Hottentots and Bushmen. Indeed, the Dutch complained, when fifty years later they founded their settlement here, that they had to use the English language, if they wished to make themselves understood by the natives; and that these latter were much more ready to barter their sheep and produce with the English than with them. In these circumstances it is not surprising that the desirability of acquiring permanently so useful a place should have occurred to the English sailors; and in 1620 two of them, Captains Andrew Shillinge and Humphrey Fitz-Herbert, being in Table Bay with four East Indiamen, hoisted the English flag on Signal Hill, and proclaimed English sovereignty over the Cape Peninsula and the adjacent mainland in the name of King James I. Their action, however, was disallowed both by the Company and King James; and so it fell to the Dutch to have the honour of planting a European population at the foot of Table Mountain.

The first Dutch fleet to reach the East Indies was that which sailed for Bantam in 1595 under Admiral Houtman. The Dutch East India Company, which was a union of smaller and competing associations of merchants, was incorporated two years later than the English Company. The General Directory, called the Chamber of XVII, met at various places in Holland, and the chief seat of its operations in the East was Jakatra in the island of Java, to which in 1621 the name " Batavia " was given. When the company was fully organised, its oversea administration consisted of a Governor-General and Council of India at Batavia, and a number of settlements or stations which were subject to the Indian Government, but administered directly by admirals, governors or commanders, assisted in each case by a local " Council of Policy." The immediate origin of the Company's station at the Cape is to be found in an incident scarcely less tragic than the death of D'Almeida. In 1848 the *Haarlem*, one of its finest ships, was wrecked on the shores of Table Bay, where the crew remained for five months before they were taken off by the homeward bound fleet. On their return to Holland two of the shipwrecked crew presented a " remonstrance " to the Chamber of XVII, setting out the advantages which the place presented for the establishment of a permanent post.

After considerable hesitation and delay the Directory were persuaded to carry out the proposal. The execution of their plans was entrusted to Jan Antony Van Riebeck, a surgeon in the company's service. His instructions were, to construct a wooden building for the accommodation of invalided sailors and soldiers ; to build a fort capable of holding a garrison of seventy or eighty men ; to make a garden where vegetables could be grown—a matter of great importance in days when scurvy was a terrible scourge—to treat the natives " kindly " ; and to keep a diary for the information of the Directory in Holland. The expedition, consisting solely of persons in the civil or military employment of the Company, who with a few female relatives numbered less than 200 in all, set sail in three ships from the Texel on December 24th,

1651. By the morning of Sunday, April 7th, 1652, the whole company, with the building materials and four culverins for the fort, had arrived safely in Table Bay.

Five years later the Company took a step which was destined to convert the naval station at the Cape into a European colony. Nine soldiers and sailors were discharged and placed upon small farms of twenty-six acres on the banks of the Liesbeck at Rondebosch. The terms upon which these original " free burghers " took their holdings are characteristic of the principles and methods pursued by the Dutch East India Company during the entire period of its rule in South Africa (1652–1795). In addition to the land, the settlers were provided with tools, seeds, and stores, and exempted from taxation for a period of years. In lieu of rent and interest upon the capital thus advanced, they undertook to bring all their produce to the Company's warehouses, there to be purchased by the Company at prices fixed from time to time by its officers. If any produce remained over and above the amount required by the Company, the settlers were free to dispose of it to any foreign ships which might come to anchor in Table Bay ; provided, however, that they did not go on board these ships until three days after their arrival, by which time the Company's agents would have themselves sold to the foreign merchantmen whatever produce they had for sale. In the case of the cattle trade the restrictions were even more stringent. The settlers were only allowed to purchase cattle from the Hottentots at the prices fixed by the Company, and were forbidden to sell the cattle, when ready for market, to any purchaser except the Company.

In Java, the most valuable of the Dutch East India Company's possessions, this system, afterwards known as the " culture " system, was pursued on a large scale and with excellent financial results. Here a great revenue was derived from the two monopolies exercised by the

Company, after they had acquired a virtual possession of the soil of the island from the native princes. These monopolies were (1) that of producing all the most valuable crops, (2) that of trading in all productions whatsoever. It was upon this system that the Cape Colony was founded. The Government of the colony were agents of the Company, the colony was their estate, and the colonists were persons employed to cultivate the soil, on perfectly well understood terms of remuneration, for the benefit of the shareholders in Holland. Such a system was suitable enough for the development of the resources of a tropical island like Java, with an abundant and semi-servile population, but it was too rigidly commercial to satisfy the elementary conditions necessary for the social and political well-being of a European community, however small. Before, however, we consider the effects which this system produced, it is necessary to trace in outline the growth of the European population to which it thus came to be applied.

The Liesbeck settlement had another result which must be noticed. The occupation of land outside the actual station on Table Bay at once led to a conflict between the Dutch and the Hottentots. The account of the affair which Van Riebeck wrote, under date July 29th, 1659, to the Governor-General and Council, Batavia, contains an interesting statement of the Hottentots' point of view. After remarking that the "Hottentots had been at work again," and that the Fiscal Gabbema had caught two of them, of whom he killed one and took the other prisoner, he continues :

> The said prisoner, who was one of the Caepmans, and spoke tolerable Dutch, being asked why they did us this injury, declared . . . because they saw that we were breaking up the best land and grass, where their cattle were accustomed to graze, trying to establish ourselves everywhere with houses and farms, as if we were never more to remove, but designed to take for our permanent occupation more and more of their Cape country which

had belonged to them from time immemorial. Aye, so that their cattle could not get at the water without passing over the corn land, which we would not allow them to do; that they, consequently, resolved (as it was their land) to dishearten us by taking away the cattle (with which they could see that we broke up and destroyed the best land); and if that would not produce the effect—by burning our houses and corn until we were all forced to go away; that Doman had also put it into their heads that after all the houses in the country were destroyed, the fort could be easily surprised—as the earth walls were built with a slope—and then the Dutch might be forced quite to abandon the country. . . .[1]

Peace was made between Van Riebeck and the "Captain and Chief of the Caepmans" on April 6th, 1660; and a complete statement of the negotiations, as duly recorded in the diary, was then forwarded for the information of the Directors of the Company in Holland. It must be remembered that it was no part of the Company's policy to drive away the Hottentots; since they were useful in raising cattle and providing the supplies required both for the use of its own ships, and for sale to the foreigners who put in at Table Bay. The directors, therefore, replied that the discontent of the Hottentots was "neither surprising nor groundless," and recommended that in the future the land should be purchased from them, or that some other means of satisfying them should be employed. In giving this advice to Van Riebeck they were again applying to the Cape Settlement a course of action which they were pursuing in Java with excellent results. In this latter case the territorial rights of the native princes were gradually and successively bought up, until in the end the Company became possessed of the entire island—and this not by conquest but by good bargaining.

Twelve years later, when as we shall see the directors began to take a greater interest in the Cape Settlement, this policy was put into effect under the direction of a high

[1] *The Record: A Collection of Original Documents made by Donald Moodie, Lieut. R.N.* Published at Capetown in 1838.

Indian official, Aernout Van Overbeke. By deeds of cession, dated respectively April 19th, and May 5th, 1672, and " done in the Fortress, the Good Hope," the Hottentot chiefs, styled " princes," but using a cross to attest their names, granted " the whole district of the Cabo de Boa Esperance," and (on the mainland) the " whole district of the land called ' Hottentoos Holland ' " to the Dutch East India Company. The merchandise—tobacco, beads, copper, brandy, ironmongery, etc.—given to the Hottentots in consideration of these grants, was trivial in value (f.33, 17, or £2 16d. 0d., and f.81, 16, or not quite £7, as reported to the directors), but, on the other hand, the chiefs expressly retained the right of " coming with their kraals and herds of cattle, freely and without molestation " to all land outside the farms and pastures taken up by the Dutch settlers.[1] Probably the Hottentots were very little, if at all, worse off in the matter of pasturage ; while the presence of a European community, by providing them with a market in which their cattle and sheep could be exchanged for European wares, brought within their reach utilities and luxuries hitherto unknown. No subsequent deeds of cession were obtained, but the Dutch appear to have had no further trouble with the yellow-skinned natives. In point of fact, when they gradually pushed their settlements further inland, they found the country to be practically uninhabited and unoccupied : for the wandering tribes of Hottentots and Bushmen, by whom alone it was peopled, could show no evidence of settled occupation sufficient to support a claim to territorial rights.

Van Overbeke's successful negotiations with the Hottentot chiefs formed part of a general effort made by the directors at this time to secure the growth of the European population at the Cape. During the first twenty years the actual work of colonisation had progressed very

[1] These details are taken from *The Record*.

slowly. Van Riebeck, after ten years' command, had been promoted to the Governorship of Malacca. The founder of the Cape Colony is revealed to us not only as a resourceful and courageous leader but as a man of parts. His diary, which gives a vivid picture of the daily life of the little community entrusted to his charge, is a racy document, flavoured with the unconscious profanity of the period and lightened by a grim humour. He himself subsequently became Secretary to the Indian administration, and his son, born at the Cape, rose to the position of Governor-General of the Dutch Indies. Under Van Riebeck's immediate successors the Cape Station continued to increase slowly in numbers and importance, and in particular the original fort was replaced by the much stronger stone building, of which a part remains to this day and is known as "the Castle" of Capetown. But the Company had experienced great difficulty in finding persons willing to leave Holland to settle in a place so distant and isolated as the Cape of Good Hope, and in the year 1779, which marks the commencement of the period of active colonisation, the population of the settlement was returned in the census of that date as consisting of 87 freemen with 55 women and 117 children; 30 men in the Company's employment; and 133 male, and 38 female slaves, with 23 slave children.

This was the position when Simon van der Stell was appointed to the command of the Cape station, with instructions from the directors to carry out a policy of active development; and it was in the thirty years (1679–1707) covered by his administration and that of his son Adrian, that the main lines of the future progress of the colony were laid down. The directors had solved the emigration question, in part at least, by deciding to send out young women from the public orphanages in Holland, to provide the existing unmarried settlers with wives. Of these welcome emigrants one or two arrived in 1685, and from this date parties of six or eight continued to come out from time to time in the Company's ships, as occasion presented itself. The Hottentot question had been settled, as we have seen by the action of Van Overbeke; and the good effect of this arrangement was soon made apparent in the increasing numbers and prosperity of the "freemen." In 1680 a settlement to which the name Stellenbosch was given in honour of the new Governor, was founded on the mainland thirty miles away from the Cape; and a little later the first Landdrost, or District Magistrate, was appointed to hold office there.

The year 1685 was marked by an event destined to bring an even more important element than the orphan girls into the parent stock from which the South African Dutch are descended—the Revocation of the Edict of Nantes. The directors of the Company were now making a strenuous endeavour to secure fresh emigrants for the Cape. They offered an asylum to some 600 or 700 of the Valdesi, who at the time had been driven from their Alpine homes by Victor Amadeus II of Savoy; but the arrangement was not carried out in view of the preparations for the "Glorious Return" achieved under the heroic Janavello in 1689.[1] But a similar offer to the Huguenot refugees, many of whom fled to protestant Holland, was accepted gratefully; and between the years 1688 and 1690 at least 150, and probably rather more, persons of French blood were added to the permanent European population of the settlement. The Huguenot emigrants were drawn from classes somewhat higher in the social scale than those which had provided the earlier settlers.

[1] The Valdesi are a community of Christian refugees, who escaped from the plains of Lombardy to the Piedmontese Alps at the time of the invasion of Italy by the Goths. In spite of repeated and severe persecutions, they maintained the tenets and organisation of the Primitive Church through the centuries, and remain a separate ecclesiastical community to this day.

Besides the artisans, farmers, and vine-growers who formed the bulk of the various parties, there were a sprinkling of men who had once held good positions, and even a few members of such noble houses as du Plessis, de Mornay, de Villiers, and du Pré. On their arrival they were settled on the mainland, some at Stellenbosch, but the majority at Drakenstein and French Hoek. Here they proved themselves to be excellent farmers; and in particular they brought a knowledge of viticulture, and of the methods of producing wine and brandy, which was of great service to the colony.

As the Directors made no special effort to secure emigrants after the year 1688, we are now in a position to ascertain the elements out of which the Dutch Afrikander stock was formed. First, there were the soldiers, sailors, and other discharged servants of the company; second, the mainly Dutch families, or individuals, emigrated direct from Holland, of whom the majority were sent out concurrently with the French refugees; third, the Dutch orphan girls, from whom a considerable proportion of the settlers obtained their wives; and fourth, this last important and homogeneous element, the Huguenots. The annual census for the year ended December 31st, 1687, *i.e.*, that immediately preceding the arrival of the French, shows that the European population of the settlement, exclusive of the Company's establishment, then consisted of 573 persons, of whom 254 were men, 88 women, and 231 children. Assuming therefore that an equal number of Dutch were sent out at the same time as the French, these latter, numbering 150 or more, formed approximately one-sixth of the parent-stock.

This is confirmed by the census for 1691, when, according to Dr. Theal, there were in round numbers, 1,000 permanent settlers of all ages and both sexes. Of this total he estimates that two-thirds were Dutch, one-sixth French, a small fraction Swedish, Danish or Belgian, and one-seventh German, *i.e.*, Low German and practically identical with the Dutch in racial characteristics. The non-European population consisted of 50 free Asiatics and Central Africans, with their wives and 60 or 70 children ; and 386 slaves (being the property of the settlers) of whom 285 were men, 57 women, and 44 children. Further returns show that rapid progress was being made now in agriculture, stock-raising, and wine-growing.[1] In short, originally by accident but subsequently by well-directed efforts and of set purpose, the Dutch East India Company had accomplished its highest achievement—the planting of a self-sufficing European community, containing the germs of future racial and industrial growth, on the remote but temperate extremity of the great Continent of Africa.

The fact revealed by this analysis, that so large a proportion of the original Afrikander parents were men and women, who, by virtue of the circumstances of their emigration, were absolutely cut loose from the ties that bind ordinary colonists to their mother countries, leads "Olive Schreiner" to some significant reflections upon the Boer character as it developed in the nineteenth century.[2]

> The South African Boer differs from every other emigrant branch of a European people whom we can recall, either in classical or modern times, in this : That, having settled in a new land, and not having mixed with the aboriginal inhabitants nor accepted their language, he has yet severed every intellectual and emotional tie between himself and the parent lands from which he sprang. . . . He is as much severed from the lands of his ancestors and from Europe, as though 3,000, instead of 200, years had elapsed since he left it.

Of the orphan girls she writes :

> The ships that bore these women to South Africa carried them

---

[1] *History and Ethnography of South Africa before* 1795, by George McCall Theal, Litt.D., vol. ii (1909), pp. 370–1.

[2] Mrs. Cronwright Schreiner, in the *Fortnightly Review* for April, 1896.

> towards the first "Good Hope" that ever dawned on their lives; and the day in which they landed at Table Bay and first trod on African soil, was also the first in which they became women, desired and sought after, and not mere numbers in a printed list. In the arms of the rough soldiers and sailors who welcomed them, they found the first home they had known! . . . to such women it was almost inevitable that, from the moment they landed, South Africa should be "home," and Europe be blotted out. . . .

And persecution for religion produced the same obliteration of the home tie in the case of the Huguenot.

> As he entered Table Bay, and for the first time the superb front of Table Mountain broke upon him, he saw in it his first token from his covenant-keeping God—"The land that I shall give thee!" And the beautiful valleys of Stellenbosch, French-Hoek, and the Paarl, in which he settled, were to him no mere terrestrial territories on which to plant and sow: they were the direct gifts of his God; the answers to prayer. . . . The vines and fig-trees which he planted, and under which he sat, were not merely the result of his labour; they were the trees which aforetime he had seen in visions while he wandered a homeless stranger in Europe. . . . To this man, France was dead from the moment he set foot on South African soil, and South Africa became his.

The importance of gaining a full insight into this aspect of the character of the Dutch Afrikander, must be my excuse for a third quotation from the same very interesting article.

> He was produced—as are all suddenly developed, marked, and permanent varieties in the human or animal world—by the close interbreeding of a very small number of progenitors. The handful of soldiers and sailors who first landed, a few agriculturists and their families, the band of orphaned girls, and a small body of French exiles . . . constitute the whole parent stock of the Boer people. From this small stock, by a process of breeding in and in, they have developed, there having been practically no additions made to the breed for the last 200 years, the comparatively large numbers to which they have attained having entirely to be accounted for by the fact of their very early marriages and prolific rate of increase.

The circumstances which determine the origin of the Boers, or South African Dutch, being now before the reader, a few salient facts will suffice to tell him what is essential in the history of the century intervening between the period of the Van der Stells and the temporary occupation of the Cape Colony by the English in 1795.

The Huguenot emigrants, in accepting the offer of an asylum at the Cape, had taken an oath of allegiance to the Dutch East India Company, which placed themselves and their future lives unreservedly under its control. It ran :

> Je promets et jure. . . .
> Et que j'observeray et executeray fidellement, *et de point en point,* toutes les lois et ordonnances, faites ou a faire tant par Messieurs les Directeurs, par le Gouverneur-General et par les Conseillers, que par le Gouverneur ou Commandant du lieu de la residence, et de me gouverner et comporter en toutes choses comme un bon et fidelle sujet, Ainsi Dieu m' aide.[1]

In the face of this promise of unlimited obedience they were unable to complain when the harsh but salutary decision of the directors to amalgamate them absolutely with the Dutch majority was put into effect. On their arrival Van der Stell was ordered to mingle them with the Dutch settlers, and to put the French children to learn Dutch ; and all efforts to preserve their national identity and language were vigorously repressed as "French impertinences." In 1709 the use of the French language in official communications was declared illegal ; in 1724 the Bible was read in French in their churches for the last time ; and by the middle of the eighteenth century they had lost all knowledge of their mother tongue, and in this and other respects the process of racial amalgamation had been completed.

The only point in which the original Dutch community may have suffered is that of language. According to "Olive Schreiner" so vigorous and abrupt were the methods by which the French language was suppressed, that the new language was imperfectly learnt by the Huguenot refugees ; and this imperfect knowledge of the

[1] *The Record.*

Dutch of Holland was communicated by them to the whole body of settlers. The result was the evolution of a patois in the "Taal," the clipt baby tongue which has come to be the national language of the Boer. But whether this be its origin or not,[1] there is no question as to the nature of the "Taal," or the effect which it has exercised upon the Dutch Afrikander mind.

"The verb 'to be,' instead of being conjugated as in the Dutch of Holland and in analogy with all civilised European languages," says "Olive Schreiner," "runs thus: Ik is, Je is, Hij is, Ons is, Yulle is, Hulle is,—which would answer in English to—'I is,' 'thou is,' 'he is,' 'us is,' 'you is,' 'they is.' And not only so, but of the commonest pronouns many are corrupted out of all resemblance to their originals. Of nouns and other words of Dutch extraction, most are so clipped as to be scarcely recognisable. A few words are from Malay[2] and other native sources; but so sparse is the vocabulary and so broken are its forms, that it is impossible in the Taal to express a subtle emotion, or abstract conception, or a wide generalisation; and a man seeking to render a scientific, philosophic, or poetical work in the Taal, would find his task impossible."[3]

By thus substituting for the mother tongues of Holland and France a patois limited to a few hundred words, the European population of the Cape cut themselves off from intellectual fellowship with the civilised world. When,

[1] Two other origins, or contributing influences, are mentioned by Mrs. Cronwright Schreiner: (1) The original soldiers and sailors "being largely Frisian and wholly uneducated, never spoke Dutch at all, but a dialect." (2) It was a sort of "pigeon" Dutch, *i.e.*, it arose out of "intercourse between the Dutchman and his slaves, and the aboriginal races of the country" (*Fortnightly Review*: *Ibid*).

[2] The Malays were imported by the Dutch East India Co. to provide labour for the settlers. Their descendants form the Malay community of to-day in the Cape Peninsula.

[3] *Ibid.*

therefore, England became responsible at the beginning of the nineteenth century for the government of the Cape Colony, she found what was in effect a seventeenth century community; a people, that is, with the intellectual ideas and moral standards of a hundred years ago. And that section of the Dutch population which left the Cape Colony in 1835 to find new homes beyond the Orange River—the Boers of the nineteenth century—protracted the severance for almost another century. And so in that common life of the European peoples, created by the interchange of thought through literature, "the Boer," says "Olive Schreiner" in 1896, "has had, and could have, no part. Behind him, like a bar, 200 years ago the Taal rose, higher and higher, and landlocked him in his own tiny lagoon." [1]

While the Huguenots were being absorbed by the Dutch, and the Taal was being evolved, the settlers as a whole were gradually spreading inland. They passed the first of the barrier ranges, which separate the coast lands from the interior plateaux, in 1700 at Tulbagh kloof. From this point they worked their way down the valley of the Breede River to the south coast, and in 1745 there were enough homesteads in this district to warrant the establishment of a magistracy at Swellendam. In the same year the colony was declared to extend eastward as far as the Gamtoos river, that is to say, almost to the site of the present town of Port Elizabeth. Twenty-five years later (1778) the Governor, Van Plettenberg, in the course of a tour of the colony met the Kosa chiefs in conference at a farm on the site of Somerset East. The Europeans, spreading east, had at length come into contact with the stream of the military Bantu, which, having flowed southwards between the great eastern ranges and the Indian Ocean, had now set towards the

[1] *Ibid.*

Cape, following the westward trend of the coast. It was here agreed that the course of the Fish River should be the dividing line between the two races; but the next year the Kafirs, who in the manner of savages had assumed that the readiness of Van Plettenberg to treat with them was a sign of fear, crossed the river, murdered the Hottentots, and pillaged the isolated homesteads of the Europeans. They were, however, quickly driven back by the Dutch; the Fish River was then proclaimed formally to be the eastern limit of the colony, and a few years later (1786) a magistracy was established among these border settlers at Graaf Reinet.

In the meantime, that is to say, from the middle of the century onwards, the Dutch East India Company had been declining in wealth and efficiency. An attempt made in 1791 by the stadtholder, afterwards William I of Holland, to avert its downfall by the appointment of commissioners to investigate abuses and suggest reforms, proved unsuccessful; and before the second, and (as it proved) permanent occupation of the Cape by Great Britain in 1806, the company had been abolished, and the administration of its possessions had been placed in the hands of the colonial department of the Government of Holland. The commissioners, who reached Table Bay in 1793, were apparently too anxious to proceed to Batavia to give any great attention to the Cape Colony. "The most important of their proceedings," says a contemporary writer,[1] "consisted in their proclamation amid firing of cannon and tolling of bells, that they represented the Prince of Orange, and the rest any office clerk might have done." They, accordingly, deputed their powers of inquiry and reform to Commissary Sluysken, an invalided official of the Indian Government,

[1] Christian L. Neathling: As translated by Judge Watermeyer in his *Three Lectures on the Cape of Good Hope*, Capetown, 1857.

who was on his way home to Holland, and themselves sailed for the more valuable possessions of the company in the East.

The condition of the European population of the Cape Colony at the time when the rule of the Dutch East India Company terminated is, however, sufficiently revealed by official documents and the published accounts of travellers. Among the former the most illuminating is the "Memorial" presented by the Free Burghers to the Company in 1779. The long series of grievances contains a request that the Fiscal, the chief law-officer of the local government, may be restrained from arbitrarily committing persons to prison, and from compounding crimes by private fines. Although the eighteenth century was drawing to its close it would appear that no printing press had been set up in the colony, since the Burghers pray that "authentic copies of the particular placaats and ordinances" in force at the Cape may be supplied from Holland, or that a printing press and printer should be sent out. The eighteenth article of the Memorial runs:

> We further humbly solicit that your Honours will be graciously pleased to allow to the Cape Colonists that two ships may be laden annually for the account of the Cape Citizens with such wares as shall be purchased by their appointed agents—the burgher representatives binding themselves to send back the said ships, laden for their account with Cape produce; that the same shall be consigned to the Honourable Company, to be sold by public auction as payment of the imported goods; the undersigned desiring to know, in case of this prayer being granted, what would be the amount of duty which the Company would see fit to impose on this concession of limited export and import.
>
> The Cape Burghers further implore to be allowed to have some vessels to carry the produce of the colony, after the requirements of the Company have been supplied, to India, and to receive in return wood, rice, and other articles of commerce; and also they pray for a concession of a trade in slaves with Madagascar and Zanzibar, that foreigners may not enjoy the exclusive profit of this lucrative traffic.

The reply of the Fiscal, to whom the Memorial was

referred by the directors, is a most significant statement of the principles upon which the Company had administered the colony. To the request for political rights he replies that the *status* of the settlers makes such a claim absurb.

> It would be, indeed, a serious error if a comparison were attempted between the inhabitants of a colony situated as this is, and the privileged free citizens of our great towns in the United Provinces . . . it would be a mere waste of words to dwell on the remarkable distinction to be drawn between burghers whose ancestors nobly fought for and conquered their freedom from tyranny, and from whose fortitude in the cause of liberty the very power of our Republic has sprung ; and such as are named burghers here, who have been permitted as a matter of grace to have a residence in a land of which possession has been taken by the Sovereign Power, there to gain a livelihood as tillers of the earth, tailors, and shoemakers. Here comparison is impossible.

The modest request of the burghers for limited commercial opportunities is declared to be wholly inconsistent with a knowledge of the purposes and objects for which the colony was founded by the Company.

> The burghers, whose number is at present far too great, and whom on this account it will soon be very difficult to restrain and govern with a due regard to the preservation of the interests of the State and the Honourable Company, desire to be allowed a right of trading beyond the colony. . . .
>
> The object of paramount importance in legislation for colonies should be the welfare of the parent State, of which such colony is but a subordinate part, and to which it owes its existence.
>
> No great penetration is needed to see plainly the impossibility of granting such a petition. The dangerous consequences which would result to the State in general, and, in particular, to the Honourable Company, from the concession to a colony situated midway between Europe and the Indies, of free commerce, are manifest. It would soon be no longer a subordinate colony, but an independent State.[1]

The material advance achieved by a community thus administered on the narrowest lines of the old colonial system was naturally of the humblest description. In

[1] As translated by Watermeyer. *Ibid.*

the course of a century the number of the European population had grown from 1,000 to 20,000. The slave population, arising out of the Central Africans and Malays imported by the Company together with the offspring of Hottentots and half-castes, was considerably larger, and the "native," or Hottentot, population was somewhat smaller than the European. In Capetown itself and in the Cape Peninsula, the officials of the Company, the merchants, and the few successful settlers, living in the solidly-built and not ungraceful houses of which many are to be seen to-day, maintained a fair standard of comfort and civilisation. And here in the Cape Peninsula, and in the neighbourhood of Stellenbosch and the Paarl, a moderate standard of proficiency in agriculture and vine-growing had been attained. Beyond these limits the "burghers" lived for the most part in isolated homesteads, and supported themselves with difficulty by cattle-raising and hunting. Apart from the machinery existing for the conduct of the company's business, the colony was absolutely devoid both of the "plant" and the institutions common to other civilised communities of the period. The progress of the Atlantic colonies affords a standard of comparison. With a start of half a century the "plantations" of Virginia and New England had by this date—the end of the eighteenth century—grown into a nation of 4,000,000 people, and produced statesmen and soldiers not inferior to those of Western Europe. "In all things political, purely despotic; in all things commercial, purely monopolist," is the stern, but just, sentence pronounced by the late Judge Watermeyer upon the Company's administration.[1]

Deputations sent to Holland by the burghers to plead their cause before the directors in person, proved no more effective than the Memorial of 1779. When, therefore, the visit of the commissioners appointed by the Prince

[1] *Ibid.*

of Orange brought no prospect of a redress of their grievances, the inland burghers openly revolted. At Graaf-Reinet the Company's landdrost was expelled, and at Swellendam in 1795 a national assembly was convoked, with Hermanus Steyn as President, and a Free Republic was declared. It was at this crisis that the renewal of the long struggle between France and England for maritime and commercial supremacy brought a squadron of British warships to the Cape.

Just as the growth of its Indian possessions had led the Dutch East India Company to form a station at Table Bay, so now the strategic importance of this half-way house to India, as an element in the security of British India and the East India trade, made it necessary for Great Britain to prevent the Cape Colony from falling into the hands of her great rival. The circumstances were these. Republican France, after her declaration of war against Great Britain and her allies in 1793, had seized Holland, driven the Stadtholder, William of Orange, into exile in England, and established a new government, styled the Batavian Republic, in alliance with her own. This change would have placed the oversea possessions of Holland at the disposal of the French, had not Great Britain, holding the command of the seas, been able to occupy them one by one. The report that the French intended to seize the Cape had determined the British Government to send an expedition of eight ships and 4,000 men, under Admiral Elphinstone (afterwards Lord Keith), to occupy the colony in the name of the Stadtholder; and this squadron was now in False Bay, the southern, and undefended, harbour of the Cape Peninsula.

Commissary Sluysken, who had received orders from the directors of the Company to defend the colony, if he could, against both French and English, refused to recognise the Mandate of the Stadtholder, duly presented by the British Admiral, which commanded him "to admit into the Castle, as also elsewhere in the colony" the troops now sent by the King of Great Britain. But the temper of the burghers was such that it was impossible for him to offer an effective resistance to so strong a force. So far from accepting his offer of an amnesty and a free pardon, the Swellendam burghers coolly replied, that they were surprised "that the Honourable Commissioner did not respect the resolution of the National Convention, and still addressed official communications to the Landdrost, whom they had deposed": and demanded a specific recognition of their independence as a condition precedent to bearing arms in defence of the colony. Apart from the burghers, Sluysken had the company's garrison of 500 Germans and a few artillerymen. With these men he made a respectable show of resistance and then surrendered; and thus on September 16th, 1795, the English flag was broken for the first time over the Castle of Capetown.

This first and temporary occupation of the Cape by Great Britain lasted until 1803. The colony was then evacuated in accordance with the terms of the short-lived peace of Amiens, concluded between France and Great Britain in the preceding year. By this time, as already mentioned, the administration of the possessions of the Dutch East India Company had passed into the hands of the Dutch Government; and the officer appointed to take over the colony from the British authorities was a member of the newly-established Council of India, Jacob de Mist. Under British rule all vexatious restrictions upon internal and external commerce had been at once abolished; and de Mist was now instructed by the Government of Holland to carry out the reforms necessary to replace the Company's methods by a more enlightened system of administration. This work accomplished, he handed the reins of office to the new Governor, General Janssens, and sailed for Batavia.

The Treaty of Amiens barely lasted a year. On October 21st, 1805, Nelson, by destroying the combined fleets of France and Spain at Trafalgar, regained the mastery of the seas for England; and a few months later, on January 19th, 1806, the Cape was again surrendered to a British force. On this occasion, General Sir David Baird took possession of the colony on behalf of the King of Great Britain, and the title by conquest, thus obtained, was subsequently ratified by treaty and formal cession in 1814, as part of the international settlement which followed the defeat of Napoleon.[1] Since the year 1806, therefore, the English flag has remained flying over the Castle of Capetown.

[1] Of the Dutch possessions taken during the Napoleonic Wars, England retained Ceylon, a part of Dutch Guiana, and the Cape Colony; she restored Java (the most valuable of all the Dutch oversea possessions, and the Dutch "India" of to-day), and paid a sum of £6,000,000 in further compensation for the territory ceded.

CHAPTER VI

SOUTH AFRICA UNDER BRITISH RULE

THE population of the Cape Colony at the beginning of the period of British rule consisted of 26,720 persons of European descent, 17,657 Hottentots, and 29,256 slaves—in all 73,633; and its external trade was of the value of £160,000 per annum. A hundred years later the number of the European population of South Africa had risen to considerably over a million, and the value of its external trade had reached the large figure of £75,000,000 per annum—showing an amount of trade per head of European population greater than that of any other of the oversea dominions. In the meantime the area under European occupation or control had expanded northwards from the narrow limits of the Dutch Colony to the borders of the Congo Free State. These are facts which put the work of the British race in South Africa on a level with the characteristic achievements of the Victorian era in Canada, Australia, New Zealand, India and Egypt.

The merit of this record is enhanced by the circumstance that it was won in the face of obstacles so various and so stubborn, that after a century of sacrifice and effort, involving an altogether unprecedented expenditure of blood and treasure, nothing less than the highest military and civil talent of the mother country, aided by the manhood of the oversea British, has availed to keep South Africa a part of the Empire. The story of this long-protracted conflict with men and circumstances is more full of tragic episodes, and happy surprises, and withall of human interest than that of any other dominion. To tell it here with any degree of completeness would be impossible. The most that can be attempted is to indicate the cardinal events which mark the course of the main stream of South African history during the last 100 years. In order, however, that these events may be seen in true perspective, it is necessary to know something of the objects of British policy, which, often deflected by unforeseen obstacles and changing circumstance, or swayed by the clamorous demands of an Empire in five Continents, will be found, nevertheless, through all its shifts and changes to have preserved an essential continuity of aim and purpose.

The economic progress of the Colony of the Cape of Good Hope was assured from the moment that it became a part of the British commercial system, and as such the home of a British population. Its political and social advance was, however, menaced by the two conflicts—the Europeans with the natives, and the Dutch with the British—which circumstances made inevitable. The primary object of British policy was, therefore, to allay these conflicts. It recognised that the supremacy of the Europeans over the native races must be established; but it aimed at carrying out this necessary task by methods which would make its accomplishment a benefit to both parties alike. In other words, the natives were not to be destroyed or expatriated, but preserved and civilised, and ultimately fitted for a partnership in industry with the Europeans. Similarly in respect of the second conflict, while it recognised that British ideas must prevail in determining the relations of the Dutch with the British, or of both with the natives, it sought to give to the original Dutch, or Franco-Dutch, population a complete liberty to live their lives in all other respects undisturbed by any interference of the new Government. Thus the adoption of British standards of thought and action by the Dutch was to be a voluntary process, and one that would come about as a natural consequence of their intercourse with the British, when the latter should begin to emigrate to the country in appreciable numbers.

A policy framed upon these wise and humane lines—lines which were laid down at the very beginning of the British administration, and more than once subsequently enunciated by the Home Government—would have attained its purposes without a tithe of the resistance it actually encountered, if it could have been applied singly to either of these two conflicts. As it was, the nationality difficulty and the native question had to be treated in conjunction. The application of the policy of non-interference to the Dutch, admirable for the treatment of the nationality difficulty in itself, made the problem of native administration infinitely more complicated; while the handling of the native question on the generous lines laid down, by compelling the British Government to interfere with the "internal economy" of the Dutch in its most vital aspect, created a perpetual and irritating source of contention between it and them—the very evil which the non-interference policy was designed to avoid.

A further element of exceptional difficulty in the South African situation must be noted. The mere presence of the natives, in addition to furnishing a chronic source of contention between the British and the Dutch, by providing a cheap supply of manual labour blocked the way against British emigration on a large scale; since the labouring class emigrant, who was the chief means of peopling Australia and Canada, could find no market for his manual labour in South Africa. And so, by a strange irony of fate, justice to the natives deprived the British administration of the readiest and most satisfactory agency for the solution of the nationality difficulty—a rapid introduction of a British population large enough to have converted the Dutch into an appreciable, but politically impotent, minority. Nor was this all.

Apart from the check upon working class emigration thus imposed by the natives, the physical and economic conditions of the country were such that it was difficult for agricultural emigrants with small means—the class who took up holdings so successfully in Canada, New Zealand, and Australia—to make a livelihood out of the land in South Africa.

The political significance of the smallness of the British population in South Africa is so great, that it will be worth while to pause for a moment to examine these adverse conditions. In the first place, both the Bantu and the Dutch were beforehand in taking up the most fertile and easily cultivated land; and although an ample area of cultivable land remained available for settlement west of the Drakenberg, it was, generally speaking, impossible to farm this land at a profit without a large initial expenditure upon irrigation, buildings and stock. In the second place, the backward condition of agriculture, which caused South Africa to import much of its food supplies, joined with the total absence or complete inadequacy of its roads and railways prior to 1870,[1] made the cost of living unusually high to all Europeans, except the self-sufficing Dutch farmers. In short, all through the nineteenth century, South Africa kept her doors closed against the British unskilled workman and the small farmer. And even when, under the stimulus of the mineral discoveries, the industrial expansion set in, and the British began to emigrate to South Africa in appreciable numbers, these industrial arrivals did not contribute much to the solution of the nationality difficulty. Being professional men, clerks, mechanics and miners, they naturally went to the mining centres or to one or other of the few considerable towns, where the population was almost exclusively British,

[1] *i.e.*, the discovery of diamonds, which provided the Cape Colony with a revenue sufficient to build up a railway system.

and not to the country districts, where the great majority of the Dutch were to be found. And so the circumstances that kept the Dutch and British apart—the former on the land and the latter in the towns—combined with the smallness of the British population as a whole to prevent the fusion of the colonists of the two nations, until the upheaval of the great war brought them together.

The existence of these adverse economic agencies explains how it was that British policy, in spite of its wise aims, should have failed to achieve its purposes without having recourse to the arbitrament of the sword. What is surprising is, not that conflicts in arms with both the natives and the Boers should have been so frequent, but that notwithstanding these conflicts so much real progress should have been accomplished. This ultimate success is due to the fact that in applying the policy of non-intervention to the Dutch, and in establishing European control over the native races, British statesmen never lost sight of the two duties which belonged to great Britain as Paramount Power : to see that the natives received full justice at the hands of the colonists, both Dutch and British ; and to provide for the safety and good government of all the European communities in South Africa, since they rightly regarded the Boer Republics and the British colonies as interdependent members of a single political and economic system. And the essential justice and consistency of British policy nowhere appear more plainly than in the circumstance that after the great war—or rather while it was still in progress—it was able to take up the work of reconstructing the new South Africa out of the wreckage of the old, without in any respect breaking with its past traditions.[1]

With this account of the aims and difficulties of British

[1] All that was necessary was to revoke the grant of the powers enjoyed by the Boers under the Conventions.

8—(2139)

Administration to guide him, it is hoped that the reader will find a concise statement sufficient to put him in possession of the main trend of events in the period 1806–1909.

With the exception of the British officials and soldiers and a few merchants and missionaries, the Dutch inhabitants remained the sole European population at the Cape for six years after the formal cession of the colony to Great Britain in 1814. In the year 1819–20, however, some 5,000 British emigrants were sent out to Algoa Bay by the Government, and established in the country between the Bushman and Fish Rivers. These Albany settlers, as they were called, founded, or developed, Port Elizabeth and Grahamstown, and to them and their descendants is due the present predominance of the British nationality in the eastern districts of the Cape Province.

In the fourteen years prior to the Albany Settlement the industries of the colony, and in particular the wool industry, had been developed; legislation had been introduced, which, while giving personal freedom and rights of property to the Hottentots, provided for the better security of the Europeans by placing these yellow-skinned natives under the operation of special regulations, similar in character to those now in force in the Transvaal and Natal for the control of the dark-skinned Bantu; and the instruction of the native African races in the Christian faith had been commenced on energetic and permanent lines by the agents of the London Missionary Society. The change in the status of the Hottentots produced some discontent among the Europeans, and the deep-seated antipathy between the Boers, or Dutch farmers, and the Missionaries had already made itself apparent. The two together provoked the extraordinary protest of Bezuidenhout and its sequel, the "Rebellion" of Slaghter's Nek (1815–16), characterised by Cloete as "the most insane attempt ever made by a set of men to wage war against their sovereign," originating entirely in the unruly passions of a few persons, who "could not suffer themselves to be brought under the authority of the law."[1] On two occasions it had been necessary to clear the Colony of Kafir invaders by force of arms (1811–12 and 1817–18), and as early as 1812 a post called Grahamstown, with a garrison of British regulars, was established to protect the eastern border.

In 1826 the Commissioners of Inquiry, who had visited the Cape in 1823 on their return from a similar mission to New South Wales, presented their report, and certain administrative changes were introduced in pursuance of its recommendations. The judicial system was enlarged and improved; the Landdrosts and Heemraaden were replaced by Resident Magistrates and Civil Commissioners; English was ordered to be used in official communications; and in 1828 the special regulations hitherto controlling the native (Hottentot) population were rescinded, and all free coloured inhabitants of the colony were brought under the operation of the laws by which the Europeans were governed. In the meantime the numbers of the British colonists, raised by the Albany Settlement to one-eighth of the total European population, were still further augmented. While this process of Anglicising the colony was in progress, the institution of slavery was abolished within the British dominions by the Abolition Act of 1833. In the case of the Cape Colony the period during which the slaves were to remain as "apprentices" with their former masters ran from December 1st, 1833, to December 1st, 1838; and out of the £20,000,000 voted by Parliament as compensation to slave-owners, the sum of £1,247,000 was apportioned to this colony, as against the £3,000,000 of the official valuation. Thus, apart from the dislocation of industry—

[1] *Five Lectures on the Emigration of the Dutch Farmers*, Capetown, 1856. In the interests of public order, it was necessary to close the colony for a time against any further missionaries.

almost exclusively agricultural and worked by slave labour —the small and backward community [1] of 50,000 to 60,000 Europeans were involved in a direct loss of nearly £2,000,000.

On Christmas Day, 1834,—little more than a year after the Abolition Act received the royal assent—the eastern border of the colony was invaded by the Kafirs to the number of 12,000, or 15,000; the isolated farm-houses of the settlers, Dutch and British, were pillaged and burnt, their cattle driven off, and the settlers themselves in many cases ruthlessly murdered. Prompt military assistance was, of course, given by the Government; but the expulsion and punishment of the Kafirs was followed by the most unfortunate "divergence of opinion" between the Governor of the colony and the Secretary of State, to which reference has been made in a previous chapter. [2] In the same year (1835) the Boers in the eastern districts of the colony resolved to divest themselves of their allegiance to the British Government, and to seek new homes for themselves and their belongings beyond the Orange River. [3] Although the emancipation of the slaves was felt to be injurious to their material interests, the justice and necessity of the measure were generally admitted by the more advanced of the Dutch in the western districts of the colony, upon whom the monetary loss chiefly fell; and the "Great Trek," as the secession of these Dutch farmers is called, is to be attributed more to the reversal of Sir Benjamin Durban's frontier policy, together with the

[1] It must be remembered that at this period slaves were as much a form of property as houses or land; and the "general investor"—especially persons of small means who required a good return for their capital—was affected as well as the actual employers of slave labour, or the actual owners of slaves.

[2] Chap. II, p. 36.

[3] So named in honour of the Stadtholder, William of Orange, by an English explorer, Captain Gordon, on August 17th, 1779.

Hottentot legislation by which it was preceded, than to the Abolition Act.

The motives of the emigrants were stated fully in a document signed by their leader, Piet Retief, of which an exact translation was published in *The Grahamstown Journal* of February 2nd, 1837; and a letter dated from Sand River on July 21st, 1837, was addressed by him to the Governor of the Cape Colony, containing an offer of a continuance of friendly relations with the British on the understanding that the independence of himself and his fellow-emigrants was acknowledged. These documents are too long to be given here, but a more concise and probably not less genuine account of the origin of the movement is to be found in "the quaint and artless record" of Mrs. Anna Elizabeth Steenekamp, a niece of Retief, which she published in *The Cape Monthly Magazine* for September, 1876.

> The reasons for which we abandoned our lands and homesteads, our country and kindred, were the following:
>
> 1. The continual depredations and robberies of the Kafirs, and their arrogance and overbearing conduct; and the fact that, in spite of the fine promises made to us by our Government, we, nevertheless, received no compensation for the property of which we were despoiled.
>
> 2. The shameful and unjust proceedings with reference to the freedom of our slaves; and yet it is not so much their freedom that drove us to such lengths, as their being placed on an equal footing with Christians, contrary to the laws of God and the natural distinction of race and religion, so that it was intolerable for any decent Christian to bow down beneath such a yoke; wherefore we rather withdrew in order thus to preserve our doctrines in purity.[1]

The Governor of the Cape, acting under instructions from the Secretary of State, refused more than once to entertain the question of independence, and until they obtained a recognised status under the Conventions

[1] As quoted by the late Sir John Robinson in his *A Lifetime in South Africa* (London, 1900), p. 46. Sir John agrees with Cloete in ascribing the secession mainly to the inherent difference between the Dutch and British attitudes towards the natives.

(1852 and 1854), the Boers remained simply "emigrant farmers, being subjects of Her Majesty, who had made unlawful incursions into the territories of the natives." On the other hand, no attempt was made to restrain the exodus, and altogether some 10,000 men, women, and children, left the colony between the years 1835 to 1838. In the country immediately north of the Orange River the *Voertrekkers* encountered little or no resistance, but beyond the Vaal and east of the Drakenberg they barely saved themselves from destruction at the hands of the military Bantu. Subsequently, when their fighting men had grown in numbers and experience, they assumed the offensive, subdued Dingaan, the treacherous and savage "King" of the Zulus, and drove the Matabele chief, Moselekatze, the "old lion of the North," across the Limpopo, there to subjugate the peaceful Mashonas and Makalaka, and establish himself and his warriors as lords of the country since called Matabeleland. Among all their defeats and victories one day stands conspicuous. On December 16th, 1838, the voertrekkers, led by Pretorius, avenged the murder of Retief by routing 10,000 Zulus and utterly destroying the power of Dingaan; and the anniversary of this victory, called "Dingaan's Day," is the greatest event in the Boer calendar. But the Epic of the Great Trek—for the courage and endurance of the voertrekkers and their women, the wildness of the lands through which they journeyed, and the overwhelming numbers, treachery, and ferocity of their Bantu enemies, raise their doings to the height of Epic—is not to be compressed into a paragraph.[1]

The secession of the Boers, by disintegrating great masses of the Bantu population before Great Britain was

[1] The story has been related with brevity, but some precision, in the author's *History of South Africa* (Temple Encyclopædic Primers). Among the earliest settlements of the Boers were Winburg, "the place of victory," so named in honour of their first successful contest with Moselekatze, founded in the Free

ready to carry out the work of bringing them under European control, largely increased the difficulties of the administration of South Africa. In particular, the primary duty of protecting the settlers in the eastern and north-eastern borders of the Cape Colony from Kafir inroads was rendered more onerous and more costly. And it was a natural impatience on the part of the Home taxpayer at the constantly recurring Kafir wars, which, more than any other consideration, led the British Government to grant internal independence to the Boers under the terms of the two conventions, concluded respectively at Sand River in 1852, and Bloemfontein in 1854.

As the circumstances in which this change of policy was put into effect have been related before,[1] it will be sufficient to notice here that the character of these two documents, which *mutatis mutandis* are identical, shows that by this action Great Britain did not intend to divest herself of any of her essential rights as the Paramount Power in South Africa. Thus, in the Sand River Convention the Assistant-Commissioners " guaranteed in the fullest manner on the part of the British Government to the Emigrant Farmers beyond the Vaal River the right

State in 1837; Pietermaritzburg, the capital of Natal, named from the two leaders, Piet Retief and Gert Maritz, and laid out in 1839; and Lydenberg and Potchefstroom in the Transvaal. Of these, Winburg was subsequently superseded by the present capital of the Free State, Bloemfontein, which lies some 50 miles to the south. Potchefstroom, the old capital of the Boers beyond the Vaal, was similarly displaced by Pretoria, the present capital of the Transvaal, and the administrative capital of the Union. This latter town, however, was not founded until some years after the Sand River Convention. It took its name from Andries Pretorius, the voertrekker, and Commandant-General of the Transvaal Boers; and, in 1860, when the three so-called republics beyond the Vaal united under one government, it was adopted as the capital of the new state—the South African Republic. Maritzburg alone has retained its dignity, and is the capital of the Natal Province to-day.

[1] In Chap. II, p. 37.

to manage their own affairs, and to govern themselves according to their own laws, ... and that no encroachment should be made by the said Government on the territory beyond, to the north of the Vaal River . . . " In Article II, provision is made for the delimitation of boundaries in the event of disputes; in Article III alliances with the coloured natives north of the Vaal River are disclaimed on behalf of the British Government; in Article IV it is agreed that no slavery is or shall be permitted, or practised by the Emigrant Farmers; and by Article VI the Farmers are permitted to obtain arms and ammunition in the British Colonies, subject to the mutual understanding that all trade in ammunition with the natives is prohibited both by the British Government and the Emigrant Farmers on both sides of the Vaal River.

The Convention with the Boers between the Orange and Vaal Rivers was not made until two years later. Here the settlements of the emigrants had been administered directly by Great Britain since 1847, under the style of the Orange River Sovereignty; and the decision to abandon this dependency, to which the Convention of Bloemfontein gave effect, was by no means welcome to the majority of its inhabitants.

In this way, then, the settlements founded by the Emigrant Farmers west of the Drakenberg became respectively the South African Republic, or Transvaal, and the Orange Free State. Natal, on the other hand, was retained. The reasons for the distinction were these. In the first place, the earliest settlers in Natal were not the Emigrant Farmers, but some Englishmen who had established themselves lawfully at Port Natal (Durban) in 1824, under a concession obtained from the Zulu king, Tshaka. In the next, since the date at which it had been proclaimed a British colony (1843), a large proportion of the Boers had returned across the Drakenberg in disgust at the ample assignment of separate lands to the natives, and in 1848–50 some 4,000 British immigrants had been introduced. The effect of this settlement—known as the Byrne settlement from the name of its chief promoter—in conjunction with the withdrawal of the Boers was to make the population of Natal predominantly British. An even more important consideration was the circumstance that Natal was not an inland, but a maritime colony, possession of which would have enabled the Boers to have entered into an effective alliance with other European powers, and thus created a relationship entirely inconsistent with Great Britain's position as Paramount Power in South Africa.[1] There was never any question, therefore, of abandoning Natal; and in 1850, when the European community had grown to 8,500, a representative constitution was conferred upon them. The sub-tropical conditions of the colony and the early introduction of British Indian immigrants, prevented any rapid increase of the European population, and full self-government was not established until 1893. At the same time the colony always retained its British character, and it forms to-day the one province of the Union in which the British element predominates. Between the Byrne settlement and the era of industrial expansion that followed the discovery of diamonds and gold, only one other considerable accession to the European population of South Africa was made by organised immigration. In 1857 some 4,000 men who had fought in the Anglo-German Legion in the Crimean War were established on farms by Sir George Grey in British Kaffraria; and in the next year they were reinforced by the introduction of 2,000 agricultural immigrants from North Germany. These German settlers,

[1] In 1842, upon the arrival of a Dutch vessel, the *Brazilia*, the Volksraad of the Emigrant Farmers of Natal concluded a formal treaty with a Mr. Smellekamp, purporting to be an acceptance on their part of the protection of Holland. It was in view of this proceeding that the military occupation of Durban was at once resumed.

together with a certain number of farmers from the Cape Colony to whom holdings had been granted on condition of their serving in arms against the Kafirs, when called upon by the Government, colonised British Kaffraria, and laid the foundations of the subsequent development of King William's Town, and its port, East London.[1]

The reader has been made acquainted now with the chief sources from which South Africa has drawn its European population, and with the circumstances in which the four European states, to-day constituting the Union of South Africa, came into being. Rhodesia, born thirty-five years later and not as yet a member of the Union, alone remains to be accounted for.

The position created in 1854 by the application of the non-interference policy to the emigrant Boers, and the further measures by which it was accompanied, was this. The British Government remained directly responsible for the Cape and Natal, and for the small and partially civilised portion of the densely populated native areas lying between the two colonies, styled British Kaffraria. In respect of the Cape Colony, however, the administrative responsibilities of the Home Government were somewhat reduced by the establishment, in 1853, of representative institutions. Over the native territories in general, and over the Boer Republics, Great Britain retained only the rights of the Paramount Power; that is to say, the right to intervene, as, and when, the interests of South Africa as a whole should seem to make such intervention necessary.

It was believed that by thus limiting her responsibilities Great Britain would the better secure the undisturbed development of the two British Colonies upon British

[1] The opportunity for the establishment of a European population in British Kaffraria was due to the remarkable panic caused by the prophecies of Nongase (1856), which reduced the native population, by flight and starvation, from 105,000 to 38,000. British Kaffraria was incorporated into the Cape Colony in 1865.

lines; and that these colonies, when thus developed, would by the mere process of economic laws draw the inhabitants of the Boer Republics back again into the British system. In any case, the decade which saw the Crimean war and the Indian mutiny was not a time in which British statesmen could afford to risk the unnecessary employment of any of the slender military resources at their disposal for the defence of the Empire as a whole. Possibly, too, this forecast would have been realised, had it not been for the suddenness with which the then undreamt of mineral wealth of South Africa was revealed and exploited.

The events which actually happened are so recent and notorious that the mere names and dates will serve to bring them before the mind of the reader. Up to the discovery of diamonds at Kimberley in 1870 South Africa, in spite of recurring conflicts between the Europeans and the natives, continued to make steady industrial progress, but progress so slow in comparison with the rapid advance achieved in Australia, New Zealand, and Canada, that it was fitly typified by the ox-wagon, the characteristic vehicle of the country. The vine, for the growth of which the soil and climate of the Cape was especially suitable, was cultivated in the Cape Peninsula and the fertile country around Stellenbosch and the Paarl, and a fair proportion of the wine produced found a market in England. The wool industry, which had originated in the importation of merinos from New South Wales in the early years of British administration, had been developed with energy and success by the British settlers in the eastern districts of the colony. The fine-haired Angora goat had been successfully introduced from Asia Minor in 1856, and the production of mohair for the Yorkshire mills formed a lucrative industry. Ostrich farming had become an even more profitable pursuit, when, in 1869, Mr. Arthur Douglas, of Albany, had perfected his artificial incubator; and Port Elizabeth, as the market and place of export for feathers and wool, had grown to be a town of some commercial importance. In Natal the sugar industry had been founded by the new British settlers in 1850, and experiments had been made in the raising of other sub-tropical produce. The communications of the Cape Colony had been improved also. An excellent system of roads had been constructed in 1844, and in the early sixties one or two short railway lines were made.

With the establishment of the diamond mines at Kimberley the industrial development of South Africa quickened its pace; and when, fifteen years later, the greatest and most permanent goldfield, as yet known to the world, was found on the rock-strewn veld of the Witwatersrand, it finally abandoned the ox-wagon tradition, and began to move forward with the impetuosity of a locomotive. The political situation developed with equal rapidity. The proclamation of British authority over the diamond fields in 1871, was followed by territorial disputes with both the Boer Republics; and the policy of non-intervention being no longer possible, the effort was made to re-unite them to the British Colonies in a federal system with which the names of Lord Carnarvon, J. A. Froude, Sir Theophilus Shepstone, and Sir Bartle Frere, are identified (1874–80). In the endeavour to give effect to Lord Carnarvon's South Africa Act (1877), the Transvaal was annexed in 1877, and two years later the Zulu tribe which under Ketchwayo had become the most formidable military power in South Africa, was subjugated and brought under European control.

The failure of this effort to achieve federation (1880) was followed by the revolt of the Transvaal Boers, and the remarkable success of the burgher arms, joined with the risk of a general rising of the Dutch population throughout South Africa, caused the British Government to withdraw its administration from the country under the terms of the Convention of Pretoria (1881). By this instrument the suzerain rights of Great Britain over the Boer Republic were specifically maintained; but three years later it was replaced by a new Convention—the Convention of London—in which, in the words of the (then) Colonial Secretary, Lord Derby, the substance of the suzerainty was retained, but the term itself in deference to Boer susceptibilities no longer appeared. Almost before the ink of the London Convention was dry, the South African Republic (as the Transvaal for the first time was now officially designated) attempted to extend its authority over the peaceful Bechuanas beyond its western border, over whom a British protectorate had been declared. This encroachment, which would have shut off the Cape Colony from the trade route to the interior of Africa, was prevented by the despatch of an expedition under Sir Charles Warren (1884–5), and the conversion of southern Bechuanaland into a Crown colony. In view of the occupation by Germany of territory on the west coast in 1884, and the imminence of further raids into native territory on the part of the Boers, the Bechuanaland Protectorate was proclaimed (March 23rd, 1885) to extend westward to East Longitude 20° and northward to South Latitude 22°; and by this action it was made impossible for the South African Republic to extend its northern or western borders without directly violating British territorial rights.

The British Government hoped that these, and other extensions of British authority made at this period, would remove any opportunity for further conflict between the Paramount Power and the northern Republic. In the very next year, however, the proclamation of the Rand as a public goldfield (September, 1886) introduced a new and disturbing factor into the field of South African politics. With the establishment of the gold industry and the sudden growth of an industrial population, mainly British and almost as numerous as the Boers, the annual revenue of the South African Republic rose from £177,876 in 1885 to nearly £5,000,000 in 1897. The authority of the President, Paul Kruger, had become meanwhile the dominating influence in the Republic, and the aim of this remarkable man was to gain the political control of all South Africa for the colonists of Dutch descent. With this end in view, the surplus millions provided by the gold industry of the Rand were employed in the purchase of arms and munitions of war sufficient to equip not only his own burghers and those of the sister Republic, but his Afrikander adherents within the British Colonies; while money was spent freely in the subvention of friendly journals, and the furtherance of Boer diplomacy, on the Continent of Europe. To add a touch of irony to the situation, the British population, to whom this sudden wealth of the Transvaal was almost entirely due, and by whom nine-tenths of the revenue was furnished, was excluded from the rights of citizenship —rights which the British Government had intended to secure, and, as it claimed, had specifically secured in the London Convention, for them and for all future British subjects who might reside in the Republic.

While fate was placing these weapons in the hands of President Kruger a powerful defender of the interests of the British in South Africa had been revealed in the person of Cecil John Rhodes. The name of Rhodes has been enrolled among the makers of the Empire, and his words and deeds have passed into the common stock of British thought and knowledge. Only two outstanding actions of his life need be recalled here. In 1888-9 he founded the British South Africa Company under Royal Charter for the purpose of developing the vast interior stretching northwards from the Crown Colony of

Bechuanaland to Central Africa ; and in the year following the Company's pioneers effected the peaceable occupation of Mashonaland, and laid the foundation of the system of colonies now called after him, Rhodesia. In the autumn of 1905, Rhodes, being then Prime Minister of the Cape Colony, sanctioned and controlled the equipment of a force of the Company's troopers, which, together with the Reform movement on the Rand, brought about the Jameson Raid (December 29th, 1895—January 2nd, 1896). Both Rhodes himself and Doctor (now Sir Starr) Jameson have freely acknowledged that a grave political wrong was then committed, and both have since performed services for South Africa and the Empire that expiate their common fault. The military preparations of President Kruger, which had begun before the Raid took place, were now urged on with redoubled energy and determination ; and when, in 1899, Mr. Chamberlain determined to insist upon the grant of political rights to the British residents in the Transvaal, the Dutch of the two Republics and a large proportion of their kinsmen in the British Colonies were so confident of their military strength, that, rather than acknowledge the rights of Great Britain as Paramount Power, they elected to submit their cause to the arbitrament of war.

The great conflict which began with the expiry of the forty-eight hours allowed by the Transvaal ultimatum on October 11th, 1899, was terminated by the Surrender Agreement of May 31st, 1902. Under the terms of this instrument the burghers then in the field laid down their arms, and acknowledged King Edward VII to be their lawful sovereign. On the other hand, the British Government undertook to repatriate the Dutch inhabitants of the late Republics, almost all of whom were either prisoners of war or inmates of the Refugee camps, and to establish representative institutions, leading up to full local autonomy in the new colonies, so soon as circumstances should permit. The terms of Peace were honourable alike to both parties; and the long war, terrible as was the loss of life and property which it entailed, left the people of the two races with a vastly better knowledge of each other.

With the departure of the army of the Empire, led successively by Sir Redvers Buller, Lord Roberts and Lord Kitchener, the reins of South African administration fell into the strong hands of Lord Milner (1897–1905). As Governor of the Cape Colony and High Commissioner for South Africa he had earned the confidence both of the Home Government and of the British and loyal Dutch in South Africa; and while the war was in its later stages, as Governor of the New Colonies he had thought out the processes, and in part created the actual administrative machinery, of the Reconstruction. The repatriation was accomplished with such smoothness and rapidity, that within seven months of the declaration of peace Mr. Chamberlain was able to visit the new and old Colonies of South Africa, and to discuss the problems of the situation with Lord Milner on the spot (December 28th, 1902—February 25th, 1903). Just two years later (March 31st, 1905) Lord Milner resigned his offices. In less than three years of peace he had reconstructed the entire political and economic fabric of the new colonies upon a wider and more enlightened basis. In so doing he had vastly increased the material resources of their inhabitants, created a civil service at once pure and efficient, doubled the railways, built schools and public buildings, and brought the joint finances of the two colonies to a point which secured the early provision of the funds necessary to complete their equipment as civilised and progressive States. At the same time, as High Commissioner, he had skilfully promoted every form of inter-state action among the separate colonies, and thus prepared the way for the establishment of an administration common to them all.

PART II

THE GOVERNMENT

CHAPTER I

THE UNION CONSTITUTION

WHEN Lord Milner left South Africa, the Balfour Ministry had determined to establish the half-way house of Representative Government in the Transvaal, but to make no change in the existing Crown Colony administration of the Orange River Colony. In the electoral system to be created by the Lyttelton[1] Constitution two principles of capital importance had been embodied, with a view of securing absolute political equality for all European citizens. The parliamentary constituencies were to be as nearly as practicable equal not merely in point of population, but in the actual number of electors; and to prevent these originally equal constituencies from becoming unequal through the irregular movement of population, there was to be an automatic redistribution of seats, again on a basis of electors, at intervals of every few years. At the same time the qualifications for the franchise were fixed so low, that every European adult who was earning a livelihood was able to obtain a vote.

Before, however, the necessary arrangements for the first elections under this constitution had been completed, the Unionist Government went out of office, and a Liberal Government succeeded them (December 12th, 1905).

[1] Mr. Alfred Lyttelton had succeeded Mr. Chamberlain in 1903 as Secretary for the Colonies.

In respect of South Africa, the policy to which Sir Henry Campbell-Bannerman and other Liberal leaders were pledged, differed materially from that of their Unionist predecessors. They disapproved of the employment of the indentured Chinese labourers introduced in 1904 for the Rand Gold industry; and they had determined—in part because they did not want to deprive the Transvaal of the new labour supply on their own responsibility—to establish "responsible" government in both the new colonies without the intermediate stage of "representative" government. As the general election of January, 1906, gave the Liberal party a very large majority in the House of Commons, steps were taken at once by the new Government to put into effect their decisions on both points. The Lyttelton Constitution was annulled, and on December 6th, 1906, Letters Patent were issued establishing responsible government in the Transvaal, while a similar constitution was granted to the Orange River Colony in the following year.

The electoral system created by the Elgin[1] Constitutions maintained the voters' basis of the equality of constituencies and automatic redistribution; but in the Transvaal more seats were assigned to districts where the majority of the inhabitants were Dutch, and correspondingly fewer to districts where the British predominated, than would have been the case under the Lyttelton Constitution. The Transvaal elections, held in February, 1907, placed a Dutch Government with General Botha as Prime Minister in power, and in November of the same year Mr. Abraham Fischer became Prime Minister of the Orange River Colony with, of course, a relatively much larger majority of Dutch members to support him.

Two other political changes must be recalled before the actual birth of the Union can be related. Lord Milner's

[1] Lord Elgin was the new (Liberal) Secretary for the Colonies.

successor, as Governor of the New Colonies and High Commissioner, was Lord Selborne, who arrived in South Africa in May, 1905; and in February, 1908, the Progressives, who had been in office in the Cape Colony since the beginning of 1904, were defeated at the polls by the South African party, with the result that Sir Starr (then Dr.) Jameson was succeeded in the premiership of this colony by Mr. John X. Merriman.

In order to trace the agencies which brought about the rapid creation of the Union (1908–9), we must go back for a moment to the work done by Lord Milner in South Africa during the three pregnant years that followed the peace of Vereeniging (1902–5). Before he left the Transvaal, the reconstruction of the new colonies had reached a point which placed these states not merely on a level with the old colonies, but actually in advance of them, in respect of administrative efficiency and industrial development. The heads of departments and other officials whom he had enlisted contained the pick of the Civil Service of the Cape and Natal, together with a number of young and brilliant men gathered from England and the Empire at large. Among the latter were a group of university men, scornfully designated by his opponents the "Oxford Kindergarten"; and it was a few members of this group that two years after Lord Milner's departure provided the enthusiasm, constructive ability, and technical knowledge, which first started, and then carried to a successful termination, the movement for the closer union of the four self-governing colonies of South Africa. Moreover, during the (nearly) two years which intervened between Lord Milner's resignation and the establishment of Responsible Government, the agencies which he had introduced had time to mature; and the new colonies, served by the Milner officials under the sympathetic guidance of Lord Selborne, advanced still further in administrative efficiency and in the general development of their resources.[1] And in South Africa at large, with the Progressive Ministry still in office in the Cape Colony, the traditions of the Milner High Commissionership were maintained. Chief among these was the campaign against "localism" in all its forms, and the endeavour to promote concerted action among the various South African Governments upon all matters of common concern. In particular the effort to terminate the injurious competition between the various Colonial Governments for the trade of the Rand, on the basis of a partial or complete amalgamation of the competing railway systems, was continued; and continued on the lines laid down by Lord Milner in the Railway Rates Conference of February, 1905, over which he had presided.

In March, 1906, the Customs Union Convention to which the five colonies had agreed in March, 1903, was renewed for two years (*i.e.*, as from July 1st, 1906, to June 30th, 1908); but the discussions by which this action was preceded, showed that the task of reconciling the financial and industrial interests of the separate South African Governments was becoming increasingly difficult. Later on in the same year Mr. Lionel Curtis, the Assistant Colonial Secretary for Urban Affairs in the Transvaal Crown Colony administration, with the assistance of his colleagues Mr. W. L. Hichens (Treasurer), Mr. Patrick Duncan (Colonial Secretary), Mr. R. H. Brand (Secretary to the Inter-Colonial Council), and of Mr. Feetham, formerly clerk to the Municipality of

[1] So admirable was Lord Milner's Civil Service, that, on the establishment of Responsible Government, although the actual heads of departments naturally resigned to give place to ministers "responsible to Parliament," the Civil servants, as a whole, were retained by the Dutch Governments in both colonies. Not only so, but, on the establishment of the Union, Lord Milner's officials were again appointed to the most important non-political, administrative positions, *e.g.*, Sir T. R. Price became head of the combined railways, and Mr. F. B. Smith head of the Union Department of Agriculture.

Johannesburg, drew up a statement showing the urgent necessity for administrative union. Mr. Curtis, who was on the eve of resigning his posit on in view of the approaching establishment of Responsible Government in the Transvaal, was warmly encouraged by Lord Selborne; and Sir Starr Jameson, as Prime Minister of the Cape Colony, provided an opportunity for bringing the arguments of Mr. Curtis and his associates before the public of South Africa, by formally asking (November 28th) Lord Selborne, as High Commissioner, to "review the situation"—the "situation" being the apparent impossibility of settling the disputes on fiscal and railway questions then outstanding between the various South African Governments. In response to this invitation, Lord Selborne embodied the work of the Curtis group in his Federation Memorandum, which was sent to the South African Governments on January 7th, 1907, and published for the use of the general public in the following July.

This group of Crown Colony officials, steeped in the Milner traditions, provided not only the technical knowledge[1] necessary for framing the Union Constitution, but the driving power which brought the Union into being. Formed into a Committee for the promotion of closer union, they obtained valuable recruits in General J. C. Smuts, then a member of the Botha (Transvaal) Ministry; Mr. Malan, the Editor of *Ons Land*, in the Cape Colony; and Sir (then Mr.) Abe Bailey, who contributed to the cost of publishing the periodical and other literature devoted to the advocacy of their cause. By the Cape General Election of February, 1908, as

[1] Mr. Curtis, in particular, as the direct agent for the establishment of the admirable system of local self-government introduced into the Transvaal before Lord Milner resigned, possessed an ample experience of the methods of adjusting the conflicting claims of central and local authorities. But the administrative experience of his colleagues was scarcely less complete and appropriate.

already noticed, the South African party was placed in office in this colony; but the new government, of which Mr. Malan was a member, was no less favourable to closer union than its predecessor, and when the Inter-Colonial Conference on Customs and Railway Rates met in the following May, the general progress of the movement throughout South Africa at once became apparent. The business of the Conference was to provide for the renewal of the Customs Union Convention expiring on the following June 30th; and the seriousness of the position was shown by the fact that the Transvaal Government had already given notice of its intention to retire from the Customs Union. In these circumstances the representatives of the four Self-governing Colonies passed a series of resolutions, in which they pledged their respective governments to carry out immediately all preliminary measures necessary for the creation of a Central Government, and then, after the Transvaal had withdrawn its notice to retire, the Customs Convention was renewed for one year.

In pursuance of this decision the National Convention met at Durban on October 8th, and again, after an adjournment of a month, at Capetown on November 23rd. Among the delegates appointed to represent the various colonies were: For the Cape, Mr. J. X. Merriman (Prime Minister), Mr. Malan, Sir Henry (now Lord) de Villiers (President of the Convention), and Sir Starr Jameson; for Natal, Mr. Moor (Prime Minister); for the Transvaal, General Louis Botha (Prime Minister), General J. C. Smuts, Sir George Farrar, and Sir Percy Fitzpatrick; for the Orange River Colony, Mr. Fischer (Prime Minister), and ex-President Steyn; while Rhodesian interests were watched over by Sir William Milton (the Administrator), and Sir Lewis Mitchell (a director of the Chartered Company). The one conspicuous absence was that of Mr. Jan Hofmeyr, the veteran leader of the Afrikander Bond in the Cape Colony, who was understood to have refused nomination on the ground that his known preference for a federal union was in conflict with the general feeling of the majority of the Cape delegates in favour of a Central Government with more than federal powers. Mr. Brand was attached to the Transvaal Delegation as its Secretary, and Mr. Patrick Duncan as its legal adviser; while Mr. Curtis himself was engaged in the work of making the proceedings of the Convention intelligible to the general public of South Africa through the agency of the closer union organisations, the platform, and the press.

What had always been recognised as the great practical obstacle to administrative union, was the difficulty of ascertaining the revenual value of that part of the assets of the several states which consisted in their respective (Government) railway systems. But, curiously enough, when the actual moment of deliberation came, this obstacle was the first to be overcome. The difficulty was solved by following the course which Lord Milner had adopted in the New Colonies. Here the inevitable friction which would have arisen from an attempt to apportion the earnings of the railways as between the Transvaal and the Orange River Colony, had been avoided by the creation of a common authority, the Inter-colonial Council, which administered the railways of both colonies as a joint concern, and applied the revenue which it received to the maintenance of certain common services. In accordance with this precedent, the Convention at once decided to place the railways, ports, and harbours under the control of a Board of Commissioners, independent of political influences, by whom these competitive sources of revenue were to be administered in the future as one industrial undertaking, and in the common interests of the people of the Union as a whole. The next question to be settled was the nature of the Central Government.

The choice lay between the two forms of union of which the Australian and Canadian constitutions were respectively typical; federation and unification. Under the former, the Central Government would derive its authority from the component states, and transact only so much of the common business of the Union as the State Governments decided to place in its hands. Under the latter, the Central Government would itself be the originating authority, and the provincial governments would exercise only such powers as it might decide to delegate to them. This second form—unification—was strongly advocated by the Transvaal delegates, and it was adopted by the Convention. In another crucial matter—the recognition of the principle of equal rights—the Transvaal delegates were also successful in guiding the decision of the conference. It was agreed accordingly that the basis of the numerical equality of the Parliamentary Constituencies of the Union should be "electors" and not "population," and that there should be an automatic redistribution of seats every five years.

Two thorny questions were compromised. In respect of the admission of natives to the parliamentary franchise the practice of the Cape Colony was in direct conflict with that of the remaining colonies. As no agreement on the question of the admission or non-admission of natives to the Union franchise could be reached, the Convention decided that the franchise qualifications existing in the several colonies should stand as the franchise qualifications for the Union Parliament in the respective provinces of the Union. As the result of this compromise, while the native voters in the Cape Province obtained the Union franchise, practically no natives were admitted to this privilege in the remaining three provinces. In South Africa, as in Australia, it was found impossible to take any one of the capital cities of the four constituent colonies as the capital of the Union.

The compromise on this head was ingenious. Pretoria became the seat of the administration; but the Legislature was to assemble at Capetown, and the Judicature was to be installed at Bloemfontein. At the same time provision was to be made for compensating the municipalities of Maritzburg and Bloemfontein, and (if necessary) those of Capetown and Pretoria, for any diminution of prosperity due to these new arrangements.

The National Convention issued its report on February 2nd, 1909. The draft constitution of the Union was published a week later, and was considered by the respective parliaments of the four colonies, assembled in special session, at the end of March. The parliaments of the Transvaal and Orange River Colony accepted the constitution bill as drafted; that of Natal agreed to its provisions subject to the promise of the Government that the question of joining the Union should be submitted to a referendum before the colony's final acceptance was notified. In the Cape Parliament, however, a number of amendments were passed, which, if they had become a part of the measure, would have subverted the principle of "equal rights" upon which the Transvaal delegates had insisted, as being essential to the prosperity of South Africa.

On May 3rd, the Convention reassembled at Bloemfontein to discuss the amendments of the Cape Parliament. For a time the fate of the Union remained uncertain. In the end, the Transvaal delegates, while refusing to sacrifice the principle of "equal rights," agreed to certain departures from the strict application of the formula "one man, one vote," which removed all the justifiable grievances of the Cape Colony. In respect of the popular chamber, the Assembly, the system of proportional representation with multi-member constituencies was abandoned in favour of the simpler, but less exact, majority vote in single-member constituencies.

The number of "electors," as against that of "population," was retained as the basis of the numerical equality of these constituencies in the several provinces, but the division of the total number of seats in the Assembly, as between the four colonies, was to be made on a basis of the total "adult males," not the total "electors," of each colony. This gave a slight advantage to the Cape Colony and Natal, where certain (low) qualifications were necessary to obtain the franchise, as against the New Colonies in which manhood suffrage (for Europeans) had been established under the Elgin Constitutions. In the Upper Chamber, the Senate, however, there was to be an absolute equality of seats as between the several colonies, each contributing sixteen members; and the system of proportional representation, with the single transferable vote, was to be applied in the election of the elective members of this body.[1]

The Convention concluded its labours on May 11th. The Draft Act of Union, as now amended, was accepted by the parliaments of all the colonies with the exception of Natal; and here, when the Referendum had been taken, it was found that a very large majority of the electors were in favour of their colony joining the Union. The Act as passed by the South African legislatures was introduced into the House of Lords on July 22nd, and into the House of Commons on August 19th. During its passage through the British parliament an amendment was carried, specifically vesting the control of matters affecting Asiatics, as well as those affecting the native population, in the Union, as against the Provincial, authorities. With this change—a change to which no objection was offered by the South African statesmen in charge of the Draft Act—the South Africa Act received the royal assent on September 20th. A proclamation of December 2nd declared May 31st, 1910,—the eighth

[1] Half were to be nominated, and half to be elected.

anniversary of the peace of Vereeniging—to be the day on which the Union Constitution should come into operation. The elections for the Assembly were held throughout the four provinces on September 15th; and the first meeting of the Union Parliament was opened by the Duke of Connaught[1] on December 4th.

The elections had given a considerable, but not overwhelming majority of members to the South African, as against the Progressive, party, and the first Union ministry, of which General Botha was the head, was constituted as follows:

Prime Minister and Minister of Agriculture, General the Rt. Hon. L. Botha.

Minister of Railways and Harbours, the Hon. J. W. Sauer.

Minister of the Interior, Minister of Mines, and Minister of Defence, General the Hon. J. C. Smuts.

Minister of Justice, the Hon. J. B. M. Hertzog.

Minister of Education, the Hon. F. S. Malan.

Minister of Finance, the Hon. H. C. Hull.

Minister of Lands, the Hon. A. Fischer.

Minister of Native Affairs, the Hon. H. Burton.

Minister of Commerce and Industries, Colonel the Hon. G. Leuchars.

Minister of Public Works and Minister of Posts and Telegraphs, the Hon. Sir D. P. de V. Graaff.

Minister without Portfolio, Senator the Hon. Dr. C. O'Grady-Gubbins.

The Provincial Elections were held on September 15th, in the Cape and Transvaal provinces, and on October 12th, in the Free State and Natal.

This account of the birth of the Union—necessarily brief and jejune—will have revealed something of the

[1] The King had intended to open the Union Parliament (as Prince of Wales), but was prevented from doing so by the death of King Edward VII in May of the same year.

character of the national Government thus for the first time established in South Africa, and thereby have prepared the way for the study of the constitution itself which now awaits us. In framing the Act of Union the members of the National Convention had the benefit of the special knowledge possessed by the Closer Union Committee, and in particular by Mr. Curtis and his colleagues of the Crown Colony administration. They were also able to avail themselves of the results shown by the working of the federal system in varying forms in Australia, Canada, and the United States. It is not surprising, therefore, that the Union constitution, as a piece of statesmanship, should be an advance upon those of the sister dominions.

The purposes of the Union Act (South Africa Act, 1909), are excellently stated in the preamble. They are: (1) To unite the several British Colonies in South Africa under one Government in a legislative union under the Crown of Great Britain and Ireland;

(2) To provide for the union of the Colonies of the Cape of Good Hope, Natal, the Transvaal, and the Orange River Colony on the terms agreed upon by resolution of their respective Parliaments;

(3) To define the executive, legislative, and judicial powers to be exercised in the Government of the Union;

(4) To provide for the establishment of provinces with powers of legislation and administration in local and other matters specially reserved for the provincial governments;

(5) To provide for the eventual admission into the Union or transfer to the Union of such parts of South Africa as are not originally included therein.

The provisions for bringing the Union constitution into force contain only one particular which merits attention. Under Section 6, effect is given to a graceful concession to Dutch Afrikander sentiment which was made by the British delegates to the National Convention. In becoming a province of the Union the Orange River Colony reverts to the name it bore before the war; and thus the four colonies mentioned in the preamble became original provinces of the Union "under the names of Cape of Good Hope, Natal, Transvaal, and Orange Free State."

The Union Executive

The executive Government of the Union is vested in the King; and is administered, in practice, by his representative, the Governor-General, for whom an annual salary of £10,000 is at present provided. In the general discharge of his duties, the Governor-General is advised by the executive council, which holds the same relation to him as the "Cabinet" of the United Kingdom holds to the King himself. There is, however, a distinction between the powers of the "Governor-General" and the powers of the "Governor-General in Council" (*i.e.*, as acting with the advice of the Executive Council). As "Governor-General," he chooses and dismisses the Executive Councillors at his pleasure, exercises the command-in-chief of the naval and military forces within the Union, and appoints officers (not exceeding ten in number) to administer the departments of the Government, and to be "the King's ministers of State for the Union." But it is as "Governor-General in Council" that he establishes such Departments, and appoints or removes officers of the public service of the Union. The ministers become members of the Executive Council upon appointment, and as in England and the other dominions, although they are Ministers of the King, they are "responsible" for their administration of the affairs of the Union to the elected representatives of the people in Parliament. That is to say, directly any ministry ceases to command a majority in the Representative chamber, the Assembly, the Prime Minister tenders his resignation to the Governor-General, who, if he accepts it, thereupon sends for the leader of the party with the more numerous following, and requests him to undertake the formation of a new ministry. If, however, the defeated ministry believe that by a direct appeal to the electors they will recover their command of the Assembly, then (as is the practice in England) the Governor-General dissolves Parliament, a general election is held, and the ministry resign, or continue in office, according as their newly-elected supporters form a minority or a majority in the new House of Assembly.

But the Union ministry, although "responsible" to the Union Parliament and people, is not entirely in the same position as an English ministry. By making a distinction between the action of the King's representative as "Governor-General" and as "Governor-General in Council," the British Parliament has refrained from investing the Union ministry with certain executive powers which a British ministry possess in practice, though not in theory. In England the King follows the advice of his ministers. If, in any grave crisis, he does not do so, then the ministry resign; and it remains for the King to find other ministers to carry on the administration—ministers with whom he is in sympathy, and whose advice, therefore, he is able to follow. But in the Union of South Africa the King acts through his representative the "Governor-General" in independence of his Union ministers. As, however, the King follows the advice of his ministers in England, it is the British Government for the time being in office which, thus acting through the "Governor-General," in reality commands the naval and military forces, appoints and dismisses ministers, and, as we shall notice in discussing the legislative powers of the Union Government, gives or withholds its assent to the bills passed by the Union Parliament. It is by virtue of these powers, and other analogous powers exercised over other dominions of the Crown, that the British Parliament is rightly styled the "Imperial Parliament."

As, moreover, the Government of the United Kingdom derives its authority from the elected representatives of the people of the United Kingdom, it is the people of the United Kingdom who ultimately possess the control over the Government of the Union, which is embodied in the powers exercised by the Governor-General as "Governor-General," as distinct from "Governor-General in Council." Thus we have the people of one part of the Empire exercising control over the people of another part. The justification for this anomaly is to be found in another anomaly. For the time being the people of the United Kingdom provide virtually the whole of the funds necessary to maintain the sea and land forces by which the safety of the Empire, as a whole, is secured. The remedy, therefore, is clear. So soon as the Union of South Africa and the other dominions are able and willing to contribute adequately to the cost of the defence of the Empire, they will require the creation of a genuine Imperial authority to control the common affairs of the different members of the Empire, and one in which they can be directly and effectively represented. When such an authority has been constituted, the people of South Africa, or the people of any other dominion, will be able to submit to its control without any loss of national dignity; since it will be an authority of which they themselves, through their representatives, form a part.

The Union Legislature

The legislative power of the Union is vested in the Parliament of the Union, which consists of the King, the Senate, and the House of Assembly.

Of these three elements the Governor-General, as the King's representative, has power to appoint the times for the sessions of Parliament, and, by proclamation or otherwise, to prorogue Parliament, and to dissolve the Senate and the House of Assembly simultaneously, or the House of Assembly alone. The original Senate, however, cannot be dissolved within ten years of the establishment of the Union (1910). Parliament must meet in session at least once in every year, "so that a period of twelve months shall not intervene between the last sitting of Parliament in one session and its first sitting in the next session." The Act declares that Pretoria shall be the seat of Government of the Union, and Cape Town the seat of the Union legislature.

The Senate, as originally constituted under the Act of Union, consists of sixteen members from each province, eight nominated by the Governor-General in Council, and eight elected prior to the establishment of the Union by the respective parliaments of the four colonies, the two houses in each case having sat together in special session as one body; and has thus a total of sixty-four members, of whom half are nominated and half elected. The original senators, whether nominated or elected, hold their seats for ten years; but after the expiration of this period the Union Parliament is empowered to provide itself for the manner in which the Senate is to be constituted. In the absence of any such provision, however, the original method of constituting the Senate is to be maintained, with the exception that the respective provincial councils are substituted for the Colonial parliaments as the bodies to elect senators.[1] In the event of an election of senators being contested, "the election is to be according to the principle of proportional representation, each voter having one transferable vote."

[1] In the case of a vacancy occurring in the number of the *original* elected members, however, the provincial council concerned is to elect a successor in conjunction with the members of the Union House of Assembly from the province in question.

The sub-section (24 ii) dealing with the appointment of the thirty-two nominated senators provides, that in the case of each province one-half of the senators nominated "shall be selected on the ground mainly of their thorough acquaintance, by reason of their official experience or otherwise, with the reasonable wants and wishes of the coloured races in South Africa." This is noticeable as being the only provision made in the constitution for the representation of the native and coloured inhabitants of the three provinces of the Transvaal, Natal, and the Free State.[1] It is also a useful, if limited, effort to provide for the special legislative requirements of the coloured, as apart from the European, population of the Union as a whole.

The qualifications of a senator require a candidate for this office (1) to be not less than thirty years of age, (2) to be qualified to vote for the election of the members of the Union House of Assembly, (3) to have resided for five years in the Union, (4) to be a "British subject of European descent," and (5) in the case of elective senators, to be "the registered owner of immovable property within the Union of the value of not less than £500 over and above any special mortgages thereon." For the purposes of this latter qualification, "residence in," and "property situated within" a colony before its incorporation in the Union, are to be treated as residence in, and property situated within the Union.

The senate elects its President from among its members. The president has a casting vote in the case of an equality of votes; he may be removed by a vote of the Senate, or he may resign by writing addressed to the Governor-General, and he ceases to hold office if he ceases to be a senator. Twelve members are necessary to form a quorum, and all questions are decided by a majority of

[1] In the Cape Colony, it will be remembered, the coloured voters obtained the Union franchise.

votes. The method of resignation in the case of a senator is the same as in the case of the President.

THE HOUSE OF ASSEMBLY AND THE ELECTORAL SYSTEM OF THE UNION

The House of Assembly is composed of members directly chosen by the voters of the Union in their respective electoral divisions. As originally constituted, the Assembly consisted of 121 members elected from the four provinces respectively in the numbers following: from the Cape, 51; from Natal, 17; from the Transvaal, 36; and from the Free State, 17. While provision is made for increasing these numbers up to a total of 150, as may be required by the growth of population in any, or all, of the four provinces from time to time, it is expressly declared that no original province shall lose any of its seats, until either this total has been reached, or ten years have elapsed since the establishment of the Union (1910), whichever may be the longer period. The formation of the Union constituencies was effected by a joint commission of four judges, of whom one was appointed by each of the four Colonial Governments prior to the Union. This commission was empowered to divide each province into single-member constituencies in accordance with the provisions of the Act, and between the date of the passing of the Act and the establishment of the Union. The method of division was as follows. A quota, or constituency unit, was first obtained in the case of each province by dividing the total number of voters on the roll at the last registration by the number of seats assigned to the province under the Act. Thus, taking x to be the total number of voters in any one province, the quota for the Cape would be $\frac{x}{51}$, that for Natal $\frac{x}{17}$, and so on. Having thus obtained the quota for any one province, the commissioners were directed to divide the total number of voters in the province into electoral divisions containing a number of voters as nearly as might be equal to this quota. In so doing they were to give due consideration to, (*a*) community or diversity of interests; (*b*) means of communication; (*c*) physical features; (*d*) existing electoral boundaries; and (*e*) sparsity or density of population. And, to enable them to give effect to this provision, they were further empowered to depart from the quota to the extent of making any constituency to contain 15 per cent. more, or 15 per cent. less voters than the actual number of the quota.

Provision is also made in the Act for the increase and redistribution of seats in the House of Assembly, in accordance with the movement of population, both as between the provinces of the Union and as between the constituencies of any one province.

The method of readjusting the representation of the provinces is laid down in Section 34. The quota of the Union is obtained by dividing the total number of European male adults in the Union, as ascertained at the census of 1904, by the total number of members of the House of Assembly as originally constituted. For the purposes of the Act the total number of European male adults, as thus ascertained, was taken to be:

For the Cape	167,546
,, Natal	34,784
,, the Transvaal	106,493
,, the Free State	41,014
Or a total of	349,837

And as the total number of members to be elected was 121, the Union quota was:

$$\frac{349,837}{121} = 2891{\cdot}2$$

or, a little under 3,000.

The Act continues :

> In 1911, and every five years thereafter, a census of the European population of the Union shall be taken for the purposes of this Act.
>
> After any such census, the number of European male adults in each province shall be compared with the number of European male adults as ascertained at the census of 1904, and, in the case of any province where an increase is shown, as compared with the census of 1904, equal to the quota of the Union or any multiple thereof, the number of members [originally] allotted to such province . . . shall be increased by an additional member or an additional number of members equal to such multiple, as the case may be.

No province, however, is to obtain an additional seat, until its total of adult European males "exceeds the quota of the Union multiplied by the number of members" already allotted to it; and then the additional members are to be allotted only in respect of this excess. The necessity of this proviso will be seen by comparing the number of seats actually assigned to each province with the number of seats to which the provinces were respectively entitled on a basis of adult male European population only. Thus the Cape Province, with 167,546 European male adults, was allotted fifty-one seats. It was actually entitled to :

$$\frac{167546}{2891 \cdot 2} = 57$$

Natal, with 34,784 European male adults, was allotted seventeen seats. It was actually entitled to twelve. The Transvaal, with 106,493 European male adults, was allotted thirty-six seats—the number to which it was entitled. The Free State, with 41,014 European male adults, was allotted seventeen seats. It was entitled only to fourteen. In other words, while the Cape and the Transvaal—allowing for the exceptionally rapid increase of population in this latter province—were under-represented, the two smaller provinces were considerably over-represented in the original House of Assembly.

The returns of the 1911 census show the respective increases of European population in the several provinces to be as follows: The Cape, 3,436; Natal, 1,473; the Transvaal, 123,554; and the Free State, 32,756. And, as the result of them, the Transvaal will gain presumably some additional seats in the next House of Assembly.

When the number of members of the House of Assembly has been increased by the operation of the above provisions to a total of 150, no more additional seats are to be allotted, "unless and until Parliament otherwise provides." The distribution of the 150 seats among the provinces is then to be such, that the proportion between the European male adults and the number of members returned in each province is "as far as possible identical throughout the Union." This identity is, however, subject to the proviso that no original province is to have less than the total number of seats originally allotted to it.

The qualifications necessary to enable persons in the four provinces to vote for the election of members of the House of Assembly are, as we have before had occasion to notice, "the qualifications of parliamentary voters, as existing in the several colonies at the establishment of the Union." And the laws and regulations for the registration of voters, the conduct of elections, etc., of the several colonies are applied *mutatis mutandis* to the elections of members of the House of Assembly, subject to the proviso that at a General Election all polls are to be taken on the same day. No member of His Majesty's forces on full pay, however, can be registered as a Union voter.

The fact that it was only in the Cape Colony that public opinion had allowed any considerable number of coloured persons to obtain the parliamentary franchise, has made it necessary to guard the political rights of these Cape coloured voters in the Union constitution. While, therefore, the Act gives power to the Union Parliament to establish by law a Union franchise, it expressly provides that no such law shall disqualify any of the coloured voters of the Cape province from being registered as a Union voter in that province "by reason of his race or colour only," unless the bill be passed "by both Houses of Parliament sitting together, and at the third reading be agreed to by not less than two-thirds of the total number of members of both Houses." And further, if a law is thus passed, no original coloured voter, or, in the words of the Act, no person registered as a voter in any province at the passing of the law, is to be removed from the register by reason only of any disqualification based on race or colour.

The quinquennial allocation of members as between the several provinces, and the redistribution of these seats among the voters of each province, is to be carried out by a commission of three judges of the Supreme Court of South Africa, which is to be appointed by the Governor-General in Council as soon as possible after every quinquennial census has been taken. These commissions are to have the powers, and to follow the methods, of the original Joint Commission in respect of both the allocation of seats as between the several provinces and the re-division of the electoral divisions of any one province. The alterations in the number of members of the House of Assembly, and the re-divisions of the provinces into electoral divisions, thus effected by the commissions from time to time, are to come into operation "at the next General Election held after the completion of the re-division, or of any allocation consequent upon such alteration, and not earlier."

In order to complete this account of the electoral system of the Union, the following brief statement of the qualifications required of Parliamentary electors in the several provinces must be added.

In the Cape province every adult male British subject,

whether European or coloured, is entitled to be registered as an elector, provided that :

(1) He can sign his name and write his address and occupation ;

(2) Has for not less than twelve months occupied property of the value of £75, or been a joint occupier of property of higher value, the part of which occupied by him has been of the value of £75 ; or,

(3) Been in receipt of wages, or of a salary, of not less than £50 per annum, with no interval of more than one month between successive situations.

In the province of Natal every adult male British subject is entitled to be registered as an elector, provided that :

(1) He owns immovable property of the value of £50, or pays an annual rent in respect of such property of the value of £10 (whether separately, or as his share in the case of a joint tenancy) ; or,

(2) Having resided for three years in the province (or colony) earns an income equal to £8 per month, or £96 per year ; and

(3) Neither belongs to a class placed by legislation under the jurisdiction of special courts, nor is subject to special laws and tribunals.

Male adult natives who have been registered in the province (or colony) for twelve years, and been exempt from the operation of the native laws and regulations for seven years, and who possess either of the two property qualifications (as above), may petition the Governor-General for a certificate, the possession of which entitles them to be registered as Parliamentary electors.

In the Transvaal and Free State provinces all adult male European British subjects are entitled to the franchise, provided that :

They have resided for not less than six months in the province in question, either immediately before the framing of the register, or in case of temporary absence during such six months within three years of this date.

It should be added that in all four provinces persons who have recently served terms of imprisonment, or are subject to any other of the customary disabilities, are excluded from the franchise. Nor can any officer or private of the Regular Army on full pay obtain the Union franchise.

The electoral system thus constituted, although it does not secure the absolute equality of voting power as between Dutch and British, and as between the individual voters in the various constituencies, which would have been provided by the system of proportional representation (with multi-member constituencies and the single transferable vote) advocated by the Transvaal delegation and originally accepted by the National Convention, is none the less far in advance of the system in operation in the United Kingdom. Such gross violations of the principle of equal rights as the spectacle of 1,500 electors in an Irish constituency (Kilkenny City), and 35,000 electors in an English constituency (Walthamstow), both returning one member to the House of Commons, are rendered impossible by the provisions for the numerical equality of the constituencies on a voters' basis, and for the quinquennial redistribution of seats. Although the higher principle of "equal rights for all civilised men" has yet to be adopted, the principle of "equal rights for all Europeans, whether Dutch or British"—which only fifteen years ago seemed an unattainable ideal—is now securely established.

Members of the House of Representatives must have the same qualifications as senators, except that they are not required to hold immovable property or to be thirty years of age. The provisions for the election of a Speaker, his resignation or removal from office, for the resignation of members, and the like, present no unusual features. Among the ordinary disqualifications such as a dishonouring sentence, or bankruptcy, which prevent any person from becoming, or remaining, a senator or member of the Assembly, it is noticeable, however, that the disqualification of "the holding of any office of profit under the Crown within the Union" does not apply to the following: a Minister of State of the Union; a person in receipt of a pension from the Crown; or an officer or member of His Majesty's naval or military forces on retired or half pay, or an officer or member of the naval or military forces of the Union whose services are not wholly employed by the Union. Thirty members constitute a quorum in the Assembly, and the Speaker, like the President of the Senate, has a casting vote but does not otherwise take part in a division. Each House of Parliament has power to make rules and orders for the conduct of its business and proceedings; but in the event of a joint sitting of the two Houses, the Speaker of the House of Assembly presides, and the rules of the Assembly, so far as practicable, are to be followed.

The Parliament of the Union has full power to legislate (subject to the provisions of the Act) for the peace, order and good government of the Union; but the King is a part of the Union Parliament, and the power of the Houses of Parliament to legislate is limited, therefore, by the necessity of obtaining the King's assent to all bills before they can become law. For this purpose bills are presented to the Governor-General, as the King's representative, and the provisions dealing with this part of his duties are clear and precise.

When a bill is presented to him for the King's assent, the Governor-General (not the Governor-General in Council) "shall declare according to his discretion, but subject to the provisions of this Act, and to such instructions as may from time to time be given in that behalf by the King, that he assents in the King's name, or that he withholds assent, or that he reserves the Bill for the signification of the King's pleasure."

All bills must be "reserved" which deal with the matters following:

(*a*) The repeal or amendment of the section relating to the royal assent (§ 64), or of the provisions relating to the electoral system and the House of Representatives (§§ 32 to 50).

(*b*) The abolition of the Provincial Councils, or the abridgment of their powers (except as provided in the Act).

In order to prevent unnecessary friction between the Imperial and the Union Governments, the Governor-General has power to return a bill, thus presented for the King's assent, to the House in which it originated; and in doing so, he may transmit with it any amendments which he may recommend, and the House may deal with the recommendation.

A bill which is reserved for the King's pleasure, has no force, unless, and until, the Governor-General has signified that the King's assent has been given to it at some time within a year of the date on which the bill was presented.

Under the Act the equality of the English and Dutch languages as official languages of the Union is established; and all records, etc., of Parliament are to be kept, and all public notices to be issued, in both languages. So soon as any law has received the King's assent, "the clerk of the House of Assembly shall cause two fair copies of such law, one being in the English and the other in the Dutch language (one of which copies shall be signed by the Governor-General), to be enrolled of record in the office of the Registrar of the Appellate Division of the Supreme Court of South Africa."

MONEY BILLS, ETC.

Money Bills and the Relations of the Two Houses

On the two important subjects of Money Bills and Disagreements between the two Houses of Parliament careful provision is made in the Act. In respect of the former, it is declared in Section 60 that the Senate may reject but not amend Money Bills. The words of the Act are :

> (1) Bills appropriating revenue or moneys or imposing taxation shall originate only in the House of Assembly . . . .
>
> (2) The Senate may not amend any Bills so far as they impose taxation or appropriate revenue or moneys for the services of the Government.
>
> (3) The Senate may not amend any Bill so as to increase any proposed charges or burden on the people.

On the other hand, "tacking" is prevented by the limiting clause in sub-section (1) (which immediately follows the words quoted above) :

> But a Bill shall not be taken to appropriate revenue or moneys, or to impose taxation by reason only of its containing provisions for the imposition or appropriation of fines or other pecuniary penalties.

And by Section 61 :

> Any Bill which appropriates revenue or moneys for the ordinary annual services of the Government shall deal only with such appropriation.

Also, under Section 62 the approval of the Governor-General (not the Governor-General in Council) is required before any money vote can be taken in the House of Assembly :

> The House of Assembly shall not originate or pass any vote, resolution, address, or Bill for the appropriation of any part of the public revenue or of any tax or impost to any purpose, unless such appropriation has been recommended by message from the Governor-General during the Session in which such vote, resolution, address, or Bill is proposed.

Disagreements between the two Houses are to be

settled by a joint sitting; and the procedure to be followed in this event is laid down in Section 63.

If the Senate rejects, or fails to pass, " any bill dealing with the appropriation of revenue or moneys for the public service," the joint sitting of the two Houses may be convened by the Governor-General during the same session; but in the case of other bills the joint sitting is to be held in the second of two successive sessions in which the Houses have failed to agree.

The section runs:

> If the House of Assembly passes any Bill and the Senate rejects or fails to pass it, or passes it with amendments to which the House of Assembly will not agree, and if the House of Assembly in the next session again passes the Bill with or without any amendments which have been made or agreed to by the Senate, and the Senate rejects or fails to pass it or passes it with amendments to which the House of Assembly will not agree, the Governor-General may during that Session convene a joint sitting of the members of the Senate and House of Assembly. The members present at any such joint sitting may deliberate, and shall vote together upon the Bill as last proposed by the House of Assembly and upon amendments, if any, which have been made therein by one House of Parliament and not agreed to by the other; and any such amendments which are affirmed by a majority of the total number of members of the Senate and House of Assembly present at such sitting shall be taken to have been carried, and if the Bill with the amendments, if any, is affirmed by a majority of the members of the Senate and House of Assembly present at such sitting, it shall be taken to have been duly passed by both Houses of Parliament: Provided that if the Senate shall reject or fail to pass any Bill dealing with the appropriation of revenue or moneys for the public service, such joint sitting may be convened during the same session in which the Senate so rejects or fails to pass such Bill.

Each senator and member of the House of Assembly, with the exception of ministers receiving salaries under the Crown, the President of the Senate, and the Speaker of the Assembly, receives an allowance of £400 per annum; from which, however, a sum of £3 is deducted for every day of the session upon which he is absent.

CHAPTER II

THE PROVINCIAL ADMINISTRATIONS

Of the four provinces of the Union the Cape is the oldest, largest, and most populous; but the Transvaal contains the Rand, the great industrial centre of South Africa, where the most rapid increase of European population has taken place in recent years. In the Free State and Natal the European population is relatively small. In the case of the former province, this is to be attributed to the circumstance that its industries are mainly agricultural and pastoral; while in the case of Natal, it is due to the fact that the retail trade has passed largely into the hands of the British Indians, who were permitted to settle there as permanent residents prior to the Union.

While, as we have seen, under the Union constitution all matters of national, or general, concern are placed in the control of the Central Government, each province is provided with a separate administration for the management of its local, or special, affairs. The Provincial Administrations, as constituted by the Union Act, are composed in each case of an Administrator, an elective Council, and an Executive Committee.

The Administrator is appointed by the Governor-General in Council; that is, by the Union Government for the time being. His appointment runs for five years, and he cannot be removed from his office except by the same authority, and then only "for cause assigned," which must be at once communicated to both Houses of Parliament. The salaries of the Provincial Administrators are fixed and provided by Parliament, and cannot be reduced during their respective terms of office. In each province the Administrator is chairman of the

Executive Committee, in which he has a casting as well as an original vote; and if he is not a member of the Provincial Council, he is entitled to take part in its proceedings, though he may not vote in any division. He convenes and prorogues the Provincial Council, and all executive acts relative to the affairs of the province are done in his name. Just as the Governor-General of the Union acts in certain matters under the instructions of the Imperial Government, and in independence of the Union Ministry, so the administrator of a province has power under the constitution to act on behalf of the Governor-General in Council (that is, the Union Government) in all matters outside the sphere of the provincial council, when required to do so, without reference to the other members of the Executive Committee of the province.

The Provincial Councils consist of the same number of members as are elected in the respective provinces for the House of Assembly; and the councillors are elected by the parliamentary voters in the parliamentary constituencies of the Union. As, however, no Provincial Council can consist of less than twenty-five members, in the case of the two lesser provinces of Natal and the Free State it has been necessary for the present to form separate and more numerous Provincial Council constituencies, which are delimited by the commissions provided for the division, and re-division, of the parliamentary constituencies. The Provincial Councillors are elected for three years, and the council is not subject to dissolution except by effluxion of time. Any person who is entitled to the Union, and therefore to the Provincial, franchise is qualified to be a councillor; and the provisions for the conduct of elections, the disqualification of councillors, etc., are *mutatis mutandis* those which are laid down in the Act in respect of members of the House of Assembly. Any councillor, however, who becomes a member of either House of Parliament thereby ceases to be a member of the Provincial Council. The dates of the elections, and the times of the sessions of the council, which must meet once at least in each year, are fixed by proclamation of the Administrator. In each council a chairman is elected from among the councillors, and rules for the conduct of business, which are subject to the approval of the Governor-General in Council, are framed. Members of the Provincial Councils enjoy the same privileges of free speech as members of Parliament, and no councillor is liable to any action or proceeding in any court by reason of his speech or vote in the council. They are also paid allowances, the amount of which is determined, however, not by the council, but by the Governor-General in Council.

The Executive Committees consist of four persons, elected by the respective Provincial Councils at their first meeting, and the Administrator as chairman. The members of the committees are not necessarily members of the respective councils, but if they are, they retain their seats; on the other hand, any member of an Executive Committee who is not a member of the council of his province, has the right to take part in the proceedings, but not to vote, in the council. The remuneration of the members of the Executive Committees is determined by the respective Provincial Councils, subject to the approval of the Governor-General in Council; and it is noticeable that the system of proportional representation is applied to the election of these Executive Committees, as well as to the election of senators.

In all matters in respect of which the Provincial Councils are competent to make ordinances, the Executive Committees are the successors of the Governors and responsible ministries of the former colonies; and all "powers, authorities and functions" relative to such matters, which formerly belonged to the Colonial Governments, are now vested in the respective committees.

It is the business of the Executive Committee in each province "to carry on the administration of provincial affairs on behalf of the Provincial Council," and it is empowered to appoint officers to carry out the services for which it is responsible, and to make and enforce regulations for the organisation and discipline of the Provincial Civil Service, as distinct from the "officers assigned to the province by the Governor-General in Council." Such appointments, however, are subject to the provisions of any law passed by the Union Parliament for regulating the conditions of appointment, tenure of office, retirement and superannuation of Civil Servants.

The matters in respect of which the Provincial Councils are competent to legislate (by ordinance)—matters which collectively indicate the administrative sphere of the Executive Committees—are stated in the Act to be as follows:

(1) Direct taxation within the province, in order to raise a revenue for provincial purposes:

(2) The borrowing of money on the sole credit of the province with the consent of the Governor-General in Council, and in accordance with regulations to be framed by Parliament:

(3) Education, other than higher education, for a period of five years and thereafter until Parliament otherwise provides:

(4) Agriculture to the extent and subject to the conditions to be defined by Parliament:

(5) The establishment, maintenance, and management of hospitals and charitable institutions.

(6) Municipal institutions, divisional councils, and other local institutions of a similar nature:

(7) Local works and undertakings within the Province, other than railways and harbours and other than such works as extend beyond the borders of the province, and subject to the power of Parliament to declare any work a national work, and to provide for its construction by arrangement with the provincial council or otherwise:

(8) Roads, outspans, ponts, and bridges, other than bridges connecting two provinces:

(9) Markets and pounds:

(10) Fish and game preservation:

(11) The imposition of punishment by fine, penalty, or imprisonment for enforcing any law or any ordinance of the province made in relation to any matter coming within any of the classes of subjects enumerated in this Section [85]:

(12) Generally all matters which, in the opinion of the Governor-General in Council, are of a merely local or private nature in the province:

(13) All other subjects in respect of which Parliament shall by any law delegate the power of making ordinances to the provincial council.

In addition to these matters, the report work in Private Bill Legislation has been handed over to a large extent to the Provincial Councils with the object of economising the time and labour of Parliament. Under Section 88 of the Act, the Provincial Council of the province to which any matter proper to be dealt with by Private Bill Legislation relates, may take evidence by select committee or otherwise for and against the bill, and "upon receipt of a report from such council, together with the evidence upon which it is founded, Parliament may pass such Act without further evidence being taken in support thereof." A Provincial Council is also empowered to recommend to Parliament the passing of any law relating to any matter in respect of which it is not competent to legislate.

All ordinances of the Provincial Councils require the assent of the Governor-General in Council to bring them into effect; and any ordinance has effect in the province "as long and as far only as it is not repugnant to any Act of Parliament." The Governor-General in Council may assent, withhold assent, or "reserve for further consideration" an ordinance; but he must declare his intention to take one or other of these three courses within one month of the presentation of the ordinance by the Administrator. Upon his assent being obtained the ordinance is promulgated by the Administrator of the Province, and two copies of it—one in English and one in Dutch—are enrolled in the Registrar's office of the

Appellate Division of the Supreme Court. As in the case of an Act of Parliament, one of these copies must be signed by the Governor-General.

The services for which the provincial administrations are responsible are provided for by ordinances of the councils making the necessary appropriations from the provincial revenue funds constituted in each province. No appropriation, however, whether general or specific, from these funds can be made by any council without a recommendation of the Administrator of the province; and no money can be paid out of them except in accordance with an appropriation, and under warrant signed by the Administrator, and countersigned by the auditor of the province. The revenue fund of a province consists of all revenues raised by, or accruing to, the provincial Council, and all moneys paid over to it by the Governor-General in Council (whether for the general purposes of the provincial administration, or for specific objects).

Careful provision is made to secure the correct management and disposal of the provincial funds. For this purpose the Act requires that there shall be in each province an auditor of accounts, who is appointed and paid by the Governor-General in Council, and removable only by the same authority for "cause assigned," which must be communicated at once to both Houses of Parliament. It is the duty of this officer to examine and audit the accounts of the province to which he is assigned, and no warrant signed by the administrator authorising the issue of money from the provincial fund has effect unless it is countersigned by him, as auditor.

The seats of the Provincial Governments are the respective capitals of the former colonies; Cape Town, Maritzburg, Pretoria and Bloemfontein.

The following table shows the area, population, revenue and expenditure, and public debts of the four constituent colonies at the time of the establishment of the Union.

	Area in sq. miles.	Population (1904).			Revenue	Expenditure (1908–9) in millions.	Public Debt (1910).
		European.	Other than European.	Total.	m. £.	m. £.	m. £.
Cape of Good Hope ..	276,995	579,741	1,830,063	2,409,804	7·3	7·6	52·5
Natal	35,371	97,109	1,011,645	1,108,754	3·5	3·5	22·8
Transvaal	110,426	297,277	972,674	1,269,951	5·7	6·0	32·3
Orange River Colony (Free State Province)	50,392	142,679	244,636	387,315	·9	·9	8·9
Services common to Transvaal and Orange River Colony (not included in above)	—	—	—	—	4·6	2·9	—
Total	473,562	1,116,806	4,059,018	5,175,824	22·2	21·1	116·7

CHAPTER III

FINANCIAL AND ADMINISTRATIVE REORGANISATION

At the time of the Union the combined revenues of the four constituent colonies amounted to rather more than £22,000,000, and their combined public debts were approximately £116,000,000. The South Africa Act left the task of completing the administrative machinery of the Union to be accomplished by the Central Government which it brought into being. But both the necessary readjustments themselves, and the methods and principles to be followed in carrying them out, were indicated with the precision required to give effect to the agreements on the various questions involved, at which the representatives of the four colonies had arrived in the National Convention; and due provision was made for carrying on the government in the period intervening between the establishment of the Union and the completion of its administrative system.

In considering these readjustments it must be remembered that the great material advantage which the establishment of a Central Government promised to the people of the four colonies, was the large reduction of taxation to be expected from the economies effected through the substitution of one, for four administrative systems. The three main readjustments were, accordingly, changes directly planned to bring about the realisation of this promised advantage. They were:

(1) The gradual elimination of the hitherto necessary, but economically bad, practice of using the State railways as agencies of taxation.

(2) The unification of the Government Departments and Civil Services of the four colonies.

(3) The division of the revenues of the four colonies other than their railway and harbour revenues, as between the Union Government and the four provincial administrations.

The first of these administrative processes is required by the Act to be accomplished within four years from May 31st, 1910. In the meantime the railways, ports, and harbours were placed at once under the management of a Board of three commissioners, with a Minister of the Union Government as Chairman, and all revenues derived from these sources are paid into a separate fund—the Railway and Harbour Fund. During the gradual lowering of the railway rates from the original revenue-producing standard to one sufficient to make the earnings merely cover the necessary expenses of working and maintenance, the Union Government is empowered by the Act to draw upon the surplus revenue of the Railway and Harbour Fund to the extent required to supplement the Consolidated Revenue Fund in providing for the general service of the Union.

The provisions for the accomplishment of this great administrative reform are an integral part of the financial system of the Union, as laid down in the constitution.

All revenues, from whatever source arising, over which the several colonies have at the establishment of the Union power of appropriation, shall vest in the Governor-General in Council. There shall be formed a Railway and Harbour Fund, into which shall be paid all revenues raised or received by the Governor-General in Council from the administration of the railways, ports, and harbours, and such fund shall be appropriated by Parliament to the purposes of the railways, ports, and harbours in the manner prescribed by this Act. There shall also be formed a Consolidated Revenue Fund, into which shall be paid all other revenues raised or received by the Governor-General in Council, and such fund shall be appropriated by Parliament for the purposes of the Union in the manner prescribed by this Act, and subject to the charges imposed thereby. (Sec. 117.)

All ports, harbours, and railways belonging to the several colonies are vested in the Governor-General in

Council, *i.e.*, the Union Government; and no works of the sort may be constructed without the sanction of Parliament. (Sec. 125.)

The Commissioners of the Railway and Harbour Board are appointed by the Union Government, and receive salaries, fixed by Parliament, which may not be reduced during their respective terms of office. The appointment runs for five years, but a commissioner may be re-appointed. No commissioner can be removed before the five years have expired except by the Union Government; and then only for "cause assigned," the particulars of which must be communicated at once to both Houses of Parliament. (Sec. 126.)

The independence of the commissioners being thus secured, the Act proceeds to state the broad principles upon which the future administration of the railways, ports, and harbours of the Union is to be based.

The railways, ports, and harbours of the Union shall be administered on business principles, due regard being had to agricultural and industrial development within the Union, and promotion, by means of cheap transport, of the settlement of an agricultural and industrial population in the inland portions of all provinces of the Union.

So far as may be, the total earnings shall be not more than are sufficient to meet the necessary outlays for working, maintenance, betterment, depreciation, and the payment of interest due on capital not being capital contributed out of railway or harbour revenue, and not including any sums payable out of the Consolidated Revenue Fund in accordance with the provisions [for the reimbursement of the Railway Fund for losses due to works constructed contrary to the advice of the Board, and gratuitous, or partly gratuitous, services or facilities]. The amount of interest due on such capital invested shall be paid over from the Railway and Harbour Fund into the Consolidated Revenue Fund.

The Governor-General in Council shall give effect to the provisions of this section as soon as and at such time as the necessary administrative and financial arrangements can be made, but in any case shall give full effect to them before the expiration of four years from the establishment of the Union. During such period if the revenues accruing to the Consolidated

Revenue Fund are insufficient to provide for the general service of the Union, and if the earnings accruing to the Railway and Harbour Fund are in excess of the outlays specified herein [for working, maintenance, etc.], Parliament may by law appropriate such excess or any part thereof towards the general expenditure of the Union, and all sums so appropriated shall be paid over to the Consolidated Revenue Fund. (Sec. 127.)

This power of the Union Government to use the railway profits for the present to supplement the regular revenue is not to be allowed, however, to interfere with the administration of the railways upon the enlightened principles laid down in the section.

Notwithstanding anything to the contrary in the last preceding section, the Board may establish a fund out of railway and harbour revenue to be used for maintaining, as far as may be, uniformity of rates, notwithstanding fluctuations in traffic. (Sec. 128.)

And :

All balances standing to the credit of any fund established in any of the colonies for railway or harbour purposes at the establishment of the Union shall be under the sole control and management of the Board, and shall be deemed to have been appropriated by Parliament for the respective purposes for which they have been provided. (Sec. 129.)

(2) The provisions for the unification of the administrative machinery and *personnel* of the four Colonies include the immediate appointment of a Public Service Commission "to make recommendations for such reorganisation and readjustment of the departments of the public service as may be necessary," and the subsequent appointment of a permanent Public Service Commission "with such powers and duties relating to the appointment, discipline, retirement, and superannuation of public officers as Parliament shall determine." After the first Commission has reported, the Union Government are to assign to each province the Civil Servants necessary for the proper discharge of the services reserved or delegated to the provincial administrations; and these Civil Servants thereupon become "officers of

the provinces." Pending the report of the Commission, the necessary officials were to be placed by the Union Government at the disposal of the Provinces. All services and departments, with their *personnel,* placed under the control of the Railway and Harbour Board, were, however, expressly excluded from the operation of this reorganisation commission.

The Act contains provisions intended to prevent the readjustment of the public services of the four colonies from injuriously affecting the rights of individuals. Any officer being in the public service of any of the colonies at the time of the establishment of the Union, but neither retained in the service of the Union, nor assigned to that of a province, is to receive the same pension, gratuity, etc., as he would have received in the like circumstances if the Union had not been established. All officers retained in the Union service, or assigned to the provincial services, are secured in all their existing and accruing rights, and their times of retirement, pensions and allowances, are not to be affected by the change. It is also specifically laid down that no official in the service of the colonies at the establishment of the Union was to be dismissed "by reason of his want of knowledge of either the English or Dutch language." A special provision is made for permanent officers of the legislatures of the several colonies. Any such officer who is neither retained in the Union service, nor provided for by the legislature of his province, is "entitled to such pension, gratuity, or compensation as Parliament may determine."

At the time of writing the Reorganisation Commission has presented several reports, to which some reference will be made in subsequent chapters; and the work of reorganising and readjusting the various departments of the Union and Provincial administrations is well advanced. The process has given rise to considerable dissatisfaction among Civil Servants whose services are no longer required. This, perhaps, was only to be expected; but the extent of the economies effected would appear also to be disappointing. Some time must elapse, however, before any reliable conclusions can be formed upon these questions.

(3) The difficult task of apportioning the revenues of the four constituent colonies (other than those derived from railways and harbours) as between the Central Government and the provincial administrations was entrusted by the Act to a Financial Relations Commission. This important body was appointed as soon as possible after the establishment of the Union, and consisted (as provided) of one representative from each province, with an officer from the Imperial Service as President. Pending the completion of the work of the commission, the Union Government was directed to pay annually out of the Consolidated Revenue Fund to the administrator of each province:

(*a*) An amount equal to the sum provided in the estimates for education, other than higher education, in respect of the financial year, 1908-9, as voted by the Legislature of the corresponding colony during the year 1908;

(*b*) Such further sums as the Governor-General in Council may consider necessary for the due performance of the services and duties assigned to the provinces respectively.

The members of the Financial Relations Commission were unable to arrive at a unanimous conclusion, and presented both majority and minority reports, which were laid before the Union Parliament early in February, 1912. The Majority Report was signed by Sir George Murray,[1] the Chairman, and by Sir Percival Laurence and Mr. Patrick Duncan, the respective representatives of the Cape and Transvaal provinces; while the Minority

[1] Sir George Herbert Murray, P.C., G.C.B., I.S.O.

Report contained the opinions formed by Sir Thomas Hyslop and Mr. Wessels, the respective representatives of Natal and the Free State. The Majority Report recommends that half the expenditure actually incurred by the provincial administrations should be provided by "block grant" from the Union Government; and the remaining half raised by direct taxation "imposed and collected" by the several administrations. The sources of provincial revenue proposed are: (1) School fees (already charged in the Cape, Natal, and Free State provinces, but not in primary schools in the Transvaal); (2) Hospital fees (including one-half of the yield of the Native Pass Fees in the Transvaal—about £340,000—which formed originally a hospital fund); (3) Licence Duties; (4) Transfer Duties (on fixed property); and (5) Rates on owners and occupiers of fixed property. The method of direct taxation suggested is that adopted in the Cape Province, where funds for educational purposes and for the maintenance of roads and bridges are raised by the School Boards and Divisional Councils; the revenue of these authorities being derived from a rate on the capital value of all property, with certain exceptions, within their respective areas. The extension of the Cape system to the remaining provinces of the Union is advocated on two grounds. In the first place, it would promote economy in the provincial administrations, since each province would be separately responsible for the provision of funds to meet half of the provincial expenditure; and in the second, the Cape would be relieved of the injustice of being the only province to pay local taxes over and above the Union taxes. The recommendation, if adopted, would impose no fresh burden upon the Cape; but the rates on fixed property—assuming the Cape system to be followed—required in the Transvaal, Natal and the Free State would be respectively seven-sixteenths, three-eighths, and less than one-eighth of a penny in the pound. The effect of these proposals upon the Union Budget is thus exhibited in the Report :

	£
The Present amount of Provincial Subsidies (including supplementary grants)	3,297,133
Future subsidy (one-half of above)	1,552,229
Saving	1,744,904
Revenue surrendered	1,058,332
	686,572
Compensation to Municipalities in Natal on account of Licence Duties surrendered, say,	15,000
Net saving	£671,572

The alternative proposal embodied in the Minority Report is based upon the opinion that the whole cost of provincial administration, exclusive of the revenue derived from school fees, hospital fees, licences (including dog tax), and pass fees, should be met by a grant from the Union Exchequer. The system proposed is that adopted in Australia, where 25s. per head of population is paid by the Commonwealth Government to the several State Governments. The Minority Report recommends, therefore, that the Union Government should make a grant to each province of £1 per head of European, and 4s. per head of Native or coloured population.

These Reports were submitted to the consideration of a conference of the provincial authorities called together by the Union Government, and the decision of this latter was announced by Mr. Hull, the Union Treasurer, in the course of his Budget speech on March 22nd, 1912. The Majority Report, with the principle of rendering the provinces directly liable for half of the expenditure of their respective provincial administrations, was to be adopted; but its recommendations were to be modified in certain respects calculated to satisfy the objections to which the Minority Report gave expression. For ten years, therefore, special grants of £80,000 and £67,000 per year were to be made by the Union Exchequer to the respective administrations of the Natal and Free State provinces, over and above the pound for pound grant; and the Natal administration was also to have the yield of the Indian Pass Fees. In the case of the Transvaal, the Union Exchequer was to retain the Totalisator revenue, but to surrender not half, but the whole, of the Native Pass Fees to the provincial administration.

Apart from these three conspicuous administrative changes of system and *personnel*, the Act provides for certain lesser readjustments the guiding principle of which is in each case laid down.

Ultimately the same laws are to run throughout the Union in respect of all national concerns; but while the work of harmonising the various laws of the several colonies is in progress, the laws in force in each colony at the date of the establishment of the Union are to continue in force in the respective provinces. In other words, the people of the provinces remain under their old laws, until these latter are repealed or amended by the Union Parliament, or by the Provincial Councils in cases where they relate to matters within the sphere of the Councils. In the same way although the Act declares that "there shall be free trade throughout the Union," until provision is made by the Union Parliament to give effect to this declaration, the old duties of custom and excise remain leviable in the several provinces.

There were, however, two departments of the State in respect of which no period of readjustment was contemplated by the Act—The administration of justice, and the control and administration of native affairs and matters affecting Asiatics within the Union. The national judicature, which the constitution brought into existence with the Union, will form the subject of the following chapter. Provision for the immediate devolution upon the Union Government of the affairs comprised in the second department was made in Section 147 of the Act, and the provincial authorities were thus at once relieved of all responsibility in respect of a sphere of administration which presents many difficulties. The section is important in itself, and contains, as we have noticed before, the one amendment introduced by the Imperial Parliament into the Draft Act as passed by the South African Legislatures.[1] It runs:

> The control and administration of native affairs and of matters specially or differentially affecting Asiatics throughout the Union shall vest in the Governor-General in Council, who shall exercise all special powers in regard to native administration hitherto vested in the Governors of the colonies or exercised by them as supreme chiefs, and any lands vested in the Governor or Governor and Executive Council of any colony for the purpose of reserves for native locations shall vest in the Governor-General in Council, who shall exercise all special powers in relation to such reserves as may hitherto have been exerciseable by any such Governor or Governor and Executive Council, and no lands set aside for the occupation of natives which cannot at the establishment of the Union be alienated except by an Act of the Colonial Legislature shall be alienated or in any way diverted from the purposes for which they are set apart except under the authority of an Act of Parliament.

In addition to the financial and administrative arrangements already noticed, the Act contains provisions under which all assets and all liabilities of the four Colonial Governments became at once assets and liabilities of the Union Government, and all officers in the public service of the colonies officers of the Union. Similarly all rights and obligations under any conventions or agreements binding on any of the colonies devolved upon the Union; and in particular the railway agreement made on the eve of the Union (Feb. 2nd, 1909) between the Transvaal, Cape, and Natal Governments, under which Durban was to receive 30 per cent., the Cape ports not less than 15

[1] See p. 128.

per cent., and Delagoa Bay from 50 to 55 per cent., of the oversea traffic to the Rand,[1] was, "as far as practicable, to be given effect to by the Government of the Union."

The Act also makes provision for the alteration of the original provinces of the Union, and for the admission into the Union of Rhodesia and of the native territories at present administered directly by the Imperial Government. The Parliament of the Union can itself carry out the first of these changes upon the petition of the several provincial councils of the provinces whose boundaries are affected. The admission of Rhodesia and the transfer of the Native Territories may be effected by "the King, with the advice of the Privy Council," upon the presentation of addresses from the Union Parliament. In the case of Rhodesia the terms and conditions (as to representation and otherwise) are to be those expressed in the addresses and approved by the King: but the terms and conditions upon which the Union Government may undertake the Government of any Native Territories which the Imperial Government may decide to transfer to the Union, are already laid down and set out in the Schedule to the Act.

The Union Parliament has a general power to repeal or alter the constitution as set out in the South Africa Act (1909), but the exercise of this power is altogether withheld in respect of certain provisions, and made conditional in respect of certain other provisions, of the Act. No provision for the operation of which a definite period of time is prescribed may be altered or repealed until the specified period has elapsed. Thus no alternation can be made in the period, viz., four years, within which the

[1] This confirmed the change in the Railway provisions of the *modus vivendi*, or agreement between the Transvaal and Mozambique Governments of 1901, previously arranged between the Transvaal and Mozambique Governments.

railways must be converted from partial agencies of taxation to an industrial service administered by the State. The following sections can only be repealed or altered by a vote of not less than two-thirds of the total number of members of both Houses of Parliament in a joint sitting.

(1) Section 152, dealing with the amendment or alteration of the Act.

(2) Sections 33 and 34, which provide for the number of members of the House of Assembly, and the division of seats as between the several provinces.

(3) Section 35, which safeguards the Cape coloured voters.

(4) Section 137, establishing the equality of English and Dutch as the official languages of the Union.

CHAPTER IV

THE SUPREME COURT OF SOUTH AFRICA

For some time prior to the establishment of the Union the need of an Appeal Court for South Africa had been recognised, and some efforts had been made to bring it into existence by agreement between the several South African Governments. What was wanted—and wanted especially by the business community—was a court strong enough to take the bulk of the appeals from the Supreme Courts of the separate colonies, which, in the absence of such an institution, had to be decided by the lengthy and expensive process of appeal to the Judicial Committee of the Privy Council in England. With the establishment of the Union this want was satisfied. For while, as we have seen, many of the benefits promised by the constitution, and notably the reduction of railway rates, could not be realised until sufficient time had elapsed to allow the necessary administrative readjustments to be completed, the people of the four provinces were enabled at once to enjoy the advantages of a common judicature.

The sections of the Act which deal with this element of the Union Constitution provide for the establishment of a Supreme Court of South Africa, with an Appellate Division, and Provincial and Local Divisions.

The Supreme Court consists of the Chief Justice,[1] two Judges of Appeal, and the Judges of the Provincial and Local Divisions.

The Court of Appeal, or Appellate Division of the Supreme Court, is composed of the Chief Justice, the two

[1] Lord de Villiers (Sir Henry de Villiers, Chief Justice of the Cape Colony, raised to the peerage as Baron de Villiers) was appointed first Chief Justice of South Africa.

ordinary Judges of Appeal, and two additional Judges of Appeal, who are assigned from the Provincial or Local Divisions to sit on the Appellate Bench when the full court of five is required. The Appeal Court sits at Bloemfontein, but may sit elsewhere in the Union to meet the convenience of suitors. It is composed of five judges of the Appellate Division in the case of appeals from a court consisting of two or more judges, and of three judges in hearing appeals from a single judge; and no judge may take part in the hearing of an appeal against a judgment given in a case heard before him. The process of the Appellate Division runs throughout the Union, and its judgments or orders are executed as though they were original judgments or orders of the respective Provincial Divisions of the Supreme Court.

The Provincial Divisions of the Supreme Court respectively correspond to the several supreme Courts of the Cape, Natal, and the Transvaal, and the High Court of the Orange River Colony.

The Local Divisions of the Supreme Court correspond similarly to the Court of the Eastern Districts of the Cape Colony, the High Court of Griqualand (the Diamond Fields), the High Court of the Witwatersrand, and the several circuit courts. The Provincial and Local Divisions, designated generically as "Superior Courts," have, in addition to the original jurisdiction exercised in each case by the corresponding courts of the colonies at the establishment of the Union, jurisdiction in all matters:

(*a*) In which the Government of the Union, or a person suing or being sued on behalf of it, is a party:

(*b*) In which the validity of any provincial ordinance shall come into question.

Further, unless and until Parliament shall otherwise provide, the Superior Courts have *mutatis mutandis* the same jurisdiction in matters affecting the validity of elections of members of the House of Assembly and Provincial Councils, as the corresponding courts of the colonies had at the establishment of the Union in regard to parliamentary elections in such colonies respectively.

Upon the establishment of the Union the judges of the several Supreme Courts of the Cape, Natal, and the Transvaal and of the High Court of the Orange River Colony, became judges of the Supreme Court of South Africa, and were assigned to its divisions in the respective provinces. The Chief Justices of the three colonies with Supreme Courts became Judges-President of these divisions in their respective provinces, but retain so long as they hold this office the title of Chief Justice of their respective provinces.

The Chief Justice and all judges of the Supreme Court are appointed by the Union Government; their salaries are fixed and provided by Parliament, and cannot be reduced during their continuance in office, and neither the Chief Justice nor any judge can be removed except by the Union Government on an address from both Houses of Parliament in the same session, praying for such removal on the ground of misbehaviour or incapacity.

The Union Parliament has power to reduce the number of judges in the event of a vacancy occurring in any division other than the Appellate Division, if such a reduction is to the public advantage.

Appeals from the Superior Courts in civil cases are taken to the Appellate Division of the Supreme Court, except in cases of orders or judgments given by a single judge, upon applications by way of motion or petition or on summons for provisional sentence or judgments as to costs only, which by law are left to the discretion of the court. Appeals in these latter cases and in criminal cases go to the respective Provincial Divisions. From the judgment of a Provincial Division given on appeal a further appeal lies to the Appellate Division, but only upon special leave obtained from the Appellate Division.

Appeals from the court of a Resident Magistrate or any other Inferior Court in both civil and criminal cases are taken to the respective Divisions of the Supreme Court; and there is a further right of appeal, but only upon special leave obtained, to the Appellate Division.

There is an appeal to the Privy Council from the Appellate Division of the Supreme Court, if special leave be obtained from the Privy Council; but the Union Parliament has power to make laws limiting the matters in respect of which such special leave may be asked. Bills, however, containing any such limitation are reserved by the Governor-General for the signification of His Majesty's pleasure. The right to appeal to the Privy Council in respect of judgments given by the Appellate Division under, or in virtue of, the Colonial Courts of Admiralty Act, 1890, remains unaffected by the above provisions.

Civil suits may be transferred from one Provincial or Local Division to another, if it can be shown that there is good ground for such transference; and such suits are proceeded with as though they had been originally commenced in the Divisions to which they are respectively transferred.

Advocates and attorneys admitted to practise in any Superior Court of any of the colonies at the establishment of the Union, are entitled to practise in the corresponding Divisions of the Supreme Court; and advocates and attorneys entitled to practise before any Provincial Division of the Supreme Court may practise before the Appellate Division. The laws regulating the admission of advocates and attorneys to practise before any Superior Court of any of the colonies apply *mutatis mutandis* to the admission of advocates and attorneys to practise in the corresponding Division of the Supreme Court of South Africa.

To sum up: the provision made for the administration of justice to the European population of the Union, and to that part of its native and coloured population which has acquired a European manner of life, consists of:

(1) A Court of Appeal for South Africa seated at Bloemfontein,

(2) Provincial Divisions of the Supreme Court at Capetown, Pretoria, Maritzburg and Bloemfontein, and Local Divisions in the three chief industrial centres; the Rand, Kimberley, and the eastern districts of the Cape Province,

(3) Circuit Courts held by the Provincial Divisions in the lesser towns of the four provinces, and

(4) The Courts held by the Resident Magistrates with other inferior courts, in which small civil causes and police cases are heard, and which, roughly speaking, perform the work done in England by the County Courts, Petty and Quarter Sessions, and the Courts of Stipendiary Magistrates.

PART III

RHODESIA & THE NATIVE TERRITORIES OF THE HIGH COMMISSION

CHAPTER I

RHODESIA

RHODESIA remains for the present outside the Union. As we have seen, however, provision for its future admission is made in the Constitution, and its representatives took part in the proceedings of the National Convention. There is every reason to believe, therefore, that, when the right time comes, Rhodesia will unite herself to the sister states of South Africa ; but in any case she and they are so closely connected economically and geographically, that no description of the Union of South Africa can be complete, unless it includes some account of this colony. And the same remark applies to the three Native Territories which have been left under the direct control of the Imperial Government, and which, therefore, are described separately in a subsequent chapter. Although not as yet a part of the Union, they, Rhodesia, and the Union are all alike members of the South Africa that owes allegiance to the British Crown.

The territories of the Chartered Company, of which Southern Rhodesia forms the most important portion, extend over an area of 439,575 square miles, and are, therefore, nearly four times as large as the British Isles. They are bounded on the south by the Bechuanaland Protectorate and the Union, on the west by German South-West Africa, on the east by Portuguese East Africa, and on the north by the Congo State and German East Africa.

This great inheritance, wrested from barbarism for the Anglo-Saxon race by the genius of Cecil Rhodes, has been happily described by a French observer [1] as "the pick of Central Africa on both sides of the Zambezi." Little more than twenty years ago—before Mashonaland was occupied in 1890 by the Pioneer Force of the British South Africa Company—it was known vaguely to a few intrepid explorers and a handful of Big Game hunters. Only one or two white men had gazed upon its greatest natural marvel, the Victoria Falls ; and its present principal town was the kraal of the savage Matabele chief, Lobengula. To-day, the traveller reaches a large and comfortable hotel within an easy walk of these same Falls, by a journey from Southampton of less than twenty-one days, during the last four of which (Tuesday to Saturday) the Zambezi Express has carried him luxuriously, and without a break, from Capetown over 1,641 miles of railway. From the garden of the Victoria Falls Hotel he looks out upon a sight scarcely less significant than the clouds of spray that mark the line of the as yet unseen cataracts. It is the majestic span of the great bridge which carries the railway over the Zambezi on its way to the present terminus of the Cape-to-Cairo transcontinental line in the Congo Free State, 100 miles beyond the northern border of Rhodesia. To-day, again, three miles from the European town of Buluwayo the traveller will find in the beautiful gardens of the official residence of the Administrator the ancient tree, with wide-spreading twisted roots, under which Lobengula held *indabas*, [2] and bade his *impis* go forth to rob and massacre the peaceful Mashonas.

These strange contrasts tell something of the rapidity with which the British South Africa Company is developing the vast territories granted to it by the Imperial

[1] M. Lionel Décle, who was sent to Central Africa on a scientific mission by the French Government.

[2] Councils of lesser chiefs.

By permission of *W. Basil Worsfold*

THE VICTORIA FALLS STATION

Government. None the less Rhodesia has had its "dark days"; and we must glance back at these periods of war and industrial difficulty before we consider the country as it is now.

The actual occupation of Mashonaland was accomplished, in 1890, by a Pioneer Force of 200 Europeans and 150 native labourers, and a Police Force of 500 men, "without the loss of a single life, and without the necessity of firing a shot." Forts with intermediate stations were erected at Tuli, Victoria, Charter and Salisbury, and garrisoned by the Company's police. A serviceable road, 400 miles long, was constructed by the Pioneers, who then dispersed to prospect for gold, and the administration of the infant colony of some 1,000 Europeans was assumed by Mr. A. R. Colquhoun, the Company's first Administrator. The time occupied in the process was just two months and a half—from June 28th, 1890, when the expedition left its camp on the Macloutsie River in Bechuanaland, to September 12th, when the site of the present town of Salisbury was reached. The cost of the expedition, exclusive of grants of land and mineral rights to the Pioneer settlers, was £89,285 10s. 0d. All this was done with the sanction and approval of Lobengula, who was supreme chief over the Matebele and Mashona peoples.

A year later the European settlers were in great straits, and the financial outlook of the Company looked almost hopeless. The 1,500 settlers, said Rhodes in his speech at the annual meeting of the Company held on November 29th, 1892, "went to work to find their reefs, but they were removed 1,700 miles from the coast, and their food cost them £70 a ton." It was at this time, when fever was prevalent owing to the unusually heavy rains of the season 1890–1, food was at famine prices owing to the enormous cost of carriage, and the Company was spending £250,000 a year on police and administration, against a revenue of practically nothing, that Dr. (now Sir Starr) Jameson was persuaded by Rhodes to give up his practice at Kimberley and take command of the Settlement. "If you will give me £3,000 a month, I can pull through," he said, when Rhodes "talked matters over with him."

Dr. Jameson proved equal to the task. The costly police force was cut down to forty men, and the settlers themselves undertook the military duties necessary for their security. The frontier dispute with the Portuguese had been settled by the Imperial Government, and the Company was on the most friendly terms with Lobengula. "He receives," said[1] Rhodes, "a globular sum of £100 a month in sovereigns, and he looks forward with great satisfaction to the day when he will receive them. I have not the least fear of any trouble in the future from Lobengula."

But Rhodes forgot that the savage associates peace with weakness. Dr. Jameson pulled the settlers and the Company through; but when Lobengula saw that the white men, in going about their business—prospecting for minerals, building houses and making roads—scrupulously refrained from committing any act of violence or injustice against himself and his people, he drew the conclusion, inevitable to the mind of the military Bantu, that this peaceful disposition was the result of fear. On July 18th, 1903, in the course of Lobengula's annual raid upon the Mashonas, a Matabele *impi* about 300 strong, a detachment of a much larger force, entered the township of Victoria, and when ordered by Dr. Jameson to retire, refused to leave the neighbourhood. This was the commencement of the Matabele war. A year later Rhodes said of it: "We either had to have that war, or leave the country. I do not blame the Matabele: their system was a military system; they once a year

[1] November 29th, 1892.

raided the surrounding people, and such a system was impossible for our development."

Two columns were despatched by the Chartered Company from Forts Charter and Victoria, so soon as Dr. Jameson had been authorised by the High Commissioner "to take all steps he considered necessary to provide for the safety of the lives and property of the settlers under his administration." They consisted of 1,227 men, of whom 672 were Europeans, and joining hands at Indaima's Mount on October 16th, 1903, advanced in a south-westerly direction over the open plateau towards Buluwayo. In the meantime a force of Bechuanaland Border Police, and a small column from Tuli, under command of Major (now Sir Hamilton) Goold-Adams, co-operated with Dr. Jameson's columns by advancing from the South. After hard fighting on the Shagani river, and again on the Imbembezi, when they successively defeated 5,000 and 7,000 Matabele, the Chartered Company's force, led by Major P. W. Forbes, and accompanied by Major Sir John Willoughby, the Company's senior military officer, and Dr. Jameson, occupied Buluwayo on the 4th November. A month later Major Alan Wilson, Captain Borrow, and thirty-three others, while in pursuit of Lobengula, were surrounded by the Matabele in overwhelming numbers, and met their death in the heroic manner which is fitly commemorated by the stately monument that stands close to Rhodes' grave in the Matopos. On December 22nd, the war was over. Lobengula had perished, and the cruel military supremacy exercised by the Matabele over the Mashonas had come to an end. The Company's columns were disbanded, and the country was thrown open for prospecting and settlement. The death roll of the Mashonaland force was eighty, of whom forty-nine were Europeans; and forty-five men, of whom twelve were Europeans, had been wounded. In the Southern Force out of a total of 445 Europeans four men had been killed, and ten wounded. The expenditure on the Matabele war appears in the Company's books as £113,488 2s. 11d.

These rapid and successful operations put the Chartered Company in possession of the whole of the central plateau from Bechuanaland to the Zambezi; and the work of industrial and administrative development proceeded for the next two years without a check. The forts and settlements had been linked together by the telegraph early in 1892; and thus the infant community had been placed in rapid communication with the rest of South Africa and England from the first moment of its existence.[1] The railway was driven vigorously northward from the Cape Colony by Rhodes, the Managing Director of the Chartered Company in South Africa, who knew well that no great progress could be made, until the settlers could get their food, mining machinery, and building materials of all kinds, quickly and cheaply by rail from the ports. Moreover he had already conceived the magnificent project of an African Transcontinental line, that should unite Capetown with Cairo—a project then deemed audacious, but now on the eve of realisation. Rhodes, then, was eager to furnish the Company's territory with railways on both of these grounds; but the railways, like the development of the country as a whole, were commercial undertakings, and the interests of the shareholders had to be considered. This latter necessity made the advance of the main line from the Cape Colony at first comparatively slow; since a line carried for hundreds of miles through the desert regions of Bechuanaland could not be expected to pay its way, still less to earn dividends, for some time to come. None the less, when a great necessity arose, Rhodes

[1] The telegraph was carried from Mafeking to Fort Victoria in December, 1891, and it reached Salisbury (819 miles from Mafeking) in February, 1892.

By permission of *W. Basil Worsfold*

THE ALAN WILSON MEMORIAL

used his own private resources to drive the iron road through to Buluwayo.

Before, however, recording the material development achieved up to the end of 1895, it will be convenient to consider for a moment the purposes for which the Chartered Company was founded, and the methods by which Rhodes proposed, in his own words, "to combine the commercial with the imaginative." In other words, how it was proposed to add these vast regions to the British Empire without the cost of sixpence to the British taxpayer, and at the same time to give the shareholders of the Company a return upon the capital which they had subscribed?

In the proposals for the formation of the British South Africa Company, submitted to the Imperial Government on April 30th, 1889, the objects of the Company were stated to be:

1. To extend northward the railway and telegraph systems in the direction of the Zambezi:

2. To encourage emigration and colonisation:

3. To promote trade and commerce: and

4. To develop and work mineral and other concessions under the management of one powerful organisation, thereby obviating conflicts and complications between the various interests that had been acquired within those regions, and leaving to the native chiefs and their subjects the rights reserved to them under the several concessions.[1]

The petition for the grant of a Royal Charter was supported by the plea, that the undertaking "could not be considered as likely to be remunerative for some time," and that "the sanction and moral support" of the Imperial Government was "necessary to the due fulfilment" of the objects of the Company. The Charter

[1] Chief among these was Lobengula's concession (obtained on October 30th, 1888, by Mr. Rochfort Maguire and two others) of the right to work minerals within his territory.

was granted on October 29th, 1889; and within twelve months of this date the occupation of Mashonaland was peacefully accomplished.

For the realisation of purposes of such magnitude large sums of money were necessary. How was a commercial return on the capital expenditure to be obtained? Rhodes himself answered this question. At a meeting of the shareholders of the Company held on January 18th, 1905, he said:

> I have once said before that out of licences and the usual sources of revenue for a government you cannot expect to pay dividends. The people would get annoyed if you did; they do not like to see licences spent in dividends—these are assets which are to pay for any public works and for good government. We must, therefore, look to our minerals to give us a return on our capital, which, you must remember, is £2,000,000.[1]

In the earlier speech[2] he said:

> My experience of the past is that, just as *qua* Government so *qua* a Company we cannot expect to do more than balance revenue and expenditure from land, customs, and assisting in other matters connected with developing the general natural resources of the country. Therefore, when we created the Charter, we had to consider by what means a return could be given to the shareholders, and I remember thinking out the various ways of making a return to those who had risked their capital in the undertaking. It has always struck me that if it were possible for the Government of a country to share in the discovery of the minerals, a very fair return would accrue. For instance, I have been a miner at Kimberley, on the discovery of the Diamond Fields, and I was allowed to mark out one claim. It has always struck me afterwards, when I had become engaged in the politics of the country, that if I had been allowed to mark out five claims, no one would have been hurt if I had pegged out two and a half for myself and two and a half for the Government. The same thought had occurred to me when I went up to Witwatersrand and saw that marvellous goldfield, where the terms were that they could each mark out one claim. It had occurred to me that, supposing the law had been that each of them could mark out ten claims—five for themselves and five

[1] The capital stands in the last (March 31st, 1911) balance sheet at £9,000,000.

[2] November 19th, 1892.

for the Government—it would not hurt the prospectors, and it would have meant wealth to the Government of the country. The only objection to the idea was that it was a perfectly new one. At any rate, we thought we would try it in Mashonaland, and it was the law of the Company that 50 per cent. of the vendor scrip went to the Charter.

Although this large share in the profits of the mines to which the Company was thus originally entitled, has been greatly reduced, the principle here laid down by Rhodes has been maintained. The Company has appropriated the ordinary revenue of the colony to the purposes of administration, and has sought to find a return for the shareholders' capital in the profits of industrial undertakings—notably the construction and working of railways, the development and sale of town and agricultural lands, and royalties on minerals—as distinct from revenue. In this way, by rigidly separating the function of government from its functions as a great commercial enterprise, the Company, as Rhodes saw, whatever return it made to the shareholders, great or little, as the case might be, would avoid the mistake of imposing any burdensome or unusual taxation upon the settlers.

The faithfulness with which the directors of the British South Africa Company have adhered to the principle laid down by Rhodes, is shown by the circumstance that, although no dividends have been paid as yet, it is only within the last few years that the annual revenue of Southern Rhodesia, the most advanced portion of the Company's territories, has become sufficient to defray the annual costs of administration. On the other hand, the Company, as a commercial undertaking, will benefit ultimately by the prosperity of the community which it has brought into being. Whether it retains the function of government, or resigns it to the Imperial or Union authorities, the commercial assets, whose value has increased in proportion as the colony has grown in population and wealth, will remain the property of the shareholders. It is from these commercial assets, thus enhanced in value, that a return should be obtained ultimately upon the vast sums disbursed by the Company in the administration and industrial development of Rhodesia. And it is to be hoped that when the settlers have reached this stage of assured prosperity, the shareholders of the Company, who have borne a humble but indispensable part in the making of the country, will at length receive the return upon their capital for which they have waited so long and patiently.

Rhodesia, then, like the Indian Empire, has been founded not by the nation, but by the energy, ability, and monetary resources of private individuals. Their primary aim, like that of the men who in the "spacious times" of Queen Elizabeth adventured their lives and property in the discovery and development of unknown lands, was to secure the profits to be gained by opening up new fields for industry and commerce in South-Central Africa. In the pursuit of this primary aim Rhodes and Dr. Jameson were interrupted by the Matabele war; but the event, unexpected and unwelcome at the moment, proved in the end to be as salutary as it was necessary. Before the war the material operations preliminary to the establishment of fresh settlements, and to the industrial development of the country, had made some progress, and directly it was over they were resumed with increased activity.

The first necessity was the provision of means of communication and transport. Telegraphic communication, as we have seen, was secured for the Mashonaland settlement contemporaneously with the occupation of the country; and one of Rhodes' earliest acts, upon the grant of the Charter, was to arrange with the Cape Government, in which he was then Prime Minister, to allow the Company to carry the railway northward from

Kimberley to Vryburg in Bechuanaland, on the understanding that, when completed, the new line should form a part of the Cape Government railways. This section, 127 miles long, was begun in November, 1889, and opened for traffic on December 3rd, 1890. The second advance, from Vryburg to Mafeking, a distance of ninety-six miles, was begun early in 1893 and completed on October 3rd, 1904; Palapwe, 263 miles from Mafeking, was reached early in 1897. Over the 229 miles of country which still separated the rail head from Buluwayo—by this time a town with a considerable European population—the iron road was carried in the astonishingly short time of *five months*;[1] and on November 4th, 1897, the fourth anniversary of the capture of Lobengula's chief kraal, the first train, bringing the High Commissioner, Lord (then Sir Alfred) Milner, to perform the opening ceremony, steamed into the station after its run of 1,360 miles from Capetown.

In the meantime a second line of railway had been built to give Rhodesia access to the sea at Beira, the Portuguese port on the east coast of Africa. The first section of this line, running for seventy-five miles across the Tsetse Fly belt,[2] was built as early as October, 1893; and the Portuguese and Rhodesian frontier was reached three years later. It was subsequently extended to Umtali and Salisbury, and it has for some years past formed a part of the considerable system of the Beira and Mashonaland Railway Company, which to-day connects Buluwayo, Rhodesia, and South Africa in general with the port of Beira.

Apart from this large construction of railways the progress, both administrative and industrial, which was

[1] At the rate of 687 miles a year.

[2] This insect being fatal to horses and cattle (although it does not attack human beings) had made transport between Beira and Mashonaland practically impossible prior to the construction of the railway.

made in the two years following the Matabele war, was remarkable. The internal communications had been greatly improved by the making of roads, and the completion of the telegraphic and postal systems. Among the four or five centres of European population, Buluwayo, the youngest, had grown most rapidly. In less than a year it had been converted from the kraal of a savage chief into a European town of 2,000 inhabitants, with brick buildings and two weekly newspapers; while in the surrounding country there was a considerable population engaged in the business of gold-mining. At Salisbury, Buluwayo, and other towns public buildings had been erected, and all the administrative machinery of a civilised state had been established, and was running smoothly. Apart from Matabeleland, gold mining was in progress at Salisbury, Mazoe, Umtali, and in other districts of Mashonaland; and some of these Mashonaland mines had begun to obtain creditable results from the crushing mills. The advance secured by the end of 1905, and the bright prospects of the immediate future, are vividly reflected in Mr. F. C. Selous' account of the condition of affairs which he found in Buluwayo upon his return to Rhodesia.

"In short, at this time," he wrote,[1] "everything was apparently *couleur de rose* in Matabeleland. Properties, whether farm lands, building sites in town, or mining claims, went up to very high values, whilst almost everyone believed that within a year Buluwayo would contain a population of 5,000 souls, and that the town itself would receive a plentiful supply of water from the reservoirs already in course of construction, and be lighted by electric light. In fact all was mirth and joy and hope in the future; for what was to hinder the ever-increasing

[1] *Sunshine and Storm in Rhodesia.* Mr. F. C. Selous, the famous hunter of big game, had acted as guide to the Pioneer Expedition in 1890.

prosperity of the country? Much good work had already been done on many of the reefs, and on the whole the promise was distinctly good. Then, again, after a probation of eighteen months, the country had been pronounced favourably upon by Dutch and Colonial farmers, especially for cattle-ranching, whilst many predicted that much of the high veld would carry sheep."

Dr. Jameson, then, in the three years of his administration had not merely "pulled through"; he had largely increased the resources of the Company's territories, and brought comfort and prosperity within reach of the settlers.

In the next year, however, the brightness of this prospect was suddenly darkened by the clouds of disaster. The Jameson Raid (December 29th, 1895—January 2nd, 1896) not only disorganised the local administration and endangered the continued existence of the Chartered Company, but it denuded Rhodesia of the white police upon whose presence and efficiency the scattered European population depended for protection against any hostile movement among the Bantu masses. The insurrection of the natives, which, commencing on March 20th, 1906, with the murderous attacks of the Matabele upon the settlers living in isolated homesteads and mining camps, spread afterwards to the Mashonas, was not subdued until the end of the year. The originating cause of the rising, in which many settlers lost their lives, and much property was destroyed, was not, in all probability, the mere absence of the white police, but the visitation of the rinderpest, or Zambezi cattle fever, in itself a disaster of sufficient magnitude, which had reached the neighbourhood of Buluwayo on March 5th. Upon the appearance of this terrible scourge, special powers were at once obtained from the High Commissioner, and in order to arrest the spread of the disease, a wholesale destruction of the cattle belonging to the natives took

13—(8139)

place in the infected districts. How necessary this was may be seen from the sequel. In spite of the vigorous enforcement of the regulations for the destruction or isolation of affected cattle in Rhodesia, the disease made its way to Palapwe in Bechuanaland by March 11th; and although £50,000 was spent in the Protectorate by the Imperial Government in the endeavour to arrest its progress, it rapidly overran the Cape Colony and Orange Free State. After the construction of a fence across the colony, following the line of the Orange River and guarded by mounted police, and other measures had failed to protect the farmers, the Cape Government, as a last resource, commissioned Dr. Koch, the German scientist, to visit South Africa and investigate the disease, in the hope that some remedy might be discovered

Other causes contributed to make the natives in Rhodesia discontented at this time. For some of these, such as the apparent confiscation of the "royal" cattle [1] and the abuses committed by the native police, the Company's administration was in part responsible; but others were physical disasters, such as a plague of locusts and drought, which, like the outbreak of rinderpest, only the malice of the "witch-doctors" and disaffected chiefs could attribute to the presence of the white men in the country. But it is unlikely that all these causes combined would have brought the natives to the point of open rebellion, if it had not been for the opportunity

[1] The Company, having succeeded by right of conquest to the ownership of Lobengula's cattle, took possession of a part only of the royal herds, and having branded the rest (90,000 head) with the Company's mark, left them to be tended and pastured by the natives as before. The officials then called up these cattle as they were required from time to time. The natives, being unable to distinguish between ownership and agistment, resented the process as being a confiscation of cattle they had come to regard as their own property. Ultimately the Company took two-fifths of all the royal cattle, and gave the remaining three-fifths (70,000) head to the natives as their absolute property.

presented by Dr. Jameson's withdrawal of so large a proportion of the white police.

Mr. Selous' narrative of his own experience affords significant evidence on this question, and is in other respects valuable; since it serves to emphasise a fact which must never be forgotten by the statesmen of South Africa, that the supremacy of the white man depends in the last resort upon force. The murder of native policemen on the night of March 20th, was, according to Mr. Selous, the first overt act of rebellion on the part of the Matabele. Returning to his homestead, Essex Vale, on mid-day of March 24th, he found that some natives from a neighbouring village had been over to borrow some axes. Settlers were accustomed to render such small services to their native neighbours, and the request had, therefore, caused no surprise to Mrs. Selous.

"After sundown," he continues,[1] "some of these same men brought the usual evening's milk, and my wife and I chatted with them for some time. We spoke about the recent murders on the Umzingwani, and the conduct of Umzobo and Umfondisi, and my wife asked me to say that she thought they had acted very foolishly, as the white men would punish them. At this they laughed, and one of them said significantly, 'How can the white men punish them? Where are the white police? There are none left in the country.'"

Mr. Selous tells us further that, apart from such isolated acts of lawlessness as may be found in any similar community, young or old, there was no ill-treatment of the natives by the white settlers. In his own case, he and his wife had lived on the best of terms with the natives in their immediate neighbourhood; but he cannot attribute his—and her—escape to this circumstance. "Why no attempt was made to murder us on that Tuesday," he writes, "will always remain a mystery

[1] *Sunshine and Storm in Rhodesia.*

to me." The assumption that he had just returned from summoning assistance, was, he believes, the real reason that made them stay their hands. That they were not by any means well-disposed was shown by the fact that they went off that night with the cattle placed under their charge, and probably assisted in the murder of Messrs. Foster, Eagleson, and Anderson.

Thus again in 1896, as before in 1893, the settlers in Rhodesia, like every other European community in South Africa, were compelled to demonstrate their superiority to the natives by force of arms. Thanks to the swift messages of the telegraph, assistance was promptly rendered by the Imperial authorities from Mafeking—600 miles from Buluwayo, but at the time the nearest railway terminus; and when, by the end of the year, the insurrection had been put down, it was found that, after the first treacherous murders of the isolated whites, comparatively few Europeans had been called upon to sacrifice their lives. The troops employed consisted of a small number of British regulars, of irregular corps raised in Kimberley and Johannesburg, and of the hastily raised citizen forces of the settlers; and among these latter many Dutch fought side by side with their British comrades. They were placed under the command of an Imperial officer, General Sir Frederick Carrington; but the Company defrayed the expenses of the entire military operations. "It is the ambition of the Chartered Company and of the people of this country to secure for England the peaceful possession of Rhodesia at their own cost, and without calling upon the taxpayers of England or of the Cape Colony for the contribution of a single sixpence," was the reply of Lord Grey, who had succeeded Dr. Jameson as the Company's administrator, to the offer of assistance from the Cape Government.

Not only was this ambition realised, but ample compensation was made to the settlers for all losses caused by the insurrection, and by the enforced destruction of cattle under the rinderpest regulations. At the same time a new and better system of native administration was introduced. Following the Natal precedent, the authority of the chiefs, and the tribal divisions and organisation of the Bantu population, were restored so far as was compatible with the due maintenance of the undisputed supremacy of the Government. The chiefs or *indunas*, under this system, are salaried officials of the Government; and, as such, they are responsible in the first instance for the maintenance of order among their people. But in each (native) division there is a European official, the Native Commissioner, to whom the chief pays the taxes due to the Government, and refers for advice and direction in all matters affecting the interests of the European inhabitants.

Certain changes in the administrative powers and *personnel* of the Chartered Company, which were the immediate result of the Raid, must be briefly mentioned. The command of all armed forces, police, volunteers, and native levies, was at once withdrawn from the Company by the Imperial Government, and placed in the hands of an Imperial officer, styled "Commandant-General of the local forces in the Bechuanaland Protectorate and in the territories south of the Zambezi under the direct administration of the British South Africa Company, and Deputy-Commissioner of the last mentioned territories." In the Bechuanaland Protectorate all administrative powers of the Company were cancelled, and it was left with its commercial rights only. In Rhodesia the powers of the Company's Administrator were henceforward to be purely civil, and in the exercise of these civil powers he was to be subordinate in matters of policy to the Deputy-Commissioner, the local representative of the Imperial Government. Four days after the surrender of Dr. Jameson's force to the Boers, Rhodes resigned the premiership of the Cape Colony, and shortly afterwards he sailed to England to give an account of his actions to Mr. Chamberlain, and the Imperial Government. In June he and Mr. Alfred Beit resigned their seats on the Board of the Company, and Rhodes' power of attorney to represent the directors in South Africa was cancelled; and on July 30th, the Select Committee of the House of Commons was appointed to inquire into the circumstances of the Raid, and report thereon. In the meantime Lord Grey,[1] a Director of the Company, had come forward to undertake the duty of administering Rhodesia at this critical moment. The native insurrection broke out while he was on his way from England; at Capetown he accepted an offer of military assistance from the Imperial Government, and on April 28th he arrived at Buluwayo, then surrounded by 10,000 Matabele rebels, with a welcome supply of arms and ammunition. His cheery defiance of circumstances—the same spirit of high courage and lofty Imperial aims which afterwards found a wider field for its expression in Canada—was reflected in his first public speech. At the beginning of the insurrection the Government had only 379 rifles, he said; but *now* "Buluwayo was as safe as London."

To obtain the services of Lord Grey at such a time was a signal and unexpected advantage, but the "crowning mercy" which softened this year of pestilence, civil commotion, and rebellion to Rhodesia was the presence of the founder himself. When the report of the insurrection and the peril of Buluwayo was cabled to England, Rhodes had already announced his intention of making Rhodesia his future home, and of henceforward devoting his energies to the development of the country which had already come to bear his name. This news, however, hastened his departure; and travelling by the Suez Canal to Beira he actually reached Salisbury in time to

[1] Since Governor-General of Canada (1904–11).

set out on April 18th with the relief column equipped at that place, which, after relieving Gwelo and doing other useful work, arrived at Buluwayo on the 30th May.

On the occasion of this march, and subsequently during the rebellion, Rhodes showed an utter disregard of the danger to which he was exposed from the Matabele guns and assagais. But he did more than this. It was due to his courage and address that the submission of the Matabele chiefs was secured when the rebellion had reached a stage in which it threatened to develop into a long protracted and harassing guerrilla war. On August 23rd, after camping without any military protection for six weeks at the foot of the Matopo hills, he rode, with Dr. Hans Sauer, Mr. Collenbrander, and four others,[1] for four miles through line after line of glittering rocks and broken kopjes, where thousands of unseen Matabele lay securely entrenched, to an unknown meeting place to hold council with the chiefs and *indunas*. Then:

> Just four miles from camp the six reached the foot of a huge kopje, and 100 yards further on was the trysting-place. Mr. Rhodes and his companions dismounted in dignified silence, and took up their position by a large ant-heap and waited.
>
> The suspense seemed interminable, and although there was the stillness of death, the six knew well enough that the place was surrounded by hundreds of armed Matabele. Any wavering sign of fear would have been fatal, but nothing of the kind was shown.
>
> The critical moment came when Grootboom[2] advanced to the kopje to say the party were awaiting the appearance of the chiefs.
>
> Suddenly there was a gleam of dead white from the kopje, and all the chiefs filed out in a row, headed by one carrying a white flag. They drew near to the party in silence, and squatted round them in a semicircle. The Indaba lasted for five hours, all points at issue being thoroughly discussed, and full explanations proffered by the white men to allay the uneasiness which the chiefs evidently felt at certain possible consequences of the surrender.

[1] One was the representative of the Press. All took revolvers, except Rhodes, who was entirely unarmed.

[2] The guide and intermediary in the negotiations between Rhodes and the chiefs.

Then the chiefs rose, and each threw a small stick at the feet of Mr. Rhodes, indicating their willingness to surrender their guns, while another similar stick meant they were ready to hand over their assagais.

In return, Mr. Rhodes, whose coolness and dignity were never at a loss, promised that the desired abolition of the native police force should be taken into serious consideration. The chiefs then solemnly declared that hostilities should cease at once, and guaranteed the safety of the roads and of the coaches.[1]

But the advantages resulting from Rhodes' presence in the colony at this time were not confined to the assistance which he gave in the suppression of the native rebellion. Employing his large private resources to hasten forward the railway from the South, he wrought with such energy that, as we have seen, an unprecedented rate of construction was attained, and before the end of the following year (1897) Buluwayo was placed in direct railway communication with Capetown. He built for himself a spacious house in the Afrikander style on the outskirts of Buluwayo, and a comfortable homestead on the Matopos. Of these the former has long served as the official residence of the Company's Administrator; while the latter, lying near to his favourite hill of the World's End View, on the summit of which he was buried, and to the great park that he made and bequeathed to the people of the colony, is maintained as a hotel for the convenience of visitors to the colony and holiday-makers from Buluwayo. Realising that the mines alone, without agriculture, would not suffice for the prosperity of the community, he turned a valley into a lake, bred sheep, ostriches and cattle, irrigated and cultivated the land, formed nurseries of forest trees, flower gardens, and plantations of tropical plants, to show the various uses to which the soil could be put. Wherever he went, he made it his business to learn by personal converse what were the difficulties and grievances of the settlers; and

[1] *Daily Telegraph*, August 24th, 1896.

By permission of *W. Basil Worsfold*

THE GRAVE OF RHODES

thus fortified by knowledge he set himself to make better provision for their needs.

By these activities Rhodes was no less benefited than Rhodesia. He had felt deeply the public chastisement to which the Raid had exposed him; but even "unctuous rectitude," he knew, must admit that in these months of untiring effort he had done some service to the Empire. He had another source of consolation, too, in this hour of bitter defeat. His will, in which the whole scheme of the Rhodes Scholarships was set forth, and the greater portion of his wealth appropriated to the realisation of his great idea, had been made already. In his moments of depression, as Lord Rosebery has told us, Rhodes thought over this document with the comforting assurance, that when he was dead, it would show his countrymen how, at any rate, no ignoble motive had led him into his great political error. And so in Rhodesia Rhodes regained the sense of moral integrity which enabled him, when he returned to the Cape Colony in the following July (1897), to become the virtual, though not the official, head of the Progressive party—a position which he maintained until almost the very hour of his death.[1]

The facts necessary to bridge the interval between these early struggles and the assured prosperity of to-day can be told in a few sentences. In 1899, the year of the great South African war (1899–1902), Rhodesia took its place as a regular contributor to the world's gold supply with an annual output of £205,690 in value. Thanks to the masterly defence of Mafeking by Baden-Powell and the Company's excellent arrangements for the protection of the frontier, Rhodesia suffered no injury from the Burgher forces; and in 1900 the value of the gold output had risen to £308,249, approximately one-tenth of the present value. For the next few years industrial progress

[1] March 26th, 1902.

was retarded by the shortage of native African labour —an economic difficulty, directly resulting from the war, which Rhodesia shared with the rest of South Africa. None the less, during this period of retarded development the railway communications were steadily extended and improved, constitutional changes giving the settlers a voice in the administration of the colony were introduced, and agriculture as well as mining made slow but appreciable progress. In the early part of the year 1905 the writer saw the first (trial) train steam slowly across the giant framework of the Victoria Falls Bridge; and in this year, with a gold export of £1,500,000, Rhodesia may be said to have "turned the corner." From this date onwards the progressive advance of its population and industrial resources has not been interrupted.

By permission of *W. Basil Worsfold*

THE VICTORIA FALLS BRIDGE (LOOKING WEST FROM ABOVE THE PALM GROVE)

CHAPTER II

SOUTHERN RHODESIA

At the present time the British South Africa Company's territories are divided for administrative purposes into Northern[1] and Southern Rhodesia. The area of the former is 291,000 square miles; or roughly twice as great as that of the latter—148,575 square miles. They are separated by the river Zambezi, running west to east; and both are traversed from south to north by the Cape-to-Cairo Railway. Northern Rhodesia is a tropical country with 830,985 native inhabitants, and a very small European population, the number of which was returned on May 7th, 1911, as 1434. Its imports for the year 1911 were £127,664 in value, and its exports, £128,458. It has a considerable export of copper ore, and it is likely to be the seat of cotton production on a great scale. Apart from its timber and rubber, and the general suitability of its soil and climate for the growth of tropical produce, it possesses an immediately remunerative source of wealth in the numerous wild beasts, which give the country its special attraction to the sportsman, and to one of which, the elephant, it owes its once considerable export of ivory.

Interesting and commercially valuable, however, as is Northern Rhodesia, it is the lesser division of the vast territories of the Chartered Company that Englishmen speak and think of as Rhodesia. For here, south of the Zambezi, is the country in which Rhodes lived, and worked and was buried; the Rhodesia that is a part of South Africa.

The Rhodesia of to-day, then—that is Southern

[1] North-Eastern and North-Western Rhodesia were united on August 17th, 1911.

Rhodesia,—has 23,582[1] European inhabitants and a native population of 750,000; an area of 148,575 square miles; a revenue of £773,209 10s. 2d. meeting an expenditure of £663,742 14s. 4d.; [2] over 2,000 miles of railways, and an external trade of £5,053,332. [3] Her gold export of £2,624,908 in value for the year 1909 placed her fourth among the gold-producing countries of the Empire; the value of her soil for cattle-ranching and all forms of agriculture has been established, and her tobacco plantations are assured of a ready market in England and America. [4] That her claim to provide a healthy and prosperous life for men and women of the British race is not baseless, may be seen from the fact that her European population has almost doubled within the last seven years.

The colony is governed, under the authority of the Crown, by the Company's Administrator; an Executive Council consisting of the chief officials; and a Legislative Council, the members of which are partly nominated, and partly elected by the European inhabitants. The enactments of this latter body, termed Ordinances, require the assent of the High Commissioner for South Africa, as the representative of the King, before they have the force of law.

The Legislative Council, the Parliament of the colony, was originally constituted in 1898, when it consisted of seven members nominated by the Company and four elected by the settlers. In the year 1911 it consisted of seven elected, and five nominated, members. Apart from the fact that the members of the Executive are

[1] May, 1911: showing an increase of 10,906 over the numbers returned in the census of 1904.

[2] For year ended March 31st, 1911.

[3] For 1911—excluding specie and re-exported imports.

[4] Some particulars of these and other industries will be found in the later chapters, where they are treated as part of the industries of South Africa as a whole.

responsible to the Company, and ultimately to the Crown, this Council exercises much the same legislative powers as are possessed by the legislature of a self-governing colony.

While Salisbury, in Mashonaland, remains the seat of the Administration, Buluwayo, in Matabeleland, is the largest European town, and the chief commercial centre of the colony. Second to these, are the towns of Umtali in Mashonaland and Gwelo in Matabeleland. The following tables show the distribution of the European population on the night of May 7th, 1911.[1]

MASHONALAND

Fiscal Division.	Population.	Chief Town, etc., of Division.	Population.
Salisbury	6,573	Salisbury township	3,019
Umtali	2,151	Umtali township	946
Victoria	437	Victoria township	97
Melsetter	600	Melsetter village	30
Charter	701	Enkeldoorn village	102
Hartley	2,031	Gatooma village	294
Railway and Coach Passengers	50		
Total	12,543		

MATABELELAND

Fiscal Division.	Population.	Chief Town, etc., of Division.	Population.
Buluwayo	8,355	Buluwayo—Town and suburbs	4,177
Gwelo	2,095	Gwelo township	306
Gwanda	407	Gwanda village	33
Railway and Coach Passengers	180		
Total	11,039		

In forming a conception from the above tables of the size and importance of these places, it must be remembered

[1] Compiled from the Government *Gazette* of the B.S.A. Co., June 16th, 1911.

that no mention is made in them of the native population. In Rhodesia not only are the household servants almost entirely male natives, but native labour performs most of the rough work of all industries—work which in England, or any other purely white state, would be done by whites of the "wage-earning," and most numerous, classes of the community. The consideration is one that applies, though in a varying degree, to all the great centres of population in South Africa, in respect of which the number of European inhabitants alone is generally, but quite wrongly, assumed to measure their importance, and afford a basis of comparison between them and purely European towns. Of the Rhodesian towns, Salisbury and Buluwayo alone possess municipal institutions; but the local affairs of the lesser towns and villages are managed by Sanitary and Village Management Boards.

This European community as a whole, and its various centres of population taken separately, may be considered to be well provided with the necessaries and conveniences of civilised life. One or two examples will serve to show that the progress of Rhodesia in this respect has been unusually rapid. Churches, law courts, schools, railways; postal,[1] telegraphic, and telephonic communication; banks and newspapers; parks and recreation grounds—all these are expected as a matter of course in a British colony, even though it is barely twenty-one years of age. But Rhodesia, owing to the personal responsibility for its welfare which Rhodes and other founders have assumed, enjoys some advantages that are rarely to be found in so young a State. The provision for Education is a case in point. Over thirty public schools have been established by the Administration in the towns, at the larger mines, and in country districts; and there are

[1] Rhodesia, in common with other parts of South Africa, has a weekly mail to and from England.

denominational schools at Buluwayo, Salisbury and Gwelo. The bursaries and scholarships are numerous and generous. Two hundred and sixty boarding grants of £20 [1] per annum per child are distributed, of which 160 are provided by the Administration and 100 by the Beit Trustees. Each year twenty scholarships of £20 per annum for day scholars, and £40 for boarders, tenable for three years, are awarded under the terms of the Beit Bequest; and the same source provides—also each year—three Bursaries of £100 per annum tenable for three years at any of the University Colleges of South Africa. Nor need the young Rhodesian scholar, whose foot has once been planted firmly upon the academic ladder, fail to reach its highest rung. In the distribution of his scholarships, Rhodes naturally gave a preference to the community which he brought into being. And thus, as "founder's kin," Rhodesia has every year no less than three of the Rhodes Scholarships; which, it is scarcely necessary to add, are each of the annual value of £300, and tenable for three years at Rhodes' own Alma Mater—the University of Oxford. [2]

Or, to turn from moral to physical needs, the provision for medical aid is on a scarcely less ample scale. It comprises:

(1) Fully equipped hospitals at Salisbury, Umtali, Gwelo, and Hartley; a Relieving Ward at Enkeldoorn; a Native Hospital at Gatooma, and Cottage Hospitals, also fully equipped, at Victoria, Gwanda, Belingwe, Abercorn, Mazoe and Sinoia;

(2) A public Memorial Hospital at Buluwayo supported in part by voluntary contributions and in part by a grant-in-aid from the Administration;

[1] The boarding fees are not generally more than £50 per annum, *plus* the ordinary school fees for day scholars of from £3 to £6.

[2] For particulars see forward, Part V, Chap. IV, "Education," p. 472.

(3) District Surgeons stationed at Salisbury, Buluwayo, Umtali, Hartley, Belingwe, Mazoe, Malsetter, Gwelo, Gwanda, Selukwe, Inyanga, Marandellas, Abercorn, Victoria, Enkeldoorn and Inyati;

(4) Subsidised Medical officers resident at Plumtree, Rusapi, Sinoia, Gatooma and Kimberley Reefs; and

(5) A department of Public Health.

A general account of the climate of Rhodesia has been included in the description of the physical characteristics of South Africa as a whole, with which this volume opened. A few specific details, however, will not be found superfluous. The determining factor is, of course, the high average elevation of the country, the central plateau of which is raised from 3,500 to 5,000 feet above sea level. This factor preserves the people of Rhodesia, though living within the tropics, from most, though not all, of the climatic disadvantages to which European residents in other tropical countries are generally subjected. The character of the Rhodesian year can be indicated most conveniently by dividing it into only two seasons; the wet, or warm season, lasting from the end of October to the beginning of April, and the dry, or cool season, which prevails during the remaining seven months. The heat of the sun's rays in the warm season is, therefore, tempered by the prevalence of rain clouds, while in the cool season it warms the earth unchecked the whole day through. Moreover the high altitude brings healthful winds, which in the warm season prevent the moist air from becoming stagnant and oppressive, and in the dry season moderate the power of the unveiled sun in its midday glory. That the excessive moisture usually associated with tropical countries is not to be feared in Rhodesia is shown by the average rainfall. In Mashonaland it is 32 inches, or practically the same as in the United Kingdom; but in Matabeleland it is only 24 inches—one-fourth less than that of the mother country.

By permission of *W. Basil Worsfold*

A RHODESIAN HOTEL

The following tables, taken together, will give a fairly complete idea of the kind of weather which may be expected.

AVERAGE RAINFALL RECORDED OVER A PERIOD OF YEARS INCLUDING 1910

MASHONALAND

Station.	Amount in Inches.	Number of Days on which Rain fell.	Number of years covered by observations
Salisbury	32·38	87	13
Melsetter	45·11	102	12
Umtali	30·59	88	11
Victoria	25·22	60	12

MATABELELAND

Station.	Amount in Inches.	Number of Days on which Rain fell.	Number of years covered by observations
Buluwayo	23·36	78	13
Gwelo	24·98	77	10
Tuli	13·72	33	11

This and the two tables overleaf show that there are no great extremes of heat, while, as for cold, even as much as two degrees of frost at night is rare. The rainfall, if properly conserved, is sufficient, but it is certainly not excessive; and the relatively short space of time in which it falls makes the number of rainy days all the fewer. The writer's own experience of the climate of Rhodesia, in the months of April and May, at Buluwayo, the Matopos, the Victoria Falls, Salisbury and Umtali, was wholly pleasant; and although the same perfection of sky and temperature is not to be expected all the year round, it may be said safely of the climate as a whole that it is "one of the most delightful in the world."

The agricultural and industrial resources of this colony will be treated as part of the common stock of South Africa, but the mention of certain political and social characteristics which place Rhodesia somewhat apart from the provinces of the Union, will not be out of place.

Maximum Rainfall in 24 Hours during each Month of the Year ended December 31st, 1910

Station.	Jan.	Feb.	Mar.	April.	May.	June.	July.	Aug.	Sept.	Oct.	Nov.	Dec.
Mashonaland												
Salisbury	0·73	0·78	5·00	1·55	—	—	0·03	—	—	1·61	1·43	0·53
Melsetter	0·90	2·90	2·90	0·96	0·25	0·70	0·70	0·20	0·21	0·60	1·40	1·69
Umtali	0·85	0·85	2·75	0·17	0·07	0·60	0·33	0·03	0·27	1·50	1·25	0·60
Victoria	0·42	1·28	1·03	0·18	0·16	—	0·22	0·04	—	0·48	1·57	0·97
Matabeleland												
Buluwayo	1·40	0·75	2·56	0·02	0·02	0·01	0·06	—	—	0·94	0·75	0·54
Gwelo	0·54	2·62	2·30	0·32	0·01	0·05	0·05	—	—	0·30	1·29	1·15
Tuli	1·32	1·00	1·35	0·01	0·06	—	0·08	—	—	0·50	0·40	0·38

AVERAGE TEMPERATURE AS RECORDED IN THE SHADE, AT 8 A.M., DURING THE YEAR ENDED DEC. 31ST, 1910

Station.	Approximate Height above Sea Level.	Mean Maximum.	Mean Minimum.	Absolute Maximum.	Absolute Minimum.
	Feet.				
Salisbury	4,880	76·8	51·2	89·2	33·4
Rhodes Matopo Park ..	4,100	73·1	52·5	98·0	30·0
Victoria Falls	2,994	84·9	55·8	98·6	33·0
Buluwayo	4,469	76·7	53·3	90·3	34·1 [1]

[1] The above Tables are taken from the *Report on Meteorology for* 1910, of the Agricultural Department of Southern Rhodesia.

In the first place, Rhodesia has benefited by the experiences of the older European communities of South Africa, with the result that political and industrial mistakes have been avoided, and the administrative machinery has been planned upon the best lines and runs smoothly. Then she has no "nationality difficulty." The Dutch Afrikanders who have settled within her borders, are content to identify themselves in sentiment, language, and manner of life with their much more numerous British neighbours. The controversies, by which the people of the Union are chiefly agitated—controversies which arise mainly out of racial antagonism—such as the respective degrees in which Dutch and English are to be taught in the public elementary schools, or whether British emigrants should be encouraged, or warned off, are not, therefore, to be found in Rhodesia. Freedom from the bilingual incubus in itself is an appreciable economic gain. There is no need for double notices—English and Dutch—or for the double medium of instruction in the schools; since English is the only language employed in politics, business, and social intercourse.[1]

The existence of these advantages bulks largely among the considerations which govern the attitude of Rhodesia upon the question of admission to the Union. What this attitude is may be gathered from the speech delivered by the Administrator, Sir William Milton, at the opening of the session of the Legislative Council on May 17th, 1909. With reference to the Union of the four colonies, which was then on the eve of accomplishment, he said:

> Although it was not thought desirable to include Rhodesia in the proposed Union from the outset, special provision has been

[1] The "Kitchen Kafir," which is the usual language of communication between the Europeans and the Natives is easily learnt; but in the towns even this knowledge is scarcely necessary, since the Natives in domestic service and other town employment have learnt, generally, to speak English—and are proud of the accomplishment.

made for admission when the circumstances of the country permit; and I desire to repeat the hope expressed by me last year, that at no distant date it may be found possible for Southern Rhodesia to take that place in the Union which is due to it by reason of the character of its population and its varied and great resources.

The Government desires to give the assurance that no agreement in regard to Union will be accepted by it which does not take into account the views and circumstances of all classes of the community.

It is not to be expected, therefore, that Rhodesia will seek admission to the Union, so long as this step involves a sacrifice of her immunity from racial conflicts or a diminution of her administrative efficiency. Such patriotic considerations as are valid in the existing circumstances point in the same direction as her economic and political interests. The degree of influence in the common counsels to which her present population would entitle her as a Province of the Union, would be too small to enable her to assist materially in allaying the racial and social controversies which form the chief obstacle to the prosperity of the Union: while for the same reason she would run the risk of seeing the development of her European population, as a homogeneous and progressive community, retarded by uncongenial legislation or administrative neglect. In these circumstances, it would seem that Rhodesia would promote alike her individual interests, and those of South Africa as a whole, by resolving to remain outside the Union, until the growth of her population and the consolidation of her material resources are such, that they will secure for her an effective part in shaping its destinies.

But apart from these considerations, there is another element in the situation in which the Union is concerned equally with Rhodesia. Up to the present the Chartered Company, as we have before noticed, has provided the whole of the large capital expenditure necessary for the acquisition, administration, and general development of the colony. If Southern Rhodesia had been a Crown Colony, this capital expenditure (which is, of course, distinct from any sums invested by the Company in specific industrial undertakings within its territories) would have been provided in part no doubt by the British taxpayer, but most of it would have been raised by loans, the principle and interest of which would have been secured upon the assets and revenues of the Government of the colony. These loans, thus secured, would have constituted collectively the public debt of the colony, and the annual service of this debt would have been a first charge upon its revenues. Rhodesia, as it is, has no such public debt; but its equivalent exists in the proprietary rights in the country possessed by the Chartered Company under the terms and conditions of its Charter of incorporation.[1]

These rights, representing the security of the shareholders of the Company for the money which they have subscribed, are, broadly, the unalienated land and mineral rights of the colony. Moreover, of the specific industrial undertakings of the Company, one, namely the construction and working of the railways, is an undertaking which in South Africa is regarded as coming within the sphere of government. And on this head it must be added that although the Rhodesia Railways, Ltd., and the Beira and Mashonaland Railway Company, Ltd.—the two companies by which the over 2,000 miles of the Rhodesian railway system are owned and worked—are separate undertakings under separate management, in both of them a controlling interest is held by the British South Africa Company. It is obvious, therefore, that before any Government of the Union could accept the responsibility of administering Southern Rhodesia as a Province, it would be necessary that the colony should be released from any such proprietary rights of

[1] Subject to subsequent and valid modifications (if any).

the Company as would limit the power of the Union Government to raise revenue, or interfere with the performance of any of its duties or functions as defined in the Union Act. With every desire on the part of the settlers to do justice to the shareholders, and with a corresponding goodwill on the part of the directors of the Company, the process of ascertaining what capital sum is the just equivalent of the rights to be surrendered will none the less be one of considerable difficulty. And when an agreement on this head has been reached, there will remain the further question, whether the public indebtedness of the colony, as thus ascertained, will be greater than the Union Government, having regard to the taxable capacity of its inhabitants and its revenual prospects in general, would be prepared to assume.

Financial considerations, no less than political, seem, therefore, to counsel delay; since every increase of European population, and every fresh development of industry, will bring Southern Rhodesia nearer to the time when its actual and potential sources of revenue will enable it, or the Union on its behalf, to offer a fair recompense to the Company without unduly burdening either its own taxpayers, or those of the Union. In the meantime it is to the interest of all parties "to let well alone."

NOTE ON SPORT IN RHODESIA

In view of the special attractions which Rhodesia presents to the sportsman, the following brief particulars may be of interest: The writings and exploits of Mr. F. C. Selous and others have made English readers acquainted with the general character of the game, big and small, which abounds in the country, both north and south of the Zambezi. The "common" kinds include the elephant, eland, sable and roan antelopes, buffalo, kudu, and waterbuck. The rhinoceros also is plentiful in certain localities, and the lion, although like other beasts of prey, seldom seen in daylight, is to be found wherever there is big game. A permit must be obtained from the Administrator, or from the British South Africa Co., 2 London Wall Buildings, E.C., before firearms or ammunition can be taken into Rhodesia. There is no gun licence ; but a duty of 20s. for a single barrel, or 30s. for a double-barrel gun, with, in either case, 10 per cent. *ad valorem* in addition, has to be paid by any person bringing a gun into the country. There is also a duty on ammunition. An ordinary game licence costs £1, but special licences (£5 to £25 and upwards) are required to shoot particular kinds of big game. Sportsmen are recommended by the Company to make themselves acquainted, on arrival in Rhodesia, with the various game reserves which have been proclaimed ; and to obtain full information in respect of them from the Company's officials in the districts in which it is proposed to hunt. Information of a general character can be obtained from the Company's offices in London (including an excellent illustrated handbook on *Big Game Shooting in Rhodesia*).

CHAPTER III

THE NATIVE TERRITORIES

FOUR-FIFTHS of the 6,000,000 natives of British South Africa, having been incorporated prior to the Union into one or other of the four constituent colonies, are now ruled by the Union Government. Three Native Territories, however, Basutoland, the Bechuanaland Protectorate, and Swaziland, with an aggregate population of rather more than half a million persons, and a total area of rather less than 300,000 square miles, remain for the present outside the Union. It is the natural destiny of these native states—for in each of them the European population numbers only a thousand, more or less—to be " taken over " by the Union Government; and the manner in which they are to be administered in this event is set out in the schedule to the Act of Union. The general intention of the schedule is to preserve the rights and privileges of the several native populations, and, subject to the necessary adjustments, to secure a continuance of the existing systems of administration. From this point of view three of the most important sections are those which deal respectively with land tenure, the sale of intoxicating drinks, and the " pitsos," or assemblies of the tribesmen, customary among the industrial Bantu. They are:

14. It shall not be lawful to alienate any land in Basutoland or any land forming part of the native reserves in the Bechuanaland Protectorate and Swaziland from the native tribes inhabiting those territories.

15 The sale of intoxicating liquor to natives shall be prohibited in the territories, and no provision giving facilities for introducing, obtaining, or possessing such liquor in any part of the territories less stringent than those existing at the time of

the transfer [of the territories from the Imperial to the Union Government] shall be allowed.

16. The custom, where it exists, of holding "pitsos" or other recognised forms of native assembly shall be maintained in the territories.

In each case, however, there were special reasons which made it seem expedient for these territories to continue for the present under the direct administration of the Imperial Government through the High Commissioner for South Africa. In two of them, Basutoland and Bechuanaland, the natives, being industrial Bantu, are sufficiently advanced in civilisation to be allowed to pursue their own native manner of life under the immediate rule of their chiefs, who are themselves, however, guided and controlled by Imperial officers. Here it was "to let well alone," to permit existing agencies of proved efficiency to complete their work, that made a change of government seem inexpedient. In the case of Swaziland, the smallest of the three, considerations of an opposite order counselled delay. The native population, which consists of military Bantu, and is probably an offshoot from the Zulu people, has no claim to moral superiority; but the country has only just been extricated from the industrial chaos into which it was thrown by the reckless and unlimited concessions granted to Europeans by the late Chief Umbandine, after the discovery of gold in his territory in 1885. Possessing considerable deposits of gold and tin, and lying, as it does, between the Transvaal and the sea in the line of the proposed direct railway from the Rand to Delagoa Bay, Swaziland may any day become an important field for European enterprise.

Basutoland

Basutoland is an oblong stretch of mountainous country, separated on its north-west side by the Caledon River from the Free State province, and rising thence in the Maluti Mountains to the steep escarpment of the Drakenberg range, which on the north-east and south-east forms, as it were, a sunk fence between it and the provinces of Natal and the Cape. At the corner of the bend, where the Drakenberg range projects due east into Natal, Champagne Castle, the highest mountain of South Africa, rises 12,000 feet above sea level. South-west, where the Kornet Spruit and the Orange River leave their mountain cradles, the land opens to the Cape and Free State. This little country of mountain peaks and rushing torrents has been called the "Switzerland" of South Africa with a greater closeness of analogy than is customary in such descriptions: since, apart from its physical likeness, its people resemble the Swiss in their composite origin, the strenuous defence of the liberties of their mountain home, and their complete environment by powerful neighbouring states. For the rest, Basutoland has an area of 10,203 square miles, an average elevation of 6,000 feet, an ample but not excessive rainfall of between 30 and 40 inches, and a mountain air the mean maximum and minimum temperatures of which were, in 1910, 72.7° and 41.6° F. respectively.

The Basuto people, as we have seen,[1] were created by the address and policy of the wise chief, Moshesh (1815-1870), out of bands of fugitives, mainly industrial Bantu, who gathered in the mountains to escape destruction at the hands of the military tribes. As early as 1833 Moshesh admitted the representatives of the Paris Evangelical Society, and since then Basutoland has remained one of the most successful fields of missionary enterprise. After years of prosperity and many successful conflicts with native and European invaders, the Basutos were broken by four years' war with the Boers of the Free State, and only escaped from eviction and dispersal at the hands of their conquerors by finding

[1] Part I, Chap. II, p. 33.

a refuge under " the large folds of the flag of England." At this date, 1868, the most fertile half of their original territory had been wrested from them by the Boers, 2,000 of their people had been killed, and 15,000 more had fled from the country. The 100,000 or so who remained were miserable, poor and in great distress ; and Moshesh, who had performed this last and crowning service of winning for them the protection of Great Britain, was soon to die.

As British subjects, however, the Basutos, being naturally industrious and eager traders, rapidly recovered their prosperity ; and the census of 1875 showed a return of 127,707 native, and 469 European inhabitants. In 1871 the country was annexed to the Cape ; and the local government remained responsible for it until the year 1884, when it was handed back to the Imperial authorities. The circumstances which led to the abandonment of Basutoland by the Cape Colony—the only instance of an administrative failure of the kind—serve to remind us of the fact that the supremacy of the white race in South Africa has only been established quite recently, and then as the result of constantly recurring and arduous warfare. In the early years of the diamond industry, the natives who worked on the mines were allowed to purchase guns and ammunition without restriction ; and among the tribes which thus became possessed of fire-arms were the Basutos, large numbers of whom were sent to work at Kimberley by their chiefs with this very object in view. It was the epoch of that general manifestation of hostility against European authority on the part of the military Bantu, which produced the last Kafir war (1877–8) and the Zulu war (1879). In the year of this latter conflict there were disturbances in Basutoland, and in 1880 the Cape Government determined to carry out a systematic disarmament of this and of other native territories for which it was responsible. The proclamation ordering the Basutos to surrender their guns and ammunition was disobeyed by the whole tribe, with the exception of a few loyal chiefs and their immediate followers. The virtual revolt of the Basutos was followed by disturbances among the natives in Griqualand East, Pondoland and Tembuland, and the great mass of Bantu population lying between the Cape and Natal was dangerously agitated. The local forces of the Cape Government were successful in putting down these risings on its eastern border, but when they advanced into Basutoland, the strenuous resistance of the tribesmen, combined with the mountainous character of the country, defied all their efforts. After an attempt to secure a peaceful settlement through the mediation of the High Commissioner had failed, the Cape Government resolved to abandon the territory to its fate.

In these circumstances the Imperial Government, foreseeing that such a *dénouement* might produce most disastrous political effects, once more assumed a direct responsibility for the administration of Basutoland. But in doing so, they made two prudent stipulations. With the Cape Government they stipulated that the duties collected at the ports on oversea imports destined for the territory should be paid over to the administration—in other words that Basutoland should not be deprived of its customs revenue; and with the Basutos themselves, that unless they gave proof of their desire to be ruled by Imperial officers by actively co-operating with them, the new administration would be withdrawn. The first Resident-Commissioner, Sir Marshall Clarke, possessed the capacity of making prompt decisions, and the sympathy and knowledge of Bantu conditions, which are requisite for the successful handling of the South African natives. In due course he restored order, and established the simple but efficient system of administration under which Basutoland has reached its present prosperity. His successor, Sir Godfrey Lagden, was no less capable and experienced; and to him belongs the credit of restraining these mountain tribesmen, who were naturally eager to secure so favourable an opportunity of paying off old scores upon the Free State Boers, from taking part in the great South African war. Thus since 1884 Basutoland, apart from tribal quarrels, has enjoyed peace and settled government under the direct administration of the Imperial Government. In this period of rather less than thirty years the native population has doubled, the annual revenue has advanced from a few thousand pounds to £119,974 in 1909–10, and the external trade of the country has reached the respectable total of over £600,000 per annum. The whole of the revenue is spent in providing for the needs of the territory; and not only is there no public debt, but the Administration has a balance, exceeding a year's revenue in amount, standing to its credit in the bank.

The system of administration under which these excellent results have been obtained is, in its general features, the system applied throughout British South Africa to all native populations living on land reserved for their sole occupation or in native territories, whether within or without the Union. The basis of the system is the maintenance of so much of the tribal organisation and authority of the chief, and the recognition of so much of the native law and custom, as is found, in each case, to be consistent with the establishment of a civilised government under European officials. It is, therefore, an application of the wider principle of the recognition of local institutions and local laws adopted alike in the Roman and British Empires for the incorporation and government of tributary states and subject populations.

The territory of Basutoland is administered by a

Resident Commissioner, who is the deputy of the High Commissioner for South Africa, together with Assistant Commissioners and the recognised native chiefs. The Legislative authority is vested in the High Commissioner, and is exercised by Proclamations, duly issued from time to time. Apart from the body of law formed by these proclamations, the regulations framed in virtue of them, and the native customary law, the law of the Cape Colony, as it was in 1884, has effect so far as the circumstances of the territory permit. By the regulations in force Courts of Justice of the Resident Commissioner and Assistant Commissioners are established, and jurisdiction is conferred upon the recognised chiefs in civil, and within certain limits in criminal, cases between natives, in which native law may be administered. In all cases, however, there is a right of appeal to the Resident Commissioner. The customary tribal authority of the hereditary chiefs is further recognised, and their services are further utilised, for administrative purposes by entrusting to them the performance of police duties and the collection of the hut tax. For these purposes the country is divided into seven districts, which are sub-divided into wards; and each ward is presided over by a native chief. In return for their judicial and other services the chiefs receive a subsidy from the Administration.

There is a representative body, the National Council, which, like the National Assembly in Egypt, has consultative and advisory, but not legislative powers. It consists of about 100 members, of whom a part are appointed by the Administration and a part nominated by the people themselves, and it meets once a year. This useful institution, which is a recent creation, has grown out of the "pitso," or annual assemblage of all the tribesmen, which is the customary "parliament"—in the literal sense of a gathering at which each tribesman could speak what was in his mind, whether concerning a personal grievance or a question of peace or war—of the industrial Bantu. In Basutoland the "pitso," in the form of an annual, or special, assemblage of the people in the open veld, was maintained from the first by the Resident Commissioner, and it survives in the more convenient and efficient gathering which meets now in the new Council House.

In Basutoland all land is the property of the tribe not of the individual, and is occupied on tribal tenure. Plots of arable land are allotted for the time being by the chiefs among their respective tribesmen, and the grazing land of the tribe provides common pasturage for the cattle of all its members. No Europeans are allowed to reside permanently in the territory except such as have approved business, and, in the case of traders, have obtained the necessary licence to trade. These European residents cannot, of course, purchase any land, but they are allowed to occupy it; and while any house or buildings they may place upon it are erected at their own risk, in practice there is no difficulty in the transfer of such buildings by sale or otherwise to another European resident.

The cost of administration is defrayed by a hut tax of £1, the customs duties collected at the ports and rebated by the Union authorities, and by licences to trade. Of these three forms of taxation, the last is levied only upon Europeans, the hut tax is the ordinary direct tax paid by the natives of South Africa, and the customs duties are an indirect tax which falls alike upon the European and native purchaser of oversea imports. In the fiscal year 1909–10 the revenue of the territory amounted to £119,974, of which approximately two-thirds was contributed by the hut tax, and one-third by customs. The expenditure for the same year was £127,437, of which total £12,240 was expended on education. The Police Force, 292 strong, is composed of European officers with native non-commissioned officers and men. It is fully mounted and equipped, and is posted round the borders of Basutoland, principally at the various magistracies, where the Assistant-Commissioners hold their courts.[1]

The prohibition of the sale of intoxicating liquors is strictly enforced. Not only is the sale (or supply) of such liquors to natives punished by severe penalties, but no spirituous liquor is allowed to enter the territory without a permit. The prohibition of the sale of intoxicating liquors to natives, which is almost universal throughout British South Africa, is based upon the fact, established by long experience, that "the native is constitutionally incapable of being a moderate drinker, and that he must either abstain entirely or the chances are that he will drink to excess, and when in drink all the failings of his nature assert themselves."[2] The use of the native, or Kafir, beer, provided it does not contain 4 per cent. of spirit, is, however, generally permitted, and indeed recommended as a preventive against scurvy and kindred complaints; and this beverage is largely made and drunk by the Basutos.

The main industry of the territory is agriculture, and apart from the grain required for home consumption large quantities of wheat, oats, and Kafir corn, or mealies, besides wool, mohair, and hides, are exported. The country being well adapted for stock-raising, the Government is endeavouring to improve the stock of the natives by the distribution of well-bred sheep and goat-rams, and by the introduction of stallions. As South Africa, as a whole, has never yet succeeded in producing all the food supplies required for its own consumption, the Basutoland export of wheat is proportionately welcome. And no less valuable, in view of the chronic shortage of African labour, is the service which the territory performs

[1] *Union Blue Book on Native Affairs*, for 1910.
[2] *Report of Native Affairs Commission*, 1903–5.

15—(2139)

in sending out its surplus of adult males to work in the industrial centres of the Union. The following figures, taken from the report of the Native Affairs Commission of 1903–5, will show the capacity of the territory in this respect. Out of a total (native) population of 347,731 in 1904, it was estimated that 86,933 were males over fifteen years of age, of whom 49,676 were married, 37,257 unmarried, and 69,546 were between the ages of fifteen and forty; of these latter, 35,186 might be expected to be at work at any one time, and since Basutoland itself demanded only 2,000, the surplus available for labour outside the territory was over 33,000. In the year 1910, 82,000 passes were taken out for labour employment outside the territory; and of the men who thus left 32,000 went to the mines in the Free State and Transvaal.

As we have noticed before, Basutoland is one of the most successful fields of missionary enterprise. Of the 250 schools in the territory, all, with the exception of two small Government schools, are worked by various missionary societies, and the great majority of them belong to the Paris Evangelical Mission. In addition to the elementary schools a training college for native teachers at Morija, and an industrial school at Roma, have been provided. In this latter the boys are taught stone-cutting, building, carpentry, blacksmithing, tailoring, and other trades, and the girls spinning, weaving, and cooking. All these mission schools are subsidised by the Administration. In the matter of education—that is, of course, native education—Basutoland stands second to the Cape Province. In 1904, according to the Native Affairs Commission, 1903–5, there were 10,484 scholars, or 3·01 per cent. of the Basuto population, in schools, as against 4·24 of the native population in the Cape, 1·13 in the Transvaal, 1·12 in Natal, and 2·76 in the Free State; and in the same year the Administration spent £7,000 on education out of the £60,528 contributed to the revenue by the hut tax. In the year 1909–10 the average attendance at schools was 9,718 and the number of teachers 352. The enrolment of pupils in the elementary schools of the territory in December, 1909, was as given below.

Mission.	Boys.	Girls.	Total.
Paris Missionary Society	4,993	6,044	11,037
Church of England	646	844	1,490
Roman Catholic	185	667	852
Totals	5,824	7,555	13,379

The capital, Maseru, which is the residence of the Resident Commissioner and the seat of the Administration, is connected by rail with the Free State, and has a population of 1,400, of whom some 200 are Europeans. There are telegraph offices here and at Mafeting, Morija and some half-dozen less important places, and a general service of Posts has been organised. A telephone line runs from the capital to all government stations throughout the territory. The usual mode of conveyance is an ox-wagon or light cart. The roads have been made serviceable for all forms of transport, but they are subject to serious damage from the overflow drainage of the high watersheds during the periodical rains.

THE BECHUANALAND PROTECTORATE

The Bechuanaland Protectorate stretches from the northern border of the Cape Province, between German South-West Africa on the west and the Transvaal on the east, to the southern border of Rhodesia and the Zambezi. Being part of the great central plateau of South Africa, it has an average elevation of over 3,000 feet, and its general character is that of an almost waterless region of sandy wastes whose sole vegetation is the camel-thorn and the wild melon. Its area, 275,000 square miles, is twice as large as that of the Transvaal; and scattered over this vast expanse of territory a population of 140,000 natives and 1,000 Europeans is to be found in the few fertile districts, or along the line of the railway which traverses it from Mafeking in the south to the Rhodesian border on the north. The sole value of the territory lies in its geographical position. To-day it forms the highway from the Cape Province to Rhodesia, while its western deserts serve as a natural barrier between British and German territory; and at the time when, in 1884, a British protectorate was first proclaimed over it, its strategical importance was much greater. The circumstances in which this step was taken by the British Government have been related before, as a significant episode in the history of British rule in South Africa; [1] and all that need be added to the facts then placed before the reader is, that certain portions of the original Protectorate, as it remained after the southern portion, *i.e.*, the "Crown Colony of Bechuanaland," had been annexed to the Cape Colony in 1895, were subsequently transferred to Rhodesia. This arrangement was in part a compensation to the Chartered Company for the loss of its administrative powers over the Protectorate, which, as we have seen, [2] was one of the immediate results of the Jameson Raid.

The territory is governed by a Resident Commissioner, with two Assistant-Commissioners, under the direction of the High Commissioner for South Africa, and the services of the native chiefs are utilised for administrative purposes even more largely than in Basutoland. The legislative power is vested in the High Commissioner, and exercised by proclamation, but subject to this special legislation the law of the Cape Colony obtains so far as the circumstances of the country permit.

[1] Part I, Chap. II, p. 42, and Chap. VI, p. 115.
[2] Part III, Chap. I, p. 187.

The people are industrial Bantu, who live on Reserves held under communal tenure. The six principal chiefs are: Khama, Sechele, Gaseitsive, Linchwe, Mathibe, and Baitlotle, and their respective tribes, over which they rule, the Bamangwato, Bakwena, Bangwaketsi, Bakhatla, Batawana, and Bamalete. The chiefs have judicial powers in both civil and criminal suits. In cases between natives native law is administered; but there is a right of appeal to courts of the Assistant-Commissioners, and a right of final appeal to the Court of the Resident Commissioner. Cases between natives and Europeans can be tried by consent of the European suitor in the Chiefs' courts; and in this case no appeal lies from the chief's decision. Otherwise, mixed cases are taken by the Commissioners.

Of these tribes the most important are the Bamangwato and the Bakhatla, ruled respectively by Khama and Linchwe. The former of these chiefs, Khama, has been described[1] as affording an instance of the administrative ability which is sometimes to be found in persons of the Bantu race, and it is only necessary to add that he has again removed his people—on this occasion from Palapwe to Serowe, where there is a European settlement. It must also be remembered that it was among the peaceable and industrious Bechuanas that Robert Moffat, David Livingstone, and John Mackenzie obtained the most notable results of their evangelistic labours.

The insufficient rainfall and poverty of the soil make agriculture difficult, but a small quantity of grain over and above what is needed for home consumption is grown and exported to Kimberley and Johannesburg. The main industry, however, is cattle-ranching, since cattle, sheep and goats can thrive on the small bushes which cover the limestone areas of hard ground: and a considerable export of cattle, apart from surplus labour,

[1] Part I, Chap. II, p. 34.

forms practically the only contribution of the Protectorate to the general resources of South Africa. Its surplus of African labourers, according to the Native Affairs Report of 1903–5, is about 9,000; that is to say, out of a total of 12,083 adult males, who might be expected to be at work at any one time, only 3,000 are required within the territory. The total exports amounted in 1909 to £123,627 in value, and the imports to £96,096.

The seat of the Administration is placed for convenience at Mafeking, just within the Cape Province, but the principal European settlements are Gaberones, Francistown, and Serowe. The expenditure of the Administration in 1910–11 amounted to £70,875, of which some two-thirds was provided by the revenue, while the balance was made up by grants from the Imperial Exchequer. The two main sources of revenue are (1) the hut tax of £1, which is collected by the native chiefs, and amounted to close upon £30,000 in 1910-11, and (2) the customs, amounting in 1909 to £10,543. The Bechuanaland Protectorate, unlike Basutoland, has an annual deficit of some £20,000; and to this slight extent it adds to the burdens of the British taxpayer.

Swaziland

Swaziland lies between the Transvaal on the north and west, Mozambique on the east, and Natal on the east and south. It is a small territory with an area of 6,536 square miles, and a population of some 90,000 natives and about 1,000 Europeans. Westward, the territory is of the same character as the high veld of the Transvaal, being in fact a terrace raised to an average height of 5,000 feet above sea level, with a temperate climate and well watered pasturage: eastward, where the land falls to only 1,000 feet above sea level and then rises again in the Lebombo range, the country is freely clothed with tropical bush, and here in the summer, or rainy season, the climate is unhealthy for Europeans, and dangerous for horses, mules, and cattle.

The Swazis belong to the Zulu-speaking group of military Bantu, and although their country was never incorporated into the South African Republic, it had come to be administered by the Boers under Transvaal law at the time that the great South African war broke out. The circumstances which led to this result are characteristic of the difficulties by which the British Government has been beset in its endeavour to do justice alike to the Dutch and natives in South Africa. In the days of Ketchwayo's power, Sir Theophilus Shepstone earned the gratitude of the Swazis by preventing the Zulu king from "washing his spears" amongst them; and in the Zulu war they declared themselves to be the allies of the British. In the final operations of this war, the storming of the stronghold of Secocoeni, Ketchwayo's "dog," Lord Wolseley was assisted by a contingent of 8,000 Swazis. When the Transvaal was given back to the Boers, the independence of the Swazis was secured under both the Pretoria (1881) and London (1884) Conventions. The Boer farmers, however, had long been in the habit of "trekking" with their cattle into Swaziland, where, on the eastern side of the great mountain ranges, they found the grass green and luxuriant at a season when their own pastures were dry and bare. Originally this was done under grazing leases obtained from the Swazis, but subsequently the Boers established themselves permanently across the border in Swaziland, as they had attempted to do in Bechuanaland. More than this, when, in 1883, gold was discovered in the eastern Transvaal, and two years later on the Rand, Swaziland was overrun by prospectors and European adventurers, who were mainly British. The chief, Umbandine, being an independent sovereign under the Conventions, had no British resident to advise or protect him in the business transactions which then took place between himself and these adventurers. Having a taste for good living and small European luxuries, he granted "concessions" for mining, trading, and carrying on industrial undertakings of every description, within his territory in return for wine or sweetmeats, cases of gin, a trifling sum in cash, and even less solid considerations than these. Among many hundreds thus granted were concessions of the sole right to lay out townships, to practise as doctors, lawyers, surveyors, etc., to levy taxes of any nature or any amount, to advertise, to work posts, telegraphs and railways, to establish banks, music-halls, theatres, lotteries, pounds, and markets, to take photographs, and, most ingenious of all—a concession of the sole right to obtain all unallotted concessions.

The handful of Europeans then obtained a "charter" from Umbandine, in virtue of which they elected a committee, with judicial and administrative powers, and created a petty republic, called the Little Free State. Then the chief fell ill, and the committee and concessionaires had to do business with the chief's councillors, whose tenure of office was rendered insecure by their liability to be "smelt out" for witchcraft, and clubbed to death with knob-kerries, at the shortest notice. On the chief's death in 1889 the condition of Swaziland had become so scandalous as to constitute a danger to the peace of South Africa, and a joint commission, representing the South African Republic and the British Government was appointed to investigate the state of affairs and suggest a remedy. Out of this commission there resulted the Swazi Convention of 1890, under which a joint administration, consisting of three commissioners representing respectively the British Government, the Transvaal and the Swazis themselves, was established. This was the time when Rhodes was endeavouring to bring the Transvaal into closer relations with the rest of

South Africa, and the Convention, besides providing for the government of Swaziland, gave President Kruger permission to construct a railway from the Transvaal to the coast at Kosi Bay, provided that the South African Republic joined the Customs Union not later than August 8th, 1893. As the required condition was not fulfilled, the Convention was denounced. The British Government thereupon, in 1893, virtually handed Swaziland over to the Boers ; but at the same time, by annexing the strip of land lying between Swaziland and the coast, then called Tongaland, effectually prevented the President from obtaining the dream of his life—a port for his Republic. As the Joint Commission of 1889 had recognised the validity of Umbandine's concessions, in spite of the grotesque character of the majority of them, the country, when it reverted to the British Government with the annexation of the Transvaal in 1900, was found to be in a state of apparently inextricable confusion. Lord Milner, in these circumstances, wisely obtained the consent of the concession holders and other interested parties to the application of certain general principles, before he allowed a commission to undertake the task of sifting the concessions, adjudicating upon conflicting claims, and harmonising the rights of individuals with those of the native and European communities. In this way it was understood before the Concessions Commission (provided for by the Swaziland proclamation of the High Commissioner in 1904) began its labours, that all concessions which either trenched upon the proper sphere of the Government, or deprived the native population of any material resources necessary for their welfare, were to be bought out at actual cost price, or otherwise expropriated, and that in the case of grants of land at least one-third of the respective areas were to be reserved as the unalienable property of the Swazis.

In view of the establishment of Responsible Government

in the Transvaal, Swaziland was placed under the control of the High Commissioner by order-in-council of December 1st, 1906; and in March, 1907, by proclamation of the High Commissioner, provision was made for the appointment of a Resident Commissioner, with Assistant-Commissioners, and for the establishment of a Police Force and a Court of Justice. The jurisdiction of the chiefs is, however, maintained in civil causes between natives. The laws of the Transvaal, subject to the special legislation of the territory, are in force. As the paramount chief, Sobhuza, is a boy of about thirteen years, his grandmother acts as Regent, with the assistance of a Council of Chiefs.

The Concessions Commission has completed its report, and the conflicting claims of the numerous concessionaires are in process of adjustment. A special commission, appointed for the purpose, completed the demarcation of the native reserves in 1910; and the fact that the native areas, the Crown Lands, and the mineral and grazing concessions, are now well defined has stimulated agricultural and mining development. At the same time the material well-being of the native population has been improved very materially by the establishment of a settled government.

For the year ended June 30th, 1910, the revenue of the Swaziland administration amounted to £46,831, of which £25,602 was the yield of the native tax; and the expenditure was £54,217. In the same year the imports were £44,309 in value (of which £31,160 was paid for oversea merchandise), and the exports amounted to £90,348; the chief items in the latter being gold, £44,499; and tin ore, £41,768. Five gold mines and four alluvial tin mines, on which 2,071 coloured and 74 white persons were employed, were being worked in 1910.

PART IV

INDUSTRIAL DEVELOPMENT

CHAPTER I

THE LABOUR SUPPLY

SOUTH AFRICA, as we have seen, is marked off from the other dominions by the presence of its numerically predominant native population. The difference between South Africa and Canada, Australia, or New Zealand, due to this primary circumstance, is as great in the economic sphere as in the political. While these other dominions, being white countries like the United Kingdom, draw their supply of unskilled labour, with rare exceptions, from their European population, South Africa depends, and has always depended, upon coloured people for the performance of all tasks of mere manual labour. And side by side with this primary circumstance there is the no less significant fact, that the native or indigenous population of South Africa, although many times as numerous as the European Colonists, has never sufficed to provide the Europeans with all the unskilled labour that they required for their various industries.

In the first years of the foundation of the Cape Colony under the Dutch East India Company, the principle was laid down, that the white man must not do mere manual labour but leave all such tasks to the natives. Directly a serious commencement of industry was made, in the era of the Van der Stells, it was found that the indigenous coloured population could not supply the unskilled labour needed for the vineyards and the farms, and for the everyday service of the European community, and the Dutch East India Company thereupon imported

Malays from their East Indian possessions, and slaves from Central Africa, to supplement the Hottentot labour supply. In the second half of the nineteenth century, when the industrial development of the sub-continent began in earnest, the sugar and tea growers of Natal brought indentured coolies from India to work their plantations; and later on the Dutch farmers of the western districts of the Cape Colony made proposals to the Government of the day for the introduction of Chinese labour. The Diamond Mines at Kimberley were able to attract sufficient native labour; since, owing to the exceptional profits of the industry, they could afford to offer wages, which were in some cases ten times as high as those paid by the farmers—at that time practically their only competitors. But the gold mines of the Rand, where the conditions of the industry so far from permitting of any such extraordinary rates of payment, demanded a rigid economy in the costs of production, required a new method of supply. To meet the labour demands of this new and rapidly developing industry, a system of recruiting was inaugurated. Labour agents, traversing the great seats of native population in Portuguese territory as well as in the South African States, visited the kraals, and secured large numbers of able-bodied workers who otherwise would never have left their homes to seek work under a European employer. By means of this system, before the war broke out the mines of the Witwatersrand were provided with a continuous supply of some 100,000 native African labourers.

These African labourers, it must be remembered, do not work year-in and year-out like the European workman; they engage themselves for a period of months only—in the case of the gold mines from six to twelve months—and, when this period of service is completed, they return to their families. Here they remain until the desire for more money, or possibly the attraction of a town life, induces them to enlist for another term of service. To have established a source of supply sufficient to maintain the native labourers on the mines at a level of 100,000, was, therefore, a very considerable achievement; and it is one which bears testimony to the exceptional energy and skill that were possessed by the men who founded and developed the gold industry of the Rand.

After the war (1899–1902) there was a shortage of unskilled labour all over South Africa; and the gold industry in particular, which had become the mainspring of the industrial system of the sub-continent, was threatened with "stagnation." That is to say, unless an adequate supply of labour could be found, instead of developing beyond the £18,000,000 annual rate of output reached in 1899, its recovery would be arrested at a rate of £12,000,000 annual output. Such a result would not only have retarded most injuriously the political and economic reconstruction of the New Colonies under Lord Milner, but have involved the whole of South Africa in financial disaster. The inability of the gold industry to obtain African labour at this time was due to three principal causes. In the first place the body of 100,000 labourers, gradually collected in the ten years preceding the war, had been scattered and could not be immediately re-assembled. In the next, many of the natives employed by the military authorities during the war had earned large sums of money, and on this account they were more than usually reluctant to engage themselves for service on the mines. In the third place—and this was the most important cause—the declaration of peace brought with it an exceptionally large demand for unskilled labour throughout South Africa. Not only had all the material destruction of the war to be repaired, and public buildings, bridges, houses, farms, roads and railways to be rebuilt or reconstructed; but in view of the industrial expansion which was expected to follow—and did in fact follow—the establishment of British authority from "Lion's Head to Line," the several Colonial Governments undertook the construction of new railways and other public works upon a great scale. Speaking broadly, in the second year of the peace (1903) South Africa required at least a third more African labourers than she could obtain, and the gold industry of the Rand, the expansion of which was a condition precedent to the solvency and industrial prosperity of all the South African Colonies, required not less than double its then available supply of African unskilled labour. And of this wholly insufficient supply—some 60,000 in all—more than two-thirds came from Portuguese territory, and not from British South Africa at all.

It was in these circumstances that the responsible leaders of the Mining industry in the Transvaal, after fruitlessly endeavouring to obtain first African labour from all parts of the continent, and then coolie labour from India, came to the conclusion that the sole available source from which the African supply could be supplemented was China. Lord Milner, having become gradually convinced of the necessity of this course, at least as a temporary measure, gave his support to the proposal; and, by so doing, saved South Africa from an economic crisis which would have thrown the finances of the four colonies into grave disorder, and brought widespread loss and suffering upon the European communities. In the last seven months of the year 1904 the gold industry employed an average of 9,668 indentured Chinese labourers; in 1905, 39,952; in 1906, 51,427; in 1907, 49,302; in 1908, 21,027; in 1909, 6,516; and in 1910, in the early months of which year the last Chinamen were repatriated, it employed an average of 305. As the result of this accession to the labour supply, the output from the Gold Mines of the Rand rose from

£12,142,307 in 1903 to nearly £29,000,000 in 1908; and the number of white men employed increased from 13,207 in 1904 to 17,593 in 1908. In the year 1907 the Transvaal Government, under pressure from the Home Government, decided on political grounds to put an end gradually to the employment of Chinese coolies; and their decision was carried out, as shown by the above figures, before the date of the Union (May 31st, 1910). At the same time—the year of the Union—the Indian Government determined to cut off the supply of Indian indentured labour, upon which the Natal planters had hitherto relied for working their plantations. The Union of South Africa, therefore, with the exception of the Malay population in the Cape Province, the domiciled British Indian population in Natal and in the Transvaal, and some miscellaneous Asiatic labourers mainly in the ports, intends to depend at present upon its native African and other indigenous coloured population for its supply of unskilled labour.

The future alone can show whether this policy, and in particular the decision to repatriate the Chinese coolies, will be justified by events. For the present it remains only to undertake the somewhat difficult task of estimating the extent of the unskilled labour supply afforded by the coloured population of the Union, and to state the degree in which the labour demands of the various industries are being satisfied.

At the outset it is necessary to recall the two modifying circumstances to which the attention of the reader has been directed incidentally in the foregoing paragraphs. Of these the first is that the greatest industry of South Africa, the Witwatersrand Gold Industry, to-day derives half of its unskilled labour supply from countries outside of the Union; and the second, that the native and coloured labourer does not work continuously, but intermittently and for short periods. To say that the adult males of the native population, between the ages of twenty and forty, may be expected to provide the equivalent of ten years of continuous service in European employment, is probably, to make a generous estimate of their industrial capacity.

No better point of departure for an inquiry into the nature of the unskilled labour supply of South Africa can be found than that which is afforded by the following paragraphs, taken from the report of the Inter-colonial Native Affairs Commission of 1903–5. The commissioners are dealing, of course, with the position as they found it in 1904, and their population figures, which are those of the census of that year, include Southern Rhodesia as well as what is now the Union, and also the Native Territories under the Imperial Government. They write :

> The calculations which have been made show an estimated constant demand of 782,000, and an estimated continuous supply of 474,472, showing a shortfall of 307,528 labourers.
>
> Under these circumstances the question naturally arises, how in South Africa agriculture or any industry is carried on ? The answer is that when carried on at all, it is carried on under difficulties, as to which there is abundant evidence. The British South African aboriginal native has not fully met the labour requirements of the country. There is no doubt that were these natives alone to be relied upon, South African industries could at present only be worked at half power. Native labour has had to be supplemented by the employment of Africans imported from other parts of Africa, Indians and Chinese. In the latest report of the Government Mining Engineer of the Transvaal, it is mentioned that of the natives employed in mines there, only 15 per cent. were British Africans. . . .
>
> There are many causes which may be deemed to have produced this situation. . . . One half of the native population lives on reserves. The bulk of these occupy land for the most part communally and free of charge except hut or poll tax, upon which it is possible for them in some fashion or other to make a living as agriculturists or peasant proprietors, without the necessity, excepting in exceptionally bad seasons, of earning wages.
>
> With regard to those who do not live on the reserves, and who have, either from personal motives or by reason of compelling circumstances, occupied either Crown Lands or the lands of private owners upon payment of rent, it may be said that this

portion of the population has continued to be able to farm, on a small scale indeed, but with sufficient measure of return to enable them to supply their own small wants and pay such rent and taxes as have been demanded from them.

Both the above classes of natives have had access to the land on terms which have enabled them to regard work for wages as a mere supplement to their means, and not as it is regarded in the older industrial communities, namely, as the urgent condition under which the majority of mankind earn their daily bread.

The theory that the South African natives are hopelessly indolent may be dismissed as being not in accordance with the facts. Even the simple wants of the native population cannot be supplied without some degree of exertion. The population of 4,652,662 has to derive its sustenance from a soil which is not everywhere fertile, and the native agriculturist has to contend with the same drawbacks of drought and pestilence that beset the European farmer. The labour of tilling the soil, weeding, and reaping, is shared, but is by no means exclusively performed, by the native women; and the representation of the native living at his own village a lazy and luxurious life, supported by his wife or wives, is misleading. The Commission is not of opinion that polygamy has any considerable effect upon the native question in retarding the development of the native as a worker.

The main reasons for the existence of the labour difficulties may be summarised as follows:

The native populations have always been pastoral and agricultural.

The rapid increase of South African labour requirements, particularly during the last quarter of a century [*i.e.*, from 1880 to 1905], has found them to a great extent unprepared to meet the new conditions which surround them.

The normal condition of native life is that of a small cultivator and herdsman, and the circumstances of their history have never developed among them a class accustomed to, and dependent upon, continuous daily labour.

The inexpensiveness of their living, the limited nature of their wants, and the comparative absence of incentive to labour.

The terms on which they occupy land.

Given such a population, possessing easy access to the land, it would have been extraordinary if the present situation had not followed on a very rapid growth of industrial requirements.[1]

This, then, was the position in 1905, as it appeared to the most experienced and capable observers. The remarks

[1] Cd. 2,399, pars. 368–375.

here made as to the conditions and general character of the native population are as valid now as they were seven years ago; but certain causes, which will be noted subsequently, have united to make the available supply of native and coloured unskilled labour to-day much more adequate to the industrial demands of South Africa. Taking the census of 1911, we find that on May 7th, of that year the total "other than white" population of the Union was 4,680,474, and the "native" population, the largest constituent of the total, was 4,061,082. The European population at the same date was 1,278,025; and the number of "all other coloured races" apart from "natives" was 619,392. As against the census of 1904, these returns show that in the intervening seven years the native population of the Union alone had increased by 565,978, while the other coloured races had added a further 55,478. Since 1904, therefore, there has been a considerable increase by natural increment in the population from which the Union draws its supply of unskilled labour.

We have next to observe how this native population is distributed throughout the Union, and what proportion of it is available for undertaking work under European employers. These questions can only be answered in very general terms; but the best basis of information on the subject is that which is furnished by the table on p. 233. It must be noticed, however, that the figures here given are estimates only, since the table was compiled before the census of 1911 was taken; and that whereas the table puts the native population of the Union at 3,862,839, it was actually found to be 4,061,082 on May 7th, 1911. The figures in the table are, therefore, probably in all cases somewhat smaller than the actual returns would have given.

From this we see that the native population of the Union is divided broadly into three groups.

Table Showing the Distribution of the Native Population in the four Provinces of the Union

(Population is estimated at date (1911) *of publication of Blue Book)*

Provinces of the Union.	Locations.	Townships and Municipal Areas.	Private Lands.	Crown Lands.	Mission Reserve or Station.	Total Estimated Native Population.
The Cape—						
The Cape Proper	343,756	102,970	190,487	13,902	16,565	677,180
Transkeian Territories ..	846,994	8,620	41,893	—	5,251	902,758
Total for the Cape	1,190,750	111,590	232,380	13,902	21,816	1,579,938
Natal—						
Natal Proper	236,580	27,339	473,247	18,951	27,026	783,143
Zululand	190,356	—	12,851	62,859	—	266,066
Total for Natal	426,936	27,339	486,098	81,810	27,026	1,049,209
The Transvaal	299,658	25,445	550,318	59,140	17,458	952,019
Orange Free State	14,600	53,585	211,951	—	1,537	281,673
The Union	1,911,930	217,959	1,480,747	154,852	67,837	3,862,839

[*This Table is taken from the " Union Blue Book on Native Affairs for* 1910.]

1. Some 2,000,000 living on Locations (or Reserves);

2. About 1,750,000 living on Private Lands, Crown Lands, and on Mission Reserves; and

3. Some 200,000 living in towns and town areas, and presumably working for European employers.

As regards the first group, the conditions under which they live are excellently stated in the paragraphs quoted above from the Report of the Native Affairs Commission of 1903–5. They live by agriculture on lands specially reserved for their occupation by successive governments, and they have no necessity to seek other employment, except in exceptionally bad seasons.

As regards the second group, the conditions of natives living on Crown Lands and on Mission Reserves do not differ greatly from those of the first group. As tenants of Crown Lands, however, all adult males are required to pay a small sum annually by way of rent to the Government. But the bulk of the population included in this group are natives living on the land of private proprietors, and this class, amounting to 1,500,000, is almost as numerous as the first group. It is upon these native tenants, or squatters, that the European farmers of the Union depend chiefly for the labour with which they work their farms. It is a class, therefore, which holds a definite and important place in the economic system of the Union, and as such it must be described with some fullness.

In the majority of cases the native who lives on private lands gives the proprietor some two or three months' labour in the year in return for his holding. But this is by no means a universal rule, since in some cases the European owner allows the native to occupy and cultivate land in consideration of his (the owner) receiving an agreed-upon portion of the produce; while in others the native pays a regular rent in money. The general conditions of the squatting native, and the part he plays in South African agriculture, are well stated by Mr. W. Windham, who was secretary for Native Affairs in the Transvaal up to the time of the Union, in his final report dated July 15th, 1910. What Mr. Windham writes refers, of course, primarily to the Transvaal province; but with the necessary allowance for provincial variations it may be taken as applicable to the Union as a whole.

After remarking that there is a serious dearth of native labour for Agriculture in the Transvaal—a circumstance which is due chiefly to the "constant and increasing demand for unskilled labour in the gold, diamond, tin, and coal mines" of this province—he writes:[1]

In a general sense, farm labour is regulated by an oral or by a written agreement between the European proprietor and the native, who is almost invariably in the position of a tenant-at-will.

In consideration for his right to live upon a farm, the native undertakes to render certain services to the owner.

The terms of agreement may differ very widely from one another. As a rule, about two or three months' work without wage is required in lieu of rent—it being a matter of mutual arrangement between the parties when the labour shall be rendered. In other cases, work for a similar period might be provided for—at current rate of agricultural wages—the native consenting to pay rent for his tenancy. Again, there may be a compact under which the native is allowed to cultivate as much land as he can, the European owner taking a half share on the crops produced.

In all such cases the native is allowed the free use of the land for his own cultivation, grazing and dwelling.

All agreements are subject to three months' notice of termination on either side, and to a right on the part of the native to reap any of his standing crops.

Permitted as they are to cultivate land for their own use, practically all native labour tenants provide themselves with their own food, which is prepared by their women folk. The principal articles of diet are Indian corn (commonly known as "the mealie"), pumpkins, and potatoes. Sour milk is freely used by those in possession of cattle, whilst a beer made from

[1] *Report of the Transvaal Native Affairs Department, July 1st, 1909, to May 31st*, 1910 (Pretoria).

Kafir corn is consumed both as a beverage and as a nutritious food. Meals are ordinarily taken twice a day.

Farm labour is given under the circumstances already stated, or, where payment is made, it ranges between £1 and £2 10s. a month.

There are about 552,000 natives living on private farms, which is over 58 per cent. of the entire native population [of the Transvaal].

With a few exceptions, males only from 10 to 45 years of age enter service on farms.

It has already been observed that the supply is short of the demand nearly all over the Transvaal. The deficiency can only be met by satisfying the ever-increasing demands of the mining industry, which maintains a large establishment to recruit labour from every available source in South and Central Africa. The recruiting agencies, so far as the Transvaal is concerned, are governed by Labour Agents' Regulations.

Agricultural labour is drawn almost entirely from the farm resident population. Each proprietor, that is to say, draws his requirements from his own native tenantry, and can rarely obtain labour otherwise.

There are no institutions for the special education of farm labourers. At a few of the mission schools in receipt of Government grants-in-aid the pupils are being taught the rudiments of agricultural work, but only in a very small way. The training thus received has not yet reached a stage at which it can be said to be of any practical benefit to the farmer.

There is no system of insurance for labourers, except in the mines, where natives are insured by the Industry against partial or total disablement and death.

The third group consists of the 200,000 natives who are, so to speak, in regular employment in the towns. This group, together with the "coloured" population formed by the descendants of the yellow-skinned Hottentots, by half-castes, and by various minglings of races, provides the manual labour required for the municipal services, for the docks at the ports, and in general, for the needs of public and private employers in towns, including in part those of domestic service.

It will be seen, therefore, that the manual labour required for industries other than agriculture, and for the construction of any exceptional public works, must come mainly from the first group: *i.e.*, the 2,000,000 natives living on reserves in the several provinces of the Union. Always remembering that the Rand Gold Industry draws half of its supply of African labour from countries outside the Union, and that the natives engage themselves for periods of months, not for continuous employment, the following table will give a general idea of the amount of the unskilled labour supply to be expected from this source.

TABLE SHOWING THE RETURN OF PASSES ISSUED TO NATIVES TO LEAVE THEIR DISTRICTS IN SEARCH OF EMPLOYMENT DURING THE YEAR ENDED 31ST DECEMBER, 1910

(These figures refer to natives leaving to obtain work both within the province, outside the province, and outside the Union, with the exception of Natal and Zululand, where magistrates keep no record of labourers employed within the province.)

	Number of Passes issued.	Observations.
The Cape Proper ..	78,382	[24,792 left for the Rand Mines, and 11,415 for Kimberley Mines.]
Transkeian Territories .	79,377	[50,886 of these left for the Mines of the Rand.]
Natal Proper	37,428	[14,920 of these left for the Mines of the Rand.]
Zululand	14,940	[3,824 for the Rand Mines, and 1,426 for other employment on the Rand.]
Transvaal	102,492	[25,295 for the Rand Mines, and 21,937 for other employment in the Rand; 1,786 left for Kimberley Mines.]
Orange Free State ..	50,340	[2,370 for the Rand Mines; 1,396 for Kimberley Mines.]
Total	362,959	

[*This Table is taken from the "Union Blue Book on Native Affairs for* 1910."]

We can now proceed to examine the state of the African labour supply at the chief industrial centres of the Union as indicated by the latest available returns.

The Transvaal Mining Industry

The number of natives employed on mines and works in the Proclaimed Labour Districts of the Transvaal on December 31st, 1910, was 206,536, as against 183,258 employed on December 31st, 1909. The average number of natives employed throughout the year 1910 was 207,921, as against 182,357 for the preceding year 1909; thus showing an average increase for the year of upwards of 25,000.

The table on p. 239 shows the sources from which these 206,536 natives, employed on December 31st, 1910, on mines and works, were drawn; and it also shows the number and origin of the 86,615 additional natives, who were engaged at the same date in employment other than mines and works in the Proclaimed Labour Districts of the Province.

The table on p. 240 exhibits the development of the supply of native labour in the Transvaal since 1906.

It will be seen that the supply of native labour in the Proclaimed Labour Districts of the Transvaal has increased largely since the repatriation of the Chinese was begun in 1907. These latter were employed exclusively in the gold industry of the Rand; but whereas the Rand gold industry employed 84,897 African labourers in 1906, it employed 183,613 in 1910. In these four years, therefore, the Rand Gold industry has secured an addition of 100,000 native labourers, or twice the number—50,000—of the Chinese supply.

This result, that is to say the large increase of native labour thus obtained both for the Transvaal industries as a whole and for the Rand Gold industry in particular, appears at first sight to be inconsistent with the statement of the extent of the shortfall of native labour in 1904, which was put on record by the Native Affairs Commission of that date. An analysis of the causes to which the increase is to be attributed, and an examination

Table Showing Territorial Analysis of Natives Employed in Proclaimed Labour Districts of the Transvaal

Territory.	Mines and Works at 31/12/10.	Other Employment at 31/12/10.	Totals at 31/12/10.	Increase (+) or Decrease (–) as against 31/12/09.
East Coast, south of latitude 22°	78,559	3,044	81,603	+ 2,582
Mozambique	9,604	36	9,640	+ 2,750
Zambesia, Quillimane, and Tete	6,089	21	6,110	+ 2,296
Beira and Chinde	255	41	296	– 158
Nyasa	3,146	2	3,148	+ 344
Portuguese Territory	97,653	3,144	100,797	+ 7,830
Basutoland	8,210	2,565	10,775	+ 2,704
British Bechuanaland	1,191	1,075	2,266	+ 145
North-Eastern Rhodesia and Fort Jameson	124	—	124	– 19
British Nyasaland and Protectorate	3,383	326	3,709	+ 554
Cape Province	56,874	18,406	75,280	+ 10,958
German South-West Africa	350	281	631	+ 116
Natal and Zululand	14,896	21,769	36,665	+ 6,556
Orange Free State	1,102	4,998	6,100	+ 759
Southern Rhodesia	905	901	1,806	+ 63
Swaziland	2,989	1,302	4,291	+ 1,366
Transvaal	18,699	31,529	50,228	+ 4,876
Others	16	319	479	– 66
Totals	206,536	86,615	293,151	nett + 35,920

[*From the "Union Blue Book," as before.*]

TABLE SHOWING THE DEVELOPMENT OF NATIVE LABOUR IN THE TRANSVAAL FROM PRINCIPAL SOURCES OF SUPPLY.

Territory.	Average Registration.								
	1906.	1907.	Increased Percentage upon 1906.	1908.	Increased Percentage upon 1906.	1909.	Increased Percentage upon 1906.	1910.	Increased Percentage upon 1906.
Portuguese	65,512	76,255	16·3	87,579	33·6	91,157	39·1	95,625	45·9
Cape Colony	29,211	36,244	24·0	47,690	63·2	61,146	109·3	78,136	167·4
Transvaal	37,190	41,580	11·8	54,046	45·3	50,819	36·6	54,343	46·1
Natal and Zululand ..	17,929	22,572	25·8	23,059	28·6	28,545	59·2	34,155	90·5
Basutoland	9,283	9,838	5·9	9,518	2·5	9,247	—	10,092	8·7
Orange Free State ..	5,191	5,233	·7	5,648	8·7	5,492	5·7	5,871	13·0
				Other Sources.					
British Bechuanaland ..	2,479	2,354	—	2,776	—	2,839	—	2,885	—
Southern Rhodesia ..	3,219	1,921	—	2,463	—	1,823	—	1,866	—
Swaziland	2,245	2,740	—	3,830	—	3,014	—	4,101	—
Brit. Cent. Africa Prot. ..	2,264	1,092	—	775	—	1,318	—	4,101	—
German S.-W. Africa ..	561	590	—	549	—	492	—	527	—
N.-E. Rhodesia and Fort Jameson	603	302	—	246	—	173	—	135	—

[*From the "Union Blue Book," as before.*]

of the relation which it bears to the increase in the existing demand for African labour in the Transvaal, will show, however, that this is not the case. Shortly put, the explanation of the apparent inconsistency is this. In 1904 the supply of African labour was *below* the normal, and the demand was considerably *above* the normal. By 1907 both supply and demand had become fairly normal, and in these circumstances the Rand Gold industry, with its highly organised system of recruiting, was able from this time forward to increase largely its supply of native labour. By so doing it minimised, though it did not by any means wholly avoid, the injury inflicted upon it by the enforced withdrawal of the 50,000 Chinese labourers. That the gold industry was adversely affected by the repatriation of the Chinese will be generally admitted. In the first place, the 50,000 Chinese were more valuable industrially, as being more efficient, than a corresponding number of African natives; and in the second, the labour requirements of the industry were so great that it needed for its unfettered development the Chinese as well as any additional African labour which it could secure.

The increase in the Transvaal native labour supply is to be attributed to three main causes.

1. The work of repairing the material destruction of the war was practically completed, and the African labour thus employed was set free.

2. The very large reduction in the world's demand for diamonds at the time in question, due chiefly to the commercial crisis in the United States, caused the diamond mines to reduce their output, with the result that the number of natives employed by this industry was reduced by many thousands.

3. Commercial depression in South Africa lessened the general demand for native labour; and in particular the governments of the Cape and Natal, owing to the fall

in their revenues, curtailed or abandoned many of the public works upon which they had embarked immediately after the war.

In addition to the operation of these general causes, there is one special reason for the increase in the Transvaal native labour supply which is eminently satisfactory. From the last table, which shows the development of the supply in recent years, it will be seen that, while there is a general increase from most of the principal sources of supply, including Portuguese territory, the number of natives coming from the Cape Colony had increased by no less than 167·4 per cent. in 1910 as compared with 1906. This remarkable result, which is especially satisfactory as being an increase of native labour from within the Union itself, is due to the active co-operation of the Transvaal and Cape Governments with the leaders of the Rand mining industry prior to the establishment of the Union. The departure is an important one, and an account of the arrangements to which it led, will serve to illustrate how greatly the methods of recruiting native labour have improved since the pre-war days.

In September, 1909, a conference was convened at Capetown by the Cape Native Affairs Department, at which the representatives of the Chamber of Mines and the principal recruiting agencies, and the Director of the Transvaal Native Labour Bureau were present; and the system of recruitment in the Transkei and the operation of the Labour Agent Regulations were considered.

As the result of this Conference, the Transvaal Government established labour registry offices at the following stations in the Cape Colony:—Queenstown, King William's Town, Butterworth, Indwe, Umtata, Flagstaff, Bizane, Mount Frere, Umzimkulu.

The duties of the registrars are:

(*a*) To receive all native labourers proceeding to the mines in the Transvaal;

(*b*) To explain and ratify their contracts of service;

(*c*) To observe the operations of labour agents and runners;

(*d*) To report on irregularities or misconduct on their part;

(*e*) To investigate and report on complaints or grievances brought to their notice;

(*f*) To arrange for the rationing and transmission of voluntary natives. A charge of 1s. a day is made for rations, and 2s. 6d. for registration fee—these being the charges of the Transvaal Government Labour Bureau. Rail warrants are issued for railway journeys; and the charges for rations and rail warrants are recovered from the employers by the Bureau;

(*g*) Generally to act as intelligence agents in labour matters.

These officers, who were to receive salaries of £300 a year *plus* transport and travelling allowances, visited Johannesburg in January, 1910, and were made conversant with the general conditions prevailing on the mines.

At the same time the principle of limiting the amounts advanced to natives by labour agents was laid down. The limit was fixed at £5 0s. 0d. Prior to this, storekeepers, as runners, would advance sums so large to the natives, that the latter became no longer free agents, but were compelled to go out to work to satisfy their creditors.

Also, it was decided that labour agents' licences should be issued to licensees in their representative capacity, not as heretofore in their individual capacity. The result of this was to make it necessary for applications for, or renewals of, licences to be supported by the company or organisation by whom the natives were to be employed.

Natives who desired to enter voluntarily into contracts with mining companies, and not to engage through labour agents, were to be assisted to proceed to the Transvaal for this purpose. Arrangements were made under which these natives can proceed without cost to themselves to the Government Native Labour Bureau Compound at Germiston in the Rand. Here they can dispose of their labour to the best advantage, choosing any mine that they may prefer; while they receive advice and assistance from the officers of the Bureau with respect to their contracts of service.

Arrangements were also made to permit of natives contracting at the registry offices in the Cape Colony for service in particular mines, before they commenced their journey to the Transvaal.[1]

This increased supply of native labour from the Cape Province, and particularly from the Transkei, apart from the direct benefit which it brings to the Transvaal industries, will, if it is maintained, serve to lessen the present economic dependence of the Union upon Portuguese territory. It is scarcely necessary to add that this is a consideration which would assume importance in the event of any political or economic friction occurring between the Union and Mozambique Governments.[1]

On the whole, therefore, the labour outlook in the

[1] This account is based upon the Transvaal Native Affairs Report, for 1909-10. Since writing the foregoing the following telegram has appeared in *The Times* of December 8th, 1911. It forms so apposite a comment upon the text that I reproduce it.

"LISBON, *Dec. 7th.*

"The Government has lately been receiving communications from the authorities in Mozambique, in which the complaint is made that owing to the recruiting of thousands of natives yearly for the Rand, Mozambique is threatened with depopulation. Apparently only a small number of such natives return to the province [Mozambique], and these are generally physically unfit for any further work. The Minister for the Colonies has expressed in the Senate the hope that Great Britain will consent to such a modification of the Transvaal Mozambique Convention as will ensure the compulsory repatriation of Portuguese natives."

Transvaal is improving. It must be remembered, however, that while the native labour supply has been largely increased since 1906, the industrial demands of the province have also increased. Indeed it may be questioned whether the absolute increase of labour, large as it is, represents a relative increase: whether, in other words, to-day the supply is much more capable of satisfying the demand for unskilled labour, than it was (say) in 1906, when the Chinese were on the mines. In the case of the gold, diamond, and coal mines the Report of the Witwatersrand Native Labour Association, supplies a precise statement of the shortfall in 1910. Broadly speaking, as will be seen from the table given below, these, the main industries of the Transvaal, at the end of 1910 required one-third more labour than they had been able to secure: or, more precisely, they had only 71·11 of their full complement of African labourers.

SUMMARY OF NATIVES EMPLOYED AT DECEMBER 31ST, 1910

	Complement.	Natives employed 31/12/10—		Total.	Percentage of Complement.
		By/Company.	By/Contractor.		
Nine Gold Mining Groups ..	234,054	148,892	16,963	165,855	70·86
Premier Diamond Mine.. ..	14,000	9,939	—	9,939	70·99
Sundry Members..	16,413	11,898	849	12,747	77·66
	264,467	170,729	17,812	188,541	71·29
Collieries	12,377	8,354	—	8,354	67·50
Total	276,844	179,083	17,812	196,895	71·11

(*This Table is taken from the Report of the W.N.L.A.*, 1911.)

And, moreover, the corresponding figures for the year 1911 show a slight falling off in the number of "natives employed," and an appreciable increase in the "complements" of the associated gold, diamond, and coal mines. Although the annual report of the Witwatersrand Native Labour Association for 1912 has not been issued at the time of writing, it is understood that the total supply of natives at the end of 1911 was, in round numbers, 195,000, and the complement required 297,000. The percentage of the complement was, therefore, only 65¾, or, in other words, there was a shortage of 102,000 natives, being 34¼ per cent. of the number required.

As showing that this deficiency in the labour supply is in no way to be attributed to any want of effort on the part of the mining companies, it may be added that during the year, June 30th, 1909—May 31st, 1910, no less than 797 recruiting licences were issued under the conditions imposed by the Labour Agents' Regulations. The revenue produced by these licences was £3,442 5s. 0d.; and a further sum of £2,506 was collected on account of the 506 Compound Overseers' licences which were also issued. These figures bring before us the magnitude and completeness of the recruiting system by which the Transvaal industries are served.

Mining Centres other than the Transvaal

The Coal Mines of Natal

In 1910 there were employed in the Natal collieries—Zulus, 5,820; Indians, 4,143; Cape boys, 97; Shanganas, 12; making a total of 10,072 unskilled coloured labourers. The report contained in the Union Blue Book states that there was a shortage of labour on every mine, and that on some the shortage was so great as to affect seriously the output. The number of additional labourers required was 3,278. Writing on February 21st, 1911, the Government official adds: "I expect that, in the near future, further mines will be opened and worked, and that the demand for labour will accordingly increase, especially as the supply of Indians will shortly stop."

The Kimberley Diamond Mines

The Protector of Natives, Kimberley, reports that in 1910 the only mines being worked were the Kimberley,

Du Toits' Pan, Bultfontein, and Wesselton mines. The fifteen compounds under his supervision were occupied by some 13,000 natives. In this centre the labour requirements of the De Beers Company were fully satisfied by the available supply.

THE NAMAQUALAND COPPER MINES

The Resident Magistrate at Springbokfontein reports that on December 31st, 1910, the total number of natives (mainly Hottentots) employed was 1,761, but that on March 29th, 1911, it had fallen to 1,034.

OTHER MINES

In addition to the above, some thousands of native labourers—say 10,000 in all—are required for the diamond mines in the Free State, the coal mines in the Cape, and the silver, tin and copper mines in the Transvaal.

It should be added that the conditions under which natives are employed on the mines are carefully supervised by Government officials throughout the Union. On the Kimberley and Witwatersrand mines especially, and generally elsewhere, it may be said that to-day nothing has been left undone, not merely to secure the health and comfort of the native labourers, but to provide them with all reasonable conveniences. In spite of a somewhat trying winter climate, the death rate of natives employed on mines and industrial works in the Proclaimed Labour Districts of the Transvaal has been reduced from 112·5 per thousand in 1903 to 36·23 per thousand in December, 1910. Or, to take another significant item, which is provided by the Transvaal Native Affairs Report for 1909–10, in the twelve months preceding May 31st, 1910, a sum of £8,234 19s. 8d. was received by the Department from employers of labour as being due to natives injured by accidents, or to the dependents of natives who had been killed, under the compensation

17—(2139)

scheme which was voluntarily arranged and adopted by the mining companies.

Demands for Native Labour other than Mines and Agriculture

Apart from the mines, agriculture, and other miscellaneous and lesser industries under private management, there remain two considerable demands for native, and coloured, unskilled labour which come respectively (1) from the State, and (2) from the Domestic Service requirements of the European community.

Including the municipalities and other public authorities under the head of "the State," a large part of the first demand is met by the 200,000 natives and coloured persons, who, as we saw, are returned as living in locations in town areas. Bearing this consideration in mind, it will assist us in forming some idea of the extent of this demand to remember that the Union Government employs some thousands of natives as police, especially in the main centres of native population, and the Union railways have a large staff of native and coloured employés. In respect of this latter precise figures are provided by the Report of the General Manager of the Union railways, issued in 1911. From this it appears that on December 31st, 1910, the State employed on its railways 23,178 natives, coloured persons, and (in Natal) Indians; the rest and greater part of the staff, of which the total was 49,762, being Europeans.

In the case of the demand for natives for domestic service, it must be remembered that the great majority of the natives thus employed are males, not females, and the class is commonly known as "house boys." Efforts are being made to supply this obvious deficiency by teaching cooking and house-work to native girls in the mission and other schools; but for the present the coloured population of the Cape Province is practically the only source from which female servants, not being Europeans, can be obtained, and of these the supply is necessarily very limited. In the absence of any returns, it is difficult to give even an approximate statement of the number of the natives continuously required as "house-boys"; but in view of the fact that almost every European family employs at least one native, or coloured, servant, 200,000 would seem to be a reasonable estimate.

To sum up, we know from the returns given in the Union Blue Book that on December 31st, 1910, the following bodies of native, or coloured, labourers were employed :

On the Proclaimed Labour Districts in the Transvaal	293,151	
Deduct Portuguese and Central African, etc., Natives	100,000	
		193,151
On the Natal Collieries		10,072
On the Kimberley Diamond Mines		13,000
On the Namaqualand Copper Mines ..		1,761
On the Railways		27,178
Total		245,162

What number of natives and coloured persons were employed in addition to the above total of (say) 250,000 on other mines, in public services other than the railways, in the lesser industries, in shops, and in domestic service, in the absence of any available statistics must remain a matter of conjecture. The needs of agriculture are met, as we have noticed, mainly though not entirely by the partial services rendered to the farmers by the native tenants, or squatters, of whom there is a total population of 1,500,000 within the Union.

It only remains to add that, having regard to the conditions of native life, the wages paid to the African labourers, both on the mines and generally in all employments, provide an ample return for the services they render. Indeed, the altogether exceptional circumstances which keep the European community a white community without a population of white unskilled labourers, together with the extraordinary mineral wealth of the country, have brought it about that in South Africa native labour is paid twice or three times as much as it is in Central and Eastern Africa, or in the Far East.

The subjoined table will give a general idea of the rates of wages paid for native and coloured labour throughout the Union. These rates, of course, vary very considerably as between the several provinces.

(These figures must be taken as *approximate* only.)

Province and Industry.	Per Day.	Per Month.
Farm Labour—		
The Transvaal	—	From £1 to £2 10s., with or without food.
Other Provinces ..	—	10s. to 20s., with food and sleeping quarters.
On the Wine Farms ..	1s. 6d. to 2s., with allowances.	
Mine Labour—		
The Transvaal	1s. 8d. to 2s. 6d., with excellent quarters, rations, and free medical attendance.	—
Kimberley Diamond Mines ..	3s. to 3s. 6d., with free quarters. Drilling boys, 4s. 6d.	—
The Natal Collieries ..	—	From £1 10s. to £3, with quarters and food.
House Boys	—	10s. to £4, with board & lodging.
General Labourers	3s. to 3s. 6d.	

CHAPTER II

TRANSPORT AND COMMUNICATIONS

UNDER the South Africa Act (1909) the administration of the railways and harbours of the Union is vested in a Board of three commissioners, with a minister of state as chairman. The original commissioners, appointed in 1910, were Sir Thomas Rees Price, K.C.M.G., Mr. Thomas Smith McEwen, and Lt.-Col. Edward Mackenzie Greene, K.C. ; and the first chairman, as minister of railways and harbours in the first Union Government, was the Hon. J. W. Sauer.

OVERSEA COMMUNICATIONS

Communication with England and the rest of the civilised world is maintained by mail and other ocean-going steamships and by submarine cable. South Africa is linked directly with England, Europe, Australia and New Zealand by means of the various lines of steamships which visit her ports at regular intervals. The length of the voyage from London or Southampton by the West Coast, and more direct, route is about 6,200 miles to Capetown, and about 7,000 miles to Durban. The time occupied, which varies, of course, with the class of steamship chosen by the traveller, is from seventeen to twenty days to Capetown, and from twenty-one to twenty-six to Durban. There is a weekly mail, outward and homeward, between South Africa and England. Letters by the outward mail posted in London on Saturday morning usually reach Capetown on the Tuesday fortnight (seventeen days), and Johannesburg on the following Thursday or Friday. The homeward mail leaves Johannesburg on Monday, and Capetown on Wednesday.

Thus the business man in London can receive a reply from a correspondent in Johannesburg within six weeks ; and the Rand can communicate with London by letter within the same period of time.

There is also a less frequent service of steamships, British and foreign, maintained between England and the South African ports by way of the Suez Canal and the East Coast of Africa. In the case of this, the East Coast route, the outward bound passenger can either embark at London or Southampton, or travel overland to join the boat at Marseilles or Naples, and thus shorten the time to be spent on board ship. Steamers by this route pass through the Mediterranean, down the Red Sea, and after touching at the principal ports on the East Coast disembark their South African passengers at Beira, Delagoa Bay, or Durban. Of the two, the East Coast Route, with its frequent breaks at Mediterranean and East African ports, is by far the more interesting ; but, owing to the great heat of the Red Sea and the East African waters, it makes more demands upon the physical endurance of the passenger.

It may be added that a proposal is on foot for constructing a railway through the Congo Free State to connect Lobito Bay, in Portuguese West Africa, with the Cape-to-Cairo line, and thereby with the Rhodesian and Union systems. If this project is successfully carried out, and Lobito Bay takes the place of Capetown as the ocean terminus of the West Coast route, not only will the duration of the voyage to South Africa be shortened, but Johannesburg will be brought several days nearer to England.

The records of the Union Post Office show how amply communication is maintained by letter between South Africa and England, and through England with other countries. During the year 1910 the articles of mail matter, exclusive of parcels, which were sent from the

Union to England and abroad numbered 13,993,250, while the number of those received was still larger—19,948,460.

Telegraphic communication with England and abroad is provided by the submarine service of the Eastern and Western routes.

A subsidy of £13,000, shared by the United Kingdom, the Union and Southern Rhodesia, is payable when the annual value of the traffic handled by the Associated Companies does not reach £300,000. The value of the South African traffic, and the rates charged during the past eleven years are as follows:

Year.	Rate per word.	Amount.
		£
1900	4/-	457,366
1901	3/6	336,986
1902	3/-	465,791
1903	2/6	391,108
1904	,,	338,476
1905	,,	324,216
1906	,,	301,947
1907	,,	250,016
1908	,,	234,985
1909	,,	283,054
1910	,,	284,630

This table shows that the charges made for cables have been reduced since 1903 to a point which would make the service unremunerative, except for the Government subsidy. And in this connection it must be remembered that a "word," if it is a code word (as most words would be in such messages), may convey a sentence. In the seven months, May 31st to December 31st, 1910, the number of private messages handed in at the offices of the Union was 49,657. They contained collectively 518,935 words, and the charges paid on them amounted to £62,292.

The four chief ports of the Union are Capetown, Port

Elizabeth, and East London in the Cape Province, and Durban in Natal. Taking them as a whole they may be said to be well supplied with wharves, warehouses, and all modern appliances necessary for the accommodation of shipping, and the prompt and economic loading and unloading of cargoes. In the case of all these ports the docks are served by railway lines which connect them with the main railway systems.

In the waters of Table Bay, Capetown has always possessed a fine natural anchorage, and on its shores harbour works have been constructed during the last half century, upon which, up to December 31st, 1908, a total sum of £4,214,822 has been expended. The harbour works, as originally planned by the late Sir John Coode, consisted of the breakwater and the inner basin. They were named the Prince Alfred Docks from the circumstance that the Duke of Edinburgh, then Prince Alfred, tipped the first wagon load of the stone which was to form the breakwater on July 17th, 1860, and formally opened the inner basin ten years later. To the accommodation thus provided there has been added an outer harbour, formed by a second pier carried parallel to the breakwater from a point to the south of the inner basin. The outer harbour has an area of 64 acres, and provides with the inner basin a total area of 72½ acres of protected water, of a minimum depth varying from twenty to thirty-five feet. The breakwater has been completely constructed to a length of 3,640 feet, and the South Pier has been carried out some 1,700 feet. The electric cranes and "Temperley" transporters, with which the docks are equipped, are capable of handling 8,000 tons of cargo *per diem;* and spacious warehouses have been erected within which 468,516 square feet of material can be stored. Altogether there are two and a half miles of wharves and quays, and the breakwater gives shelter to vessels lying at anchor in the bay outside the actual harbour area. It should be noticed also that Capetown is suplied with a graving dock built of Paarl granite. It is 500 feet long and 68 feet wide at the entrance, which latter has a depth over the cill of 24½ feet at high water ordinary spring tides.

Algoa Bay, the harbour of Port Elizabeth, is a fine roadstead, with open but secure anchorage, which is only exposed to winds from the south-east. There is no artificial harbour; but three landing jetties, of which two are 1,160 feet, and one 1,460 feet long, have been carried so far out into the bay that their ends stand in 22 feet of water at low tide. It is alongside these jetties, which are equipped with the cranes, capstans, and wool hoists worked by the port's hydraulic plant, that vessels lie to discharge and ship their cargo. The harbour has some 35,000 square yards of warehouse area, and its present appliances and accommodation permit of 5,000 tons of cargo being handled in a day. In addition to the three landing jetties, there is an explosives jetty, 1,380 feet from the shore, which has an overhead roadway communication capable of landing 150 tons *per diem*. The jetties, warehouses, and harbour yards are lighted by electricity; and the same force, obtained from the municipal supply, is employed to work three powerful overhead cranes, or ganties—one 10-ton and two 5-ton—as well as the machinery in the workshops. Algoa Bay is the chief port of the wool industry; and it has been calculated that it is practicable for vessels to load and unload alongside the jetties, in spite of the absence of any artificial harbour, on 300 out of the 312 weekdays in the year. The cost of the two shorter jetties was £260,000, and that of the later and longer jetty, £185,000.

East London is a river harbour. The mouth of the Buffalo River, at which the town stands, was originally blocked by a "bar" of sand, which prevented the passage of large ships. Harbour works, comprising a sheltering breakwater 1,600 feet in length, and two training walls which, narrowing the entrance of the river to 250 feet, have greatly increased the force of the outflowing current, together with the continuous employment of three powerful dredgers, have now made it possible to maintain a depth of water over the bar of 19 feet at low water ordinary spring tides throughout the year. Upon these necessary but costly works, originally planned like the Prince Alfred Docks at Capetown by the firm of the late Sir John Coode, and upon the general equipment of the port the large sum of £2,084,407 had been expended up to December 31st, 1909. As the result vessels of 8,000 tons gross register now enter the river, and lie alongside the wharves, where, especially on the west bank with its recently constructed quay, they find ample accommodation and a dock equipment of the latest and most approved type. Indeed, it is claimed that the speed with which cargoes are landed and despatched from the East London docks is exceptional; since goods are either loaded direct from the ship's hold into the railway trucks, or placed in sheds which are so spacious that they permit of an unusually rapid handling of the material deposited in them.

Durban, the port of Natal, affords an even more striking example of a long and successful conflict with the forces of nature. The harbour consists of a landlocked tidal lagoon, the entrance of which, the Bluff Channel, was originally blocked by a " bar " of shifting sand. Between the years 1854 and 1881 (the year of the retrocession of the Transvaal), the average depth of the channel was 6 feet 5 inches. To-day the harbour can be entered at all times by the largest of the ocean-going steamships that visit the South African coasts. In this achievement, which affords one of the most interesting episodes in the industrial history of South Africa, a debt of £3,595,445 6s. 3d. had been incurred, up to the date of the Union, by the small European community of Natal; and this total has been subsequently increased.

The entrance and harbour channels are equipped with leading lights and light buoys, night navigation being an ordinary feature of the port work. Pilotage is compulsory. The harbour is approached through the Bluff Channel, 4,000 ft. in length, formed by a north pier and a south breakwater. These works are 800 ft. apart, except at the pier heads, where the breadth is only 600 ft. The average low water depth at the entrance is 34 ft., and on the harbour channels 30 ft. . . . The land-locked bay inside these piers extends about 3½ miles east and west, and nearly 2 miles north and south. The deep water space available for shipping covers about one-fifth of the area, and gives a low water depth of not less than 30 ft. over its greater part. This area is being constantly extended by dredging. There is a total of 10,370 ft. of concrete quay, and wooden wharf and jetty berthage, giving a low water depth alongside of from 21 to 34 ft., and the concrete quay is presently being extended to provide 38 ft. berthage.[1]

The port has nine wharf sheds with a collective capacity of 270,000 tons, and cranage and other equipment which permits of 8,000 tons of cargo being handled daily, exclusive of coal. For this latter material there is a special plant, worked by electricity and provided with storage bins of 10,000 tons capacity, by means of which vessels can be loaded with coal at the rate of 400 to 500 tons an hour. One of the special advantages possessed by Durban is its Floating Repairing Dock, which will take vessels with a dead weight of 8,500 tons, of 23 feet draught, and 475 feet in length; and to this there is attached a floating workshop, furnished with a 15-ton crane, and a patent slip for small craft up to 250 tons dead weight. The docks are lit by electricity, and have an inexhaustible supply of water which serves all berthage; and the harbour is provided with three powerful tugs, and an ample and efficient dredging fleet.

Among the lesser ports of South Africa are Port Nolloth, the port of the copper mines, and Saldanha Bay on the west coast; Mossel Bay, midway between Capetown and

[1] Colonial Office List, 1911.

Port Elizabeth, Port Alfred at the mouth of the Kowie River, and Port St. John's, the large estuary at the mouth of the St. John's River in Pondoland. Nor must it be forgotten that the Imperial Government maintains a naval station at Simonstown in False Bay, a few miles to the south of Capetown.

The debt incurred by the two coastal colonies on account of expenditure upon harbours amounted at May 30th, 1910, to the sum of £9,562,720 17s. 4d. of which £5,967,275 11s. 1d. had been spent by the Cape and £3,562,720 17s. 4d. by Natal. This debt was, of course, assumed by the Union Government, and formed part of its total public debt of £116,000,000. The largeness of this expenditure, and the consequent improvement in the accommodation and equipment of the four principal ports of South Africa, are to be attributed largely to the keen competition for the carrying trade to the Rand, which marked the six years immediately preceding the establishment of the Union. This competition was manifested not only as between the individual ports, but as between the Cape ports collectively and Durban, the port of Natal; and then further as between the British ports collectively and Delagoa Bay, a "foreign" port, but nevertheless recognised by the Transvaal, because of its proximity, as its "natural" port. The accommodation for shipping at Delagoa Bay, it is almost needless to add, has also been greatly improved in recent years; and the enlargement of the wharfage, and the increase of the railway and general facilities of the port, have been carried on under the direct inspiration of the Government of the Transvaal and to some extent with the assistance of British capital.

The Union Railways

But little more than half a century ago there were no railways in South Africa. On March 31st, 1859, the first sod was turned by Sir George Grey in the construction of the first railway, which was to run from Capetown to Wellington—a distance of fifty-eight miles. In 1862 a branch line from Salt River—the terminus of the Capetown–Wellington line—to the surburb of Wynberg was started, and a little later a railway was carried from Port Elizabeth to Uitenhage. These lines, which were originally built by private enterprise, were subsequently purchased by the Cape Government. With the addition of a two-mile line in Natal, they were all that South Africa had in the way of railways, until the discovery of the diamond mines in Griqualand West, in 1870, gave the first decided impetus to the industrial development of the sub-continent. The foundation of Kimberley and the diamond industry brought a material increase to the trade and revenues of the Cape Colony, and when, in 1873, Responsible Government was established, the new administration embarked upon a programme of railway construction. In the course of the next ten years railway lines were carried into the interior from the three principal ports, Capetown, Port Elizabeth, and East London, to meet at De Aar junction, on the southern border of the Free State; and from this point the railway was taken northwards direct to the goal of Kimberley, which was reached in 1885. In the meantime a railway had been built in Natal to connect Durban, the port, with Maritzburg, the seat of government.

A year later, in September, 1886, the Rand was declared a public goldfield by the Transvaal Government. Then followed the sudden birth of Johannesburg, and the rapid conversion of the barren uplands around it into the greatest industrial centre of South Africa. Within nine years of the proclamation of the Rand goldfield, three railway systems were competing for the profits to be earned by carrying goods and passengers between it and the coast.

The Cape Government was first in the field. By arrangements with the Free State Government and the Netherlands South Africa Railways Company, direct railway communication between the Cape ports and the Rand was established in 1892. The Delagoa Bay line of the Netherlands Company to Pretoria and Johannesburg was completed in 1894; and a year later the Natal system, which had been carried to the extreme north of the Colony by the Natal Government, was linked by the Netherlands Company with Johannesburg, and Durban was placed on a level with the Cape ports and Delagoa Bay.

In the meantime Rhodes had founded the British South Africa Company, and conceived the project of a trans-continental railway uniting Capetown with Cairo. The development of Rhodesia drew the Cape main trunk line from Kimberley to Vryburg and Mafeking, whence it reached Buluwayo, a distance of 1,362 miles from Capetown, in 1897. Altogether in the decade 1887–1897, 2,000 miles of new railways were constructed in South Africa under the immediate stimulus of gold discovery.

After the war the Rhodesian railway system was rapidly extended, and the trans-continental main line was carried over the Zambezi at the Victoria Falls, and thence through Northern Rhodesia into the Congo Free State. In the four older colonies, which now form the provinces of the Union, railway construction was undertaken with vigour as a necessary adjunct to the industrial reconstruction of South Africa. Lord Milner, in the three short years of his governorship, doubled the mileage of open lines in the Transvaal and Orange River Colony, and otherwise vastly improved their joint system, known as the Central South African Railways. More than this, he inspired and negotiated agreements with the Cape and Natal Governments and with the De Beers Company, which resulted in the construction of new lines directly calculated to link the Central South African Railways more closely with the Cape and Natal systems, and thereby made intercourse between the people of the several colonies more speedy and more convenient.

These three systems, thus extended and unified largely under Lord Milner's guidance, are now administered by a single authority, the Harbour and Railway Board, and constitute the State railways of the Union of South Africa. Taking the Report of the General Manager of Railways for South Africa, issued in 1911, we find that on December 31st, 1910, the extent of the system, which with slight exceptions includes all railways within the Union, was 7,039 miles of open lines, with more than a thousand miles of new lines under construction. Except in the suburban systems of the few large towns, the Union railways are single track lines; and the gauge is three feet six inches, which is the standard gauge for all South African lines within and without the Union.

The following table will afford a rough measure of the value of this system, considered as an important element in the national "plant" of the new dominion. The table is only a "rough" measure, because it does not distinguish single from double (or quadruple) lines, and contains no account of the variations in the number, carrying capacity, etc., etc., of the trains run upon the several systems.

TABLE SHOWING THE LENGTH OF RAILWAYS IN THE CHIEF DIVISIONS OF THE BRITISH EMPIRE

Division.	Area in Sq. Miles.	Population—		Mileage of Open Lines in 1909—	
		European.	Coloured.	State.	Private.
United Kingdom ..	121,090	45,365,599	Nil	—	23,280
Dom. of Canada ..	3,729,665	7,151,869	Negligible	1,720	23,373
Commonwealth of Australia	2,974,581	4,500,000	"	15,073	1,580
Dom. of New Zealand	104,751	1,050,000	"	2,717	29
British India (exclusive of Native States)	1,097,821	British Army and Officials, etc.	244,267,542	31,490 (mainly State, worked by private companies)	
Union of South Africa	473,184	1,278,025	4,680,474	6,705	516

The financial position of the Union Railways at the end of 1910 was excellent. According to the General Manager's Report[1] the gross takings for the year (1910) amounted to £12,056,871, being an increase of £1,899,600 over the collective takings of the three separate systems in the previous year. The excess of these gross takings over working expenses and renewals was £5,458,960; and the net profit (*i.e.*, the surplus left after the addition of £195,053 "other receipts," and the subtraction of interest and sinking fund on the loan capital) was £3,339,583. The capital expenditure from loans upon the three systems up to the date of the Union was £65,671,973 2s. 10d.; and the annual interest payable in respect of it, amounted to £2,250,754 2s. 2d. But to this sum, which formed considerably more than one-half of the total of the public debts of the four colonies assumed by the Union, must be added another £8,000,000 of capital expenditure upon the Central South African railways, which was provided out of Railway or Government revenue, and other lesser sums. Thus the total capital expenditure on the open lines of the Union railways up to the end of the year 1910 is returned as £75,100,228, and the interest payable thereon as £2,314,430; while the total capital expenditure upon open lines, lines under construction, and subsidies to private lines, amounted to £77,333,514. The percentage of gross profit on capital entitled to interest for the year was £7 12s. 3d., an increase of £1 2s. 2d. on the percentage for 1909. It is scarcely necessary to add that this is a rate of interest considerably in excess of that commonly earned by railways, whether under State or private management.

Of this successful result the General Manager writes:

The unification of the railway interests brought under one management 7,692 miles of completed line (7,207 miles within

[1] Report of the General Manager of Railways and Harbours for the year ended December 31st, 1910. Pretoria, 1911. (U. G., No. 39, 1911.)

and 485 miles outside the Union), with approximately 860 miles under construction. The railways within the Union at the 31st of December, 1910, comprised 7,041 miles of Government-owned and 545 miles of privately-owned lines, serving an area of 472,730 square miles. . . .

The main feature of the year's working, so far as the railways are concerned, has been a substantial increase in practically all classes of traffic. The anticipation and consummation of Union seem to have established greater confidence in the country, and there has been an unprecedented development of passenger, general goods, and coal traffic, which is indicative of a most gratifying extension of business throughout the Union. There also seems every prospect of the export trade in maize and wool assuming large dimensions, and there is reason to hope that meat will soon be added to the list of considerable exports. The principal harbours of the Union have shared in the general prosperity which characterised the year's working. . . . There is every reason to anticipate a bright future, but before any general advance takes place there is likely to be a lull. It would, therefore, be advisable not to be too sanguine, or to base a policy for the immediate future on the exceptional prosperity of the period covered by this report.

This statement of the financial position of the Union railways brings us at once face to face with the very important decision at which the National Convention arrived in respect of their future management. As subsequently expressed in the South Africa Act, the Convention determined that the State railways should no longer be used as an instrument of taxation, but that they should be employed in promoting directly the development of the agricultural and industrial interests of the Union as a whole. And, moreover, in view of the fact that this new policy would lessen their earnings both by the general reduction of rates and by the construction and working of unremunerative "development" lines, it was laid down that at the end of four years from the establishment of the Union, the Government should cease to draw any part of its revenue from the State railways. During this period, however, as we have seen,[1] the Union Parliament "may by law appropriate"

[1] Part II, Chap. III, p. 157.

any excess in the earnings accruing to the Railway and Harbour Fund to make up deficiencies in the consolidated Revenue Fund ; and the contribution required from these earnings by the Union Treasurer for the financial year ending March 31st, 1912, was estimated at £1,159,000. As the railways earned in 1910 a net profit of £3,339,583 —representing the railway taxation of the four colonies prior to the Union—over £2,000,000 of revenue could be abandoned by the Railway Board at once. Wisely applied this ability to reduce the railway rates should prove of great benefit to South Africa ; since the heavy cost of carriage, both by sea and land, is an appreciable element in the price of all those many " necessaries " that are imported from oversea. And, as we have noticed before, the high cost of living to Europeans, especially in the inland provinces, is a recognised economic evil, as tending to retard materially the growth of the European, as against the native population in South Africa. In itself, therefore, a reduction of railway rates is most desirable, as naturally tending to reduce the price of imported commodities of all kinds. There is a danger, however, as the experience of the Crown Colony Administration in the new colonies showed, lest a lowering of railway rates should turn out to produce not a reduction in the price of goods to the consumer, but an increase in the profits of the wholesale importer and the retail dealer.

That the Railway Board is not unconscious of this danger, may be inferred from the cautious tone in which the General Manager concludes his remarks upon the reduction of rates. After stating that substantial reductions have been made in passenger fares, and that reductions in the coal rates are under consideration, he writes :

I gladly identify myself with the policy of reducing rates where possible, but the public interest will best be served by reductions being made only after the most careful deliberation. The reduction of rates is a very popular demand, and every interest presses for first consideration ; but to comply with the many representations made would produce results which would be harmful to the general progress of the country.

There are two directions in which it would seem to be possible, however, to use this power of abandoning railway revenue in a manner directly calculated to promote the prosperity of the Union as a whole. Of these, the first is the endeavour to bring the cost of living to Europeans in the inland centres of industry down to the level of the coastal centres ; and the second, to assist the development of backward districts by the construction and working of new lines, which, although necessarily worked at a loss to the Railway Board, would ultimately increase the agricultural and industrial resources of the country. And in this connection the physical conditions of South Africa must be recalled. Owing to its vast extent the towns and villages into which its meagre population is gathered are separated by great distances. The physical structure of the country renders its rivers, which are few and comparatively insignificant, practically useless for purposes of navigation ; and prevents the construction of artificial waterways. Transport by road is costly and slow. In South Africa, therefore, railways constitute an economic factor of exceptional importance.

These conditions explain and justify the determination of the National Convention not only to place the control of this factor in the hands of an authority intended to be unaffected by the influences of party politics, but to treat transport and communication by railway as an elementary necessity of the national life, which the State must cheapen and facilitate in every way.

The wisdom of this determination is conspicuous. It remains, however, for the people of the Union to see to it, that not only is the policy of providing railway communication at cost price carried into effect, but that the independence of the railway administration is maintained unimpaired. Political interference, if it were permitted, might cause a useless line to be built across the veld to a party-leader's homestead, and then—nowhere. And—an evil worse than "log-rolling"—it might undermine the discipline, and therefore the efficiency, of the employees.

Here we touch upon the weak point in the system of the State ownership of railways. According to the report the number of persons employed on the Union railways at December 31st, 1910, was as follows:

Whites (excluding White labourers)	23,165
White labourers	3,419
Coloured and Natives	19,365
Indians	3,813
Total	49,762

And on the same date there were employed on the harbours:

Whites	1,380
Coloured and Natives	1,263
Indians	190
Total	2,833
Total for Railways and Harbours	52,595

The successful handling of a staff so numerous, and having a predominant European element, must be a difficult matter at all times in South Africa, where white men are jealously resentful of any encroachment upon what they believe to be their just rights. But if the original difficulty of the task is to be increased by political interference, it will quickly become impossible.

Here again it would appear that the Board are fully alive to the danger. Forewarned is forearmed; and we may hope that the emphatic warning of the Union General-Manager on this subject, embodied in this—

his first—report, will serve to arrest the evil, before it approaches in South Africa the serious proportions to which, apparently, it has attained in Australia. He writes:

What is most needed now is some respite during which to frame and give effect to a constructive policy. The railway staff during recent years has been much disturbed by constant investigations. In 1907–8 there was the Central South African Railways Commission, which was probably the most thorough inquiry of the kind ever made; reorganisation followed, with the tremendous detail it involved. Then came the Truter Inquiry in 1909, and a further Committee of Investigation is now to be appointed which will deal with many matters already inquired into and duplicate much of the work undertaken by the Staff Committee[1] already referred to, the main duty of which has been to make investigations in connection with the several services and devise means of assimilating the conditions. Added to all this, there is the constant political tendency on the part of those not responsible for the management of railway affairs to interfere in personal staff matters, thus rendering it almost impossible for the Government and the Administration to manage successfully the variety of interests of a service like the Union railways, and resulting in much work, of questionable benefit both to the staff and to the service, being thrust upon the responsible officers of the Administration. But South Africa is not alone in this unfortunate experience, as the following observations made by the Commissioner of Railways for Western Australia, in his report for the year ended the 30th of June, 1910, will indicate:

"The methods by which these matters (namely, certain grievances which were discussed at a meeting arranged by the Honourable the Premier with certain members of Parliament and representatives of the Railway Officers' and Amalgamated Employés' Societies) have been dealt with, and the apparent acquiescence of the Government and Parliament in their reference to political action for settlement seem to indicate a desire for the amendment of those provisions of the Government Railways Act which remove, or were intended to remove, the administration of the Department from political influence.

[1] Pending legislation for unifying the grading and rates of pay in the [formerly existing] three railway systems, "a Staff Committee, consisting of officers and employés from the different sections of the service, has been appointed to inquire into existing conditions, and thereafter make recommendations as to the grading and pay of all classes of the staff."

Unless the Act is so amended, however, its provisions should, I submit with all due respect, be observed, and while they continue as they now are I can only sincerely urge that members of the Legislature assist me in the execution of my duty as defined by the Act, especially by refraining from actions tending to lessen the authority which that responsibility necessarily entails and requires. Further, while I have no desire to place anything in the way of the Government improving the position of the employés if they consider it necessary to do so, either in pay or condition of service, it is certainly desirable, for the maintenance of the authority referred to, that such decisions of the Government should be given direct to me, so that I alone may deal with the employés concerned."

Australia is frequently quoted in matters of service conditions, and the foregoing comments are worthy of careful deliberation by those charged with the administration of the country's interests, if administration is to be effected through the constituted railway authorities. It is subversive of all discipline and good and efficient management to encourage or allow railway servants to ignore their superior officers in advancing their grievances, fancied or real. Every railway servant has the right of appeal to the Minister or the Railway Board, but all appeals should be made through the medium of the servant's superior officer. Business experience has shown that efficiency and economy can best be obtained when the authority of the senior officer is upheld as far as possible consistent with justice.

The fact that the South African railways have now reached a high state of efficiency would make it more than ever deplorable, if this insidious evil were allowed to wreck the Union system, or even to impair the great advances which have been so recently achieved. We have seen that the financial results obtained in the first year of the Union are exceptionally favourable. Nor is the system, taken as a whole, deficient in general equipment. During the years intervening between the Peace of Vereeniging and the birth of the Union the three separate State systems were greatly improved in permanent way, rolling-stock, general equipment, and efficiency of working. While, however, good results were secured in the coastal colonies, the most conspicuous progress was made in the new colonies under the Crown Colony Administration. Here, mainly owing to the excellency of Lord Milner's financial arrangements, more ample funds were available for railway development than in the Cape and Natal. In 1902 (the year of the Peace) there were on the Central South African railways only 26 locomotives of an average tractive force of 25,000 lbs. and over; in 1908, there were 137. In 1902 the number of trucks of a running capacity of 30 tons or over was 540; in 1909 it had risen to 2,024. In 1902 the total area of the covered railway shops in Pretoria was 6,000 square yards; to-day it is 45,500, and the enlarged shops are vastly improved in equipment. Before the Crown Colony régime gave place to Responsible Government (1906–7) the permanent way of the joint system of the two colonies had been completely remodelled. All bridges on the main lines had been rebuilt or adequately strengthened; where traffic was most heavy the lines had been re-graded; dangerous curves had been removed, and badly constructed portions of the line replaced by convenient deviations; and to serve the more powerful engines, and the heavier, though more economic, trains now employed, the main trunk lines had been relaid with new 80 lb. rails. At the same time the inadequate station buildings and goods siding accommodation had been enlarged, or altogether replaced; storage dams had been constructed, and the most approved appliances for watering engines installed; spacious engine sheds and mechanical coaling plants had been erected at the principal locomotive depôts; and the important stations, and most of the lines traversed by fast passenger trains, had been provided with the most recent and efficient systems of signalling and interlocking. Moreover, the financial results obtained amply justified the large expenditure required to carry out the expansion and improvement of the system. Within the period of Crown Colony rule, while the rates were lowered to an extent which represented the remission of £1,500,000 of annual taxation by railways, the earnings of the Central South African Railways were sufficient to provide year by year the interest, sinking fund, and charges payable upon the whole of the £35,000,000 Guaranteed Loan, and, in addition, to make a small contribution towards defraying the cost of other services common to the two colonies.

These details refer exclusively to the railway development which was an essential feature in the economic reconstruction of the late Republics—a reconstruction marvellous alike in its rapidity as in the variety and completeness of its operations—effected under the masterly guidance of Lord Milner. They may, however, be taken as representative of the kind of improvements which, on a lesser scale, were carried out on the Cape and Natal systems. And so, to-day, the European visitor is generally ready to admit that the shortcomings of the South African railways in the matter of speed and frequency of service are counterbalanced by the consideration which is shown for the comfort of the traveller. The length of the journeys, which is the distinctive feature of railway travelling in South Africa, and the climatic conditions of the country, have together determined the character of the carriages and the nature of the accommodation to be provided. The coaches, accordingly, are constructed on the corridor plan, and furnished with window appliances for excluding dust and sun, while freely admitting the air. The seats in every compartment can be converted readily into sleeping quarters, and all the main line trains have dining cars, where breakfast, lunch, and dinner can be obtained at prices which are not merely moderate in view of the higher cost of living in South Africa, but are actually lower than those charged on some British railways.

As Capetown is the main port for the arrival and departure of oversea passengers, it will be convenient to take the railway service as it runs, north and east, from this point, remembering, however, that there are fast through trains from all four ports to the Rand. From Capetown, then, apart from communication with the inland towns of the Cape Province, trains run twice a day to the Transvaal; once a day to the Free State, Natal, and Delagoa Bay; and three times a week to Rhodesia. In the service thus maintained one or two of the most noticeable trains may be mentioned. To begin with the longest journey, trains for the Victoria Falls leave Capetown on Tuesday and Friday about mid-day, connect with the "Zambezi Express" at Kimberley, and reach the Victoria Falls on the following Saturday and Tuesday respectively, thus taking four days to cover the 1,642 miles. The Tuesday train connects with the arrival of the outward mail boat, and enables the traveller to reach the Falls within twenty-one days of his departure from London. The corresponding southward trains meet the "Rhodesian Express" at Buluwayo on Sunday and Friday, connect with the ordinary Transvaal train at Kimberley, and arrive at Capetown on the following Wednesday and Sunday respectively, the former (leaving the Falls on the preceding Saturday) connecting with the departure of the homeward mail boat from Capetown on Wednesday. The daily train to Delagoa Bay from Capetown comes next in point of distance, and takes nearly three days for the journey (about 1,200 miles). The "Transvaal Limited" leaves Capetown on Tuesday and Saturday at 9 a.m., and runs to Johannesburg and Pretoria by the quickest route, viz., viâ Kimberley and Fourteen Streams. Corresponding coastward trains leave Pretoria and Johannesburg on Thursday and Saturday evenings, and respectively reach Capetown early on the following Saturday and Monday mornings. On Thursday morning, also, at 9 a.m., the "Orange Limited" runs from Capetown to the Transvaal by the old route through the Free State. The "Imperial Mail Train-de-Luxe," which carries passengers from Pretoria and Johannesburg direct to the docks at Capetown, follows the same route. It leaves late on Monday evening, and arrives at the docks at 2.5 p.m. on Wednesday, where its passengers have only to walk on board the homeward mail boat.

Of the trains running between Johannesburg and Capetown, five trains a week—three Capetown to Johannesburg, and two Johannesburg to Capetown—are now scheduled to perform the journey in thirty-six hours. It may be added that on the arrival of the outward mail boat at Capetown on Tuesday in each week, a special train, carrying mails only, is at once despatched from the docks; and by this means the mails are delivered with the least possible delay in all parts of South Africa.

The following tables will exhibit the greater length of the journeys which have to be taken in travelling from place to place in South Africa by a comparison of the distances separating some well-known towns respectively in South Africa and in Great Britain:

South African Distances.	Miles by Rail.
Capetown to Kimberley	647
,, Bloemfontein	750
,, Johannesburg	957
,, Pretoria	1,001
,, Buluwayo	1,362
,, Victoria Falls	1,642
,, Beira	2,037
,, Bwana M'Kubwa (Northern Rhodesia).	2,142
Johannesburg to Durban	482
,, East London	665
,, Port Elizabeth	712

British Distances.	Miles by Rail.
London (by Great Northern) to York	188
,, ,, ,, Edinboro'	393
,, ,, ,, Glasgow	440
,, ,, ,, Aberdeen	523
,, ,, ,, Inverness	558

British Distances.		Miles by Rail.
London (by Great Western) to Bristol	.. ..	118
,, ,, ,, Plymouth	.. ..	247
,, ,, ,, Penzance	.. ..	325
London (by L. & N.-W) to Birmingham	.. ..	113
,, ,, ,, Manchester	.. ..	183
,, ,, ,, Liverpool	.. ..	192

Communication by Post, Telegraph, and Telephone

On the eve of the establishment of the Union the Postmasters-General of the four colonies, having met in conference at Capetown (February 3rd–26th, 1910), prepared a Report on the assimilation of the postal rates of the several colonies, and drafted a Bill to consolidate their respective postal and telegraph laws for subsequent submission to the Union Parliament. On May 31st, 1910, the four Colonial General Post Offices were amalgamated and placed under one control, and the General Post Office of the Union was established at Capetown. The Postmasters-General of the Transvaal, the Orange River Colony, and Natal retired on pension, and Mr W. T. Hoal, the Postmaster-General of the Cape, was appointed to act as Postmaster-General for the Union. Three out of the four Post Office secretaryships were abolished, and, with the exception of a few officials who were retired, the remainder of the principal administrative officers were brought to Capetown, where duties of the same class as those which they had performed in their respective colonies were allotted to them. Such changes in the direction of the unification of the charges for postal and telegraphic services in the four provinces as could be made without fresh legislation were carried out at once; and provision for securing at the earliest possible moment the authority of Parliament for the further changes required was made. At the same time the process of reorganising the department in all its branches was begun, without waiting for the Civil Service Commission to report upon it.[1]

The revenue and expenditure of the Union Post Office for the seven months May 31st to December 31st, 1910, were as follows :

	£	£
Revenue.		
Postal	566,088	
Telegraph and Telephone	288,088	
Total		854,176
Expenditure.		
Postal, Telegraph, and Telephone ..		800,552
Surplus, or Net Revenue ..		£53,824

On December 31st, 1910, there were 2,466 post and telegraph offices open ; 752 receptacles for letters, etc., were in use ; and there were provided, in addition, 28 travelling post boxes, and 9 post boxes on Union-Castle steamers. On the same date there were 1,282 telegraph offices open to the public of the Union, of which 709 were postal, 573 railway, and 277 were worked by telephone. The length of the telegraph lines open was 12,516 miles, and that of the wires 47,421 miles. In addition to these, the department was maintaining 6,482 miles of wire in the Cape Province, and 2,547 miles in Natal, on behalf of the railway administration.

In the year direct telegraphic connection had been established with German South-west Africa, and with the Belgian Congo. In the case of the former country, communication was opened between Capetown and Warmbad through a repeating apparatus at O'okiep, and the distance was 530 miles, of which only 40 miles were in German territory. In the case of the latter,

[1] The above account is taken from the Report of the Union Postmaster-General for the year 1910. Mr. Hoal, the first acting P.M.G., died suddenly on October 18th, 1910. The Report is presented by " Jer. Wilson, Acting for the Postmaster-General," and bears date, Capetown, April 20th, 1911.

communication was opened through Broken Hill, in Northern Rhodesia ; and the offices in the Congo which accepted public telegrams were Chinsinda, Elizabethville, Mikola, and Sakania, all of which were on, and worked by, the Katanga Railways.

The capital expenditure on the telegraphs of the Union, up to December 31st, 1910, was £1,077,189. In spite of economic administration, the telegraph service is worked at a loss to the State. In the period May 31st to December 31st, 1910, the number of public messages conveyed, exclusive of Government, railway, Imperial, and military telegrams, was 2,402,027, and the receipts on account of them amounted to £152,900. It may be noticed that the Press is charged at a rate only one-fourth of that paid by the general public.

The Post Office Savings Bank of the Union is in an eminently satisfactory condition ; but the progress of this useful institution will be reserved for consideration in connection with the general banking statistics of the Union.[1]

Wireless Telegraphy

The following information on this subject, which is taken from the Report of the Union Postmaster-General for 1910, will be of interest. A radiotelegraph station was opened at Durban on June 13th, 1910, for the convenience of the shipping community and the travelling public. The station has a minimum sea range by day of 250 miles, but, at night, when signalling conditions are improved, communications are frequently exchanged with vessels at a distance of over 1,000 miles. At the time the station was opened there were only three vessels trading in South African waters which were able to exchange signals with it ; but on December 31st, 1910, there were thirty vessels among those trading with, or calling periodically at, South African ports, which were

[1] See forward, p. 384.

fitted with wireless apparatus; and other vessels were being similarly equipped.

Tenders for the erection of a wireless telegraph station in the Cape Province were called for in June, 1910, and that of the Marconi Wireless Telegraph Co. was accepted. The contract was for the erection of a 5 kilowatt station, with a guaranteed minimum sea range by day of 400 miles, and by night of from 600 to 1,600 miles. It was decided that the first station should be erected at Slangkop on the Cape Peninsula. A suitable piece of Government land was selected, and the erection of an operating house was put in hand by the Public Works Department. The station itself was expected to be in working order by the end of April, 1911.

Telephones

On December 31st, 1910, there were in the Union 103 exchanges and 614 call offices, with 10,483 exchange lines, 15,378 telephones, and 46,165 miles of wire in use. The capital expenditure on telephones up to this date was £893,239; but a further expenditure on trunk lines was contemplated, and the value of the Union system was expected to stand shortly at £1,250,000.

This convenient method of communication is being developed rapidly, and efforts are being made to cheapen, and thereby popularise, its use throughout the Union. During the last three years the capital expenditure on telephones in the four colonies has grown from £508,562 to £893,239, while the number of telephones in use has risen from 10,569 to 15,378. The financial position of the Union Department in 1910 was as follows:

Revenue	£167,271
Expenditure (including interest on loans and depreciation)..	189,049
Deficit	21,778

The subjoined tables exhibit the extent to which the telephone has been utilised in the several provinces of the Union, and in the Union itself as compared with other countries.

RATIO OF TELEPHONES TO POPULATION IN THE UNION

Cape..	1 to every	126 Europeans	or to every	512 of total population.
Transvaal	1 ,,	36 ,,	,, ,,	167 ,, ,,
Orange Free State	1 ,,	198 ,,	,, ,,	530 ,, ,,
Natal (excluding the population of Durban, and its municipal telephone system)	1 ,,	40 ,,	,, ,,	672 ,, ,,

The corresponding figures in other countries are:

U.S. of America	1	to every	11	of population
Sweden	1	,,	30	,,
Germany	1	,,	65	,,
Great Britain	1	,,	68	,,
Johannesburg	1	,,	20	of European or 40 of total population
Capetown	1	,,	29	,, ,, 49 ,, ,,

In view of the fact that the cost of telephone "plant" is greater in South Africa than in most other countries, the rates which are charged must be considered very moderate. The following information on this point is given in the Postmaster-General's Report for 1910:

The subscription rates, as reduced at October, 1910, within the Exchange Area (*i.e.*, within a radius of 25 miles from the Central Exchange at Pretoria and Johannesburg respectively), are based on the "measured service" method of charging, which was introduced in 1908. And the same system is to be adopted in Capetown and its suburbs. They are:

Business Premises (local rate)	£7 10 0	900 free calls
,, ,, (area ,,)	10 0 0	,, ,, ,,
Residences (local ,,)	5 0 0	600 ,, ,,

The charge for extra calls is 1d.; calls to exchanges from 5 to 10 miles distant count as two local calls, and to exchanges over 10 miles as three local calls. The Business (area) subscription covers 900 local and area calls, and the extra charge for either is 1d. The premises must be not more than 2 miles distant from the Exchange.

The "measured rate" system is not applied to the smaller exchanges. The fixed annual rentals charged included an unlimited number of calls, but the charges in the different provinces had not yet been assimilated. They were:

	Cape.	Transvaal.	Natal.	Free State (Harrismith).	Great Britain (Provincial)
Length of line allowed for minimum subscription in miles	1	2	2	1	1
Business Subscription	£9	£6	£10	£10	£8
Total Charge for a 2-mile line	£13			£15	£13
Domestic Subscription	£7	£5	£5	£5	£8
Total Charge for a 2-mile line	£11			£10	£13

The trunk line charges in South Africa compare with those of Australia and Great Britain as under :

Distance.	South Africa.	Australia.	Great Britain.
25 miles.	–/3	–/4	–/3
50 ,,	–/6	–/8	–/6
75 ,,	–/9	1/–	–/9
100 ,,	1/–	1/4	1/–
200 ,,	2/–	2/4	2/6
400 ,,	4/–	4/4	5/–

CHAPTER III

THE MINES

FROM an industrial point of view, South Africa has been made by its mines. The predominance of the mining industry in the economic system of the Union will be seen by a glance at the subjoined table, which reveals the fact that the gold and diamond mines provide 78·5 per cent. of the total exports from South African ports.

Exports of South African produce from South African ports during the year 1910:

Article.	Value.	Percentage of Total.
	£	
Gold, Raw	34,322,136	63·0
Diamonds	8,480,875	15·5
Other Articles (including Specie) ..	11,706,259	21·5
Total	54,509,270	100·0

(From the Union Department of Commerce and Industries.)

First diamonds, then gold, wrought mightily in the land, and together they have changed South Africa from a poor and struggling pastoral country into the industrial compeer of the great dominions, Canada and Australia. Both offered to enterprising men the prospect of the rapid acquisition of wealth, and thereby drew to South Africa both energy and capital. Of the two, gold has had the greater influence. Just as the middle of the nineteenth century saw the magnet of gold discovery at work peopling North America and Australia, so its closing years witnessed a great demonstration of the

19—(2139)

same creative force in South Africa. And here, as we know, the economic changes of the gold era gave rise to political changes that exercised a supreme influence in moulding the destiny of the country.

No material exists at present sufficient to furnish an exact and complete account of the mineral deposits of South Africa: we know, however, that they are both varied and widely distributed. Apart from diamonds and gold, iron is found in all provinces of the Union, but is not worked;[1] but in Southern Rhodesia the annual production of chrome iron ore reached 53,499 tons in quantity and £118,064 in value, in 1911. Coal is found and worked in all provinces of the Union and in Southern Rhodesia; copper in the Cape, the Transvaal, and in Northern Rhodesia; silver in the Transvaal and Southern Rhodesia; and tin in the Transvaal and in Swaziland.

Of these lesser mining industries, copper is the oldest and coal the most important. Copper mining was begun in the Cape Colony in the neighbourhood of O'okiep, in the division of Namaqualand, as early as 1852. In 1864 the annual export had risen to £100,000 in value; and in 1909 the copper production of the Cape was returned as being 95,530 tons of the approximate local value of £426,925.

The value of coal as a factor in the industrial development of a country need not be emphasised; and South Africa is fortunate, as we have seen, in being well supplied both with this useful mineral and with iron. The coal mines of the Cape Province are in the Stormberg range, and the coal, which is of good quality (giving results equivalent to two-thirds of those obtained from an equal quantity of imported Merthyr coal), is largely used for locomotives on the Union railways. The Natal deposits are in the highest and most northern of the three "terraces" into which the area of this province

[1] See forward, however, at p. 498.

is divided, and Dundee is the centre of the colliery district. The Natal coal production is mainly sent to Durban, where it is shipped for bunker purposes, or exported for consumption in the southern parts of Africa, Asia, and America, and occasionally in Australia. The Free State deposits are found in the north of the province on the Transvaal border, and the output of the collieries is consumed locally. The main coal area in the Transvaal extends from the Vaal River in the south to Middelburg in the north, and from the Rand westward to the Drakenberg Mountains ; and there are lesser areas to the south-west of the Rand. The collieries are, therefore, in close juxtaposition to the great goldfield, and they supply coal for both the industrial and the domestic needs of the Rand at moderate rates. In Southern Rhodesia the Wankie coalfields, which lie some 200 miles north-west of Buluwayo, are just beginning to make an appreciable contribution to the resources of Rhodesia.

The absolute and relative coal production of South Africa is exhibited in the two following tables :

Table showing the quantity and value of the coal produced in the four provinces of the Union and in Southern Rhodesia in 1909 :

	Tons.	Value.
		£
Cape	76,846	65,972
Natal	1,786,583	633,604
Free State	420,170	125,627
Transvaal	3,235,407	916,452
Southern Rhodesia	152,583	77,313
Total	5,671,589	£1,818,968

Table showing the quantity and value of the coal produced in South Africa, the Commonwealth of

Australia, the Dominion of New Zealand, and the Dominion of Canada in 1909:

	Tons.	Value.
		£
South Africa	5,671,589	1,818,968
Australia	8,203,221	3,120,554
New Zealand	1,911,247	1,038,742
Canada	9,296,388	5,022,000

[Compiled from the *Statistical Abstract for the British Dominions, etc.*, 1910.]

Taking the Transvaal coal industry by itself, we find that, in 1910, nineteen companies, of which four were in the Springs-Brakpan area, twelve in the Middelburg area, and three in other districts (including the Vereeniging Estates Central Colliery), produced an output of 3,970,069 tons, of the value at the pit's mouth of £986,253. In the same year, six companies distributed dividends to the total value of £167,774. The number of white employés was 469, and of coloured 8,327; and the wages bill for the year showed a total of £363,559, of which sum £56,862 was paid in salaries, £101,741 in wages to whites, and £204,956 in wages to coloured labourers.

The following table exhibits the progress of the Transvaal coal industry since 1901 (at periods of three years):

Year.	Tons.	Value at Pit's Mouth.	Value per Ton.
		£	s. d.
1901	797,144	329,113	8 3
1904	2,409,033	883,891	7 4
1907	2,883,342	775,721	5 5
1910	3,970,069	986,253	5 0

[From the *Wit. Chamber of Mines Report*, 1911.]

The two subjoined tables, which show respectively the progress of the silver and base metal mining in the Transvaal, will serve to complete this account of the lesser mining industries of South Africa.

TRANSVAAL SILVER OUTPUT since 1903—(at three-year periods):

Year.	Fine ozs.	Value.
		£
1903..	349,955	36,744
1906..	632,188	78,617
1909..	804,222	84,168
1910..	823,752	88,029

TRANSVAAL BASE METAL OUTPUT since 1908:

Year.	Tons Copper Ore.	Value.	Tons Tin Ore.	Value.
		£		£
1908.. ..	1,144	33,018	1,426	97,128
1909.. ..	1,947	53,950	2,647	227,752
1910.. ..	3,181	77,612	3,383	328,487

[From the *Chamber of Mines Report.*]

THE DIAMOND MINES

The first discovery of diamonds in South Africa was purely fortuitous, and as such it presents a sharp contrast to the laborious and persistent effort by which the existence of the famous gold-bearing conglomerates of the Rand Basin was revealed. In 1867 a hunter or trader, O'Reilly, found at a Dutch farmstead, in the Hopetown district of the Cape Colony, among a collection of river pebbles, a white stone, to which he took a fancy. It was shown to Mr. Lorenzo Boyes, the Civil Commissioner at Colesberg, proved to be a diamond, and afterwards bought by the Governor of the Cape, Sir Philip Wodehouse, for £500. Two years later, a second stone of the same character was purchased from a Griqua Hottentot by the farmer himself, Van Niekerk; and this stone, sold to Messrs. Lilienfeld, of Hopetown, for £11,200, and christened the "Star of South Africa," was ultimately purchased by the (then) Earl of Dudley for £25,000. These "finds" naturally aroused public attention, and, as the white pebbles were river stones, adventurous searchers were soon at work in the neighbourhood of the confluence of the Vaal, the Modder, and the Orange Rivers. In 1870 there were 10,000 men seeking, and finding, similar stones on the banks of the Vaal, at Klip Drift (now Barkly West). From these, the "wet diggings," as they were called, the searchers were suddenly called away, in September of the same year, by the discovery of diamonds at Dutoitspan[1] and Bultfontein farm, some 20 miles to the south-east of the river. In May of the following year the "old" De Beers mine was found, and, a little later, the Kimberley mine, known at first as the "Colesberg Kopje." All four mines lay within a circle of 3½ miles in diameter, in a barren and desolate region, where wild animals had roamed at will. Here, nevertheless, the diggers proceeded to establish their canvas tents and corrugated iron sheds, and within nine months of the first rush to Dutoitspan, the "Dry Diggings" had a population of 50,000 white and coloured persons. These pioneers of the future town of Kimberley were at first exposed to great hardships. They had no regular communication, even by road, with the nearest settled districts; water was scarce and food supplies uncertain; and their makeshift dwellings of wood or corrugated iron gave them a precarious shelter against the elements. For eight

[1] *i.e.*, Du Toit's Pan; a "pan" being a natural basin in which water collects in the rainy season.

months out of the twelve the miners had their throats choked and their eyes blinded with South African dust at its worst, and in the remaining four they were drenched with torrential rains.

It was long before the diamond fields ceased to be a mere collection of mining camps, and began to assume even the external appearance of a town. At first, no one guessed the real character of the diamondiferous deposits upon which he and his fellows had stumbled, and all alike expected that a few years' digging at most would leave the earth barren of its treasure. And so it was not thought worth while for anyone to build substantial houses or offices, and twenty years after its foundation, Kimberley remained "a straggling, haphazard collection of small, low buildings, constructed almost entirely of corrugated iron or of wood, laid out with hardly any attempt at regularity, and without the slightest trace of municipal magnificence." [1] This natural belief, however, proved to be the direct opposite of the truth; and the supply of diamonds has long been known to be practically inexhaustible. The geological facts by which the permanency of the deposits of the "blue ground," or true diamondiferous earth, is established take us back to that remote and Titanic conflict between the elemental forces of nature which gave birth to the physical South Africa of to-day. For the Kimberley mines are the pipes and craters of long-extinct volcanoes; the "blue ground" is volcanic mud forced upwards from the bowels of the earth to the surface, which at the time was covered by the waters of a great freshwater lake. The main supply of diamonds lies, therefore, not in the craters excavated in the period of "open mining," but in the oval columns of blue ground which, penetrating to unknown depths, are now reached by subterranean

[1] So the late Lord Randolph Churchill describes it (as he saw it in 1891) in *Men, Mines, and Animals in South Africa*.

workings. Indeed, so far from there being any likelihood of the supply being exhausted, it is only by rigorously restricting the output that the rarity, and, therefore, the value, of the diamond can be maintained.

It was just here that Rhodes performed, perhaps, the greatest of his many services to Kimberley, and by so doing for the first time gave public and unmistakable evidence of his exceptional ability. The diamond industry was twice threatened with overwhelming disaster. As the excavations of the open workings grew deeper and deeper, the flooding of the mines and the continuous and increasingly serious downfalls of the soft rock by which they were encased, exhausted the funds and baffled the utmost efforts of the Mining Board; and in 1883, having spent more than half a million of money in removing "reef," it abandoned its hopeless task. From the almost total paralysis which then followed, the industry was saved by the introduction of shaft-sinking and subterranean mining. The new system was first adopted by the De Beers Mining Company, with the result that the annual output from this mine rose from half a million to a million carats, while the cost of production was largely diminished. At the same time it was discovered that the round surface area of the open mine narrowed to a column, and that this column of blue ground ran perpendicularly downwards—consisted, in fact, of the contents of the pipe of an extinct volcano. The conditions of the industry were now entirely changed; but these changed conditions brought with them a new danger which threatened even more complete disaster.

In order to understand the nature of this new danger—the cheapening of the diamond by over-production under stress of competition—we must first glance at another aspect of the industry. Under the early regulations of the diamond fields no person was allowed to hold more than one "claim." In the case of the Kimberley mine a claim was 31 ft. by 31 ft., an area no larger than the floor of a good-sized room; and when a claim proved rich in diamond-bearing earth, it was subdivided among several individual miners, so that at one time this mine was divided up among 1,600 separate owners. In 1874, when the difficulties of working the claims had increased with the progress of excavation, the number of the claims tenable by a single owner was fixed at ten; and soon afterwards this limit was removed. When the huge pits grew still deeper, the work of clearing the workings of water and fallen "reef" became more onerous, and the business of raising the blue ground from the workings to the edge of the mines required mechanical appliances and an ample supply of labour. This ever-increasing complexity and cost in mining operations brought about the elimination of the individual miner, and in 1880–1 the majority of such ownerships were converted into limited liability companies. In 1883, when, as we have seen, the industry was threatened with paralysis, it was declared by the Inspector of Mines that the Kimberley Mine "could never be worked to best advantage until all the payable holdings had been amalgamated." Two years later the four mines were owned by forty-two companies and fifty-six private firms; but one, the De Beers Mine, had passed into the possession of seven companies and three private owners.

It was at this period, when, on the one hand, underground mining had increased the output and lessened the cost of production, and, on the other, the formation of companies, with ample capital and large holdings, had provided the means of working the new system to best advantage, that the industry was threatened by this new danger of over-production. How real the danger was may be seen from the fact that, while in 1882 the value of the carat was 27s. 3d., in 1887 it had fallen to

18s. 5½d. The fall was due mainly to the keen competition between the Kimberley and De Beers Mines, where, in each case, one company had become dominant by amalgamation with some of its neighbours and by buying out others. Moreover, the efforts of the rival combinations had been directed in each case by an ambitious and determined man towards the attainment of an identical aim—the amalgamation of all four mines and the consequent control of the entire diamond production of Kimberley.

Rhodes and Barnato (Barnett Isaacs) both came to the diamond fields in 1873. Seven years later Barnato formed the Barnato Diamond Mining Company, and Rhodes the De Beers Mining Company. Both men prospered, and both saw in the gradual amalgamation of the ownerships of a single mine a first and long step forward in the direction of their common goal. Barnato, by becoming a large shareholder in the Kimberley Central Company, dominated the Kimberley Mine; while the De Beers Company, under Rhodes' methods, had obtained complete control of the De Beers Mine by May, 1887. In the same year, Rhodes enlisted the interest of Lord Rothschild and Mr. Alfred Beit in his plans for the amalgamation of the four mines, and with their assistance obtained the funds necessary to enable him to bid successfully against Barnato for the outstanding properties in the Kimberley Mine. Mr. Beit, moreover, did much more than supply Rhodes with funds. He gave him much important advice, and actively co-operated with him in the final stage of his financial duel with Barnato. There followed a short period of fierce competitive buying by Rhodes and Beit on the one side, and Barnato on the other, in which the price of the Kimberley shares rose to an abnormal level. Then, finding the resources of his opponents apparently unlimited, Barnato at last came to terms with Rhodes. On March 12th, 1888, the

De Beers Mining Company became the De Beers Consolidated Mines, Ltd., and a cheque for £5,338,650, paid by the new company, bore witness to the severity of the financial conflict. In June, 1889, the Company obtained formal possession of the De Beers and Kimberley Mines, and an efficient and economic system of mining was at once introduced by Mr. Gardner Williams, the General Manager;[1] and in the year following the directors were able to inform their shareholders that their ultimate object "had at last been accomplished, and the four diamond-producing mines of De Beers, Kimberley, Dutoitspan, and Bultfontein were now practically under the control of the Company." In 1891 the amalgamation of the industry was completed by the purchase by De Beers of a fifth mine, the Wesselton (known originally as the "Premier"); and with this transaction all fear of a ruinous competition between rival producers was removed.

The part played by Rhodes himself, and the effect which his achievement produced upon his contemporaries, are well described by Lord Randolph Churchill:

> It was this great work accomplished in the teeth of unheard-of difficulties, and almost insurmountable opposition, representing the conciliation and unification of almost innumerable jarring and conflicting interests which revealed to South Africa that it possessed a public man of the first order.[2]

Nor must it be forgotten that Rhodes used the De Beers dividends to acquire for Great Britain the most fertile areas of South Central Africa, thereby doing a work which, in point of magnitude and character, might properly have been made the business of the nation, and earning for himself the title of "empire-builder."

[1] Mr. Gardner Williams retired in 1905; and was succeeded by his son, Mr. Alphæus Williams.

[2] *Ibid.*

From the time of the amalgamation onwards, Kimberley has become more and more an appanage of the great mining corporation of De Beers. While the policy of restricting the output to the current requirements of the diamond market has necessarily kept the population almost stationary in point of numbers,[1] the appearance of the town has been greatly improved in recent years. Many of the most striking improvements, and, in particular, the creation of the "garden suburb" of Kenilworth for the European employés of the Company, were carried out at the expense and under the supervision of Rhodes himself; and Kimberley, almost as much as Rhodesia, is pervaded by his spirit and genius. Rhodes took a special interest in planting trees; and during the months that Kimberley was besieged by Cronje's force, when the mines were, of course, shut down, he caused roads to be macadamised and trees planted with the double purpose of embellishing the town and providing a livelihood for the unemployed. At the head of the finest of these new roads, the "Siege Avenue," which is more than a mile in length, stands the most beautiful thing that Kimberley possesses—the monument to the citizen soldiers who lost their lives in defence of the town. Built out of red and white sandstone from Rhodesia, its Ionian columns and massive base wear not unworthily the dignity, simplicity, and unity of the Greek models whose form it borrows. At its foot, pointing to the low hills, once held by the enemy whom it helped to defy, is "Long Cecil," the gun made during the siege, under Rhodes' instructions, by Labram, the De Beers' chief engineer. On the base of the monument are two panels, of which one commemorates

[1] The population of Kimberley in 1911 was 29,519, showing a decrease of nearly 5,000 as against 1904. This decrease, however, was confined to the coloured element. The European population showed an increase of ·74 per cent.

By permission of *W. Basil Worsfold*

THE KIMBERLEY SIEGE MEMORIAL

Labram,[1] and the other records the names of the officers and privates of the irregular forces who fell during the siege; and on the side fronting the avenue Mr. Kipling's fine lines are carved upon the stone:

This for a charge to our children in sign of the price we paid;
  The price that we paid for freedom that comes unsoiled to
      your hand;
Read, revere, and uncover; here are the victors laid;
  They that died for their city, being Sons of the Land.

THE PROCESSES OF THE DIAMOND MINES

The processes by which the diamonds are won may be grouped into four main operations:

1. *Extracting the "Blue Ground."* The blue ground is raised from the underground[2] workings, where it is hewn with pick and shovel by human labour, by the familiar appliances of truck, skip, and shaft haulage. It has been calculated[3] that the material thus brought to the surface annually, roughly, 4,000,000 tons of blue ground, would form a cube which would more than hold St. Paul's Cathedral, while its ultimate residuum of rough diamonds, albeit some £4,000,000 in value, could be packed into a "good-sized chest."

2. *Disintegration.* From the head of the shaft, the blue ground is tipped into surface boxes, drawn off into trucks, and carried by endless wire-rope haulage to the "floors," or open spaces, of a total area of 17½ square miles, where it lies, exposed to the disintegrating action of air, sun, and rain, for fifteen months.

[1] Labram was killed by a Boer shell which dropped into his room as he was dressing for dinner. This chance shot was an instance of the malignity of Fate, and compares curiously with Rhodes' experience. The Sanatorium, where Rhodes was known to have taken up his quarters, was a constant target for the Boer gunners; yet Rhodes came through the siege unwounded.

[2] Of the four mines working in 1910, however, the blue ground was being removed from one mine, the Wesselton, by surface mining.

[3] *Stones of Fire; The Times*, November 5th, 1910.

3. *Washing.* When thus disintegrated, the blue ground is carried, again by mechanical haulage, and "dumped" into the puddling cylinder, from which the finer ground passes into the great rotary washing pans. From these pans a deposit, one-hundredth part of the original bulk of the blue ground, is drawn off. It is in this sordid mass, a mere collection of pebbles, bits of iron, and miscellaneous débris, dropped from the slot of the washing pan into common round tin pots, that the dull, white stones—most precious of all precious stones—are contained.

4. *Separation.* The earlier methods by which the diamonds were separated from the deposit drawn from the washing machines have recently been superseded; and the methods now pursued are thus described by a writer in *The Times*:[1]

> Every known mechanical means of separation was tried without success, until one day one of the employés, Mr. Fred. Kirsten, observed that diamonds seemed to have a peculiar affinity for oily matter. Kirsten asked to be allowed to catch the diamonds by placing a coat of lard on the surface of a shaking table which had been used for some other experiments. Complete success crowned the new experiment. The diamonds alone stuck to the grease. The other stones flowed away in the water which was passed over the table. In ingenious adaptations of this principle, the sorting of the diamonds is now done by machines, whose power of distinction is far superior to the keenest and most highly trained eye. Experiments have also been made with rubies, sapphires, and emeralds; and it has been found that stiff grease will catch these gems with the same certainty that it arrests diamonds, whilst all other minerals pass over the surface and fail to adhere. What is the cause of this amazing discrimination? The answer is yet to seek. Precious stones alone are trapped in this manner, and a piece of glass, be it never so cunningly shaped so as to imitate a diamond, will drop on the table and flow away in the tailings.

The rough diamonds thus found are then sorted, classified by reference to their size, colour, and purity, made up into parcels, and finally shipped by special

[1] *Stones of Fire; The Times*, November 5th, 1910.

By permission of *De Beers, Ltd.*

WASHING DIAMONDIFEROUS EARTH, DUTOITSPAN

registered post to the syndicate of diamond merchants in London, to whom the annual production of the De Beers Mines is now sold in advance. The delicate processes of cleaving, cutting, and polishing, by which these dull, white stones are converted into the sparkling gems that the jeweller displays, are performed almost exclusively by the lapidaries of Amsterdam, whose skill in these matters is unrivalled. Then, with the full graces of their liquid splendour revealed, the diamonds are at length ready for the jeweller; and here it is interesting to note that a single country—the United States—buys two-thirds of the diamond output of South Africa.

Illicit Diamond Buying

The circumstance that the rough diamond combines great value with extreme portability has evolved in South Africa the special statutory crime known as illicit diamond buying. Under the enactments dealing with this offence, the accused is deprived of the benefit of the ordinary legal presumption in favour of his innocence, and is liable to a long term of penal servitude, if he is unable to give a proper account of any uncut diamond found in his possession. Every dealer in diamonds is required to take out a licence, and his books are open for public inspection. All dealings in individual diamonds, or parcels of diamonds, from discovery to shipment from South Africa, must be duly registered, and the stone, or stones, properly described.

In view of the unusual character of this legislation, the provisions of the Cape Trade in Diamonds Consolidation Act (No. 48, of 1882, of the Cape Colony), which has been applied, with the necessary modifications, in other provinces of the Union where diamond mining is carried on, will be of interest.

Under this Act:

(1) It is declared unlawful for any person to have in his possession any rough or uncut diamond, unless he is able to

produce his proper permit for the same, or to account satisfactorily for, or prove his right to, the possession of the same. Only duly licensed dealers, etc., are permitted to buy, etc., any rough or uncut diamond. The penalty for contravention of this provision is a fine not exceeding £1,000, or imprisonment up to fifteen years, or both.

(2) Persons finding diamonds on private property are required to make a declaration of the fact within fourteen days to the resident magistrate of the district.

(3) All persons authorised to deal, etc., in diamonds are required to keep a register of their transactions, which record is to include (*a*) date of purchases, etc.; (*b*) name of consigner, cutter, seller, buyer, owner, etc.; (*c*) weight of each parcel; (*d*) number of stones of 10 carats and upwards in each parcel; (*e*) price paid or received; (*f*) weight of each single stone valued by buyer at over £100. A copy of this register must be forwarded every month to the chief of police, etc., and produced when required.

Though these strict regulations and heavy penalties have greatly lessened, they have not, as yet, completely suppressed, the practice of illicit diamond buying in South Africa. While, therefore, the Diamond Laws protect the Company in a large measure against dishonesty on the part of the Europeans employed in the mines, the most stringent precautions are necessary to prevent theft in the case of the native and coloured workpeople. In Kimberley, and in other diamond mines, the compounds, or native quarters, are not merely enclosed by fences and buildings, but are entirely covered with a stout wire netting, which makes egress, otherwise than through the gates, practically impossible. The native labourer in the diamond mines is, therefore, literally a prisoner during the term of his contract; and before he is allowed to leave the premises, he is stripped, searched, and subjected to medical supervision. As all alike have to submit to this treatment, none but the convicted thieves are affected by it; and the fact that this ordeal awaits them, as well as all other conditions of service, are, of course, perfectly well known to the natives before they enter upon their contracts. That the restrictions and necessary precautions essential to the industry are not found irksome, may be inferred by the circumstance that the same "boys" come back from their homes again and again to work for the De Beers Company. Also, it must be remembered that native labourers on the diamond mines receive even higher wages than the high wages which they earn on the gold mines and in other industries in South Africa.

Condition of Natives Employed by De Beers

The Report of the Protector of Natives, contained in the Union Blue Book on Native Affairs for **1910**, will afford evidence on both these points:

> There are 15 compounds under my supervision, occupied by 13,000 natives—4 mine compounds and 11 surface, all De Beers. The only mines working at present are the Kimberley, Du Toit's Pan, Bultfontein, and Wesselton.
>
> The compounds throughout are all built with a view to good ventilation and space, with bunks, and the whole kept scrupulously clean and well fumigated. Special yard boys are employed to look to the sweeping and cleaning of the enclosure and rooms, and the sanitary arrangements are as perfect as possible. Very fine hospitals, excellent medical attention, and every comfort for the sick are provided. All the natives are registered, and pay 12s. a year or 1s. a month hospital tax.
>
> Nothing can be said other than that the natives employed in these mines are well treated in every way. Not a single complaint as to ill-treatment has been made to me for at least two years. . . .
>
> The natives are contracted for four and six months at a wage of 15s. a week for general work, and for loading, 12s. per 100 loads surface and 20s. mine, so that they can earn more if they wish to exert themselves. The average wages earned are from 3s. to 3s. 6d. *per diem.* Drilling boys earn 4s. 6d.
>
> Ministers of religion of every denomination visit the compounds regularly, and hold services; from these the natives purchase Bibles and other requisites for their mental improvement. . . .

To this I may, perhaps, be allowed to add the rough notes which record some personal impressions of the compound of the Kimberley Mine, as I saw it one Sunday in **1905**:

> The compound of the Kimberley Mine is a little town in itself.

It houses 2,000 natives. It has its own shop, at which the natives purchase their own food with brass money, of special coinage, provided by the Company for the purpose. The shop appeared to contain just such provisions as would be met with in the "general" shop of an English village. Except for his "mealie-meal pap," the luxurious native, who earns easily from 30s. to £2 a week, can afford and elects to live much on the scale of the English working classes at home. He pays, however, the European penalty for European food. In other words, he learns to know the meaning of toothache, from which in his natural condition he is exempt. Therefore, the dentist's chair, which is in evidence in the dispensary. The hospital, which we were shown over by a European nurse in uniform, contains its operating room, as well as all the other equipment which a similar establishment would possess in England. Good medical attendance, and the best of food, are provided for the natives when they are ill. In reply to my inquiry, the nurse said they made good patients. In her own words, "they were not fidgety, and they did what they were told." I suggested that they were, perhaps, less sensitive to pain than Europeans, and this, she agreed, was undoubtedly the case. We were shown through the wards—large, clean, and airy rooms, and, fortunately, by no means full. Pneumonia claims the larger number of patients, whilst the accident ward was not without its inmates. Accidents, the nurse said, were generally the result of the "boys'" own carelessness, or, perhaps, more rightly, of their want of intelligence in avoiding danger. As in England, on Sundays, within certain hours, the patients are allowed visitors. I noticed more than one boy squatting in silence on the floor between two beds. These were visitors, the nurse explained, and they frequently sat thus in silence throughout the whole of their visit, neither they nor the patients having apparently anything to talk about. The walled-in compound is a busy, noisy scene on Sundays, though by no means uproariously so. Groups of boys were scattered over the large yard. Some were merely chattering; finding, apparently, no lack of topics outside the solemnity of the hospital. Other groups were busy cooking over wood fires, outside their sleeping-places. One boy was absorbed in producing sounds on a Kafir "piano," a curious flat instrument, consisting of a succession of wooden bars with spaces between. By striking these bars with a stick, he did succeed in producing sounds, if not music. Another group were dancing, that is to say, they stood in a circle chanting, moving their feet up and down, as though steadily moving the treadmill, whilst one amongst them made weird sounds on a concertina. This monotonous entertainment is kept up, we were told, from Saturday afternoon to Monday morning; as one tired boy falls out, another stepping into the ring to take his place. It is curious that these "boys"—grown men, it must be remembered—should possess so low an order of intelligence as to find a perennial source of enjoyment in such an entertainment as this, and are yet capable of becoming really useful domestic servants and workmen.

On leaving the compound, we passed a *queue* of boys waiting to pay their money into the bank; that is to say, the Company take care of it for them until they leave. The boys sign for a short term of service only—six months—but many of them spend years in the service of De Beers, and amass, what must be for them, very considerable worldly wealth. Just inside the compound some Kafir chiefs in European straw hats, and with note-books in their hand, were levying tribute in the most business-like fashion from members of their respective tribes. These taxes, we were told, were always readily paid.[1]

The subjoined table shows the numbers of both the European and coloured employés of De Beers in the years 1908 and 1909, and the rates at which they were respectively paid:

NUMBER OF EMPLOYÉS AT FOUR OPEN MINES IN KIMBERLEY DIVISION.

Mine.	White: 1908.	White: 1909.	Coloured: 1908.	Coloured: 1909.	Wages: Whites.	Wages: Coloured.
Kimberley ..	200	217	1,034	2,618	£3 to £5 10s. at all.	24/– per week & lodging at all.
De Beers ..	1,028	767	2,444	6,441		
Du Toit's Pan .	74	64	657	516		
Bultfontein ..	315	376	2,636	3,035		
Totals ..	1,617	1,424	6,771[2]	12,610		

[From the Statistical Register of the Province of the Cape of Good Hope for the year 1909. Capetown, 1910.]

We may conclude from the foregoing accounts that all reasonable measures are taken by the De Beers' authorities to ensure the comfort of the natives in their

[1] Unpublished MS.

[2] Owing mainly to the financial crisis in the United States, the diamond industry was depressed at this time, and large numbers of natives were set free for employment on the Gold Mines (see Part III, Chap. I, p. 173).

employment. Nor has the Company shown itself less solicitous for the welfare of its European employés.

Kenilworth

The most conspicuous and picturesque result of its efforts in this direction is that presented by the model village of Kenilworth, which was planned and laid out, under the supervision of Rhodes himself, in 1888. It is built on land owned by the Company, and is approached from Kimberley by a wide road, 2½ miles long, which during the siege Rhodes caused to be planted with double rows of pepper trees. The same tree, whose foliage, apart from its welcome shade, possesses health-giving properties, is freely used to line the main avenue of the village and the roads which converge upon it. Each of the neat houses has its own plot of garden, and the two sides of the avenue are screened and sheltered by the many rows of pepper trees which grow between them. Close at hand are the fruit gardens and vineyard of the Company, planned by Rhodes for the special purpose of providing the employés, both European and coloured, with a cheap and abundant supply of fruit during the hot months of the year. The long walk—a mile and 20 yards in length—in the vineyard, and the Kenilworth Gardens generally, were one of Rhodes' special interests; and here, as in the Rhodes Matopo Park, every endeavour has been made to introduce and acclimatise new varieties of fruit-bearing trees and plants, in order that the enterprise, besides fulfilling its immediate purpose, may be of utility to the general public of South Africa.

The Diamond Production of the Cape Province

The quantity and value of the annual diamond export of the Cape, from 1895 to 1909, is exhibited in the table shown on the next page.

KIMBERLEY OUTPUT

Diamonds : Weight, value, and average price per carat of rough and uncut diamonds exported from the Kimberley Division.

Year.	Weight : Carats.	Value.	Average Price Per Carat.
		£	£ s. d.
1895	3,355,863$\frac{3}{16}$	4,323,308	1 5 9$\frac{3}{16}$
1896	3,283,439$\frac{3}{8}$	4,195,651	1 5 6$\frac{37}{64}$
1897	3,220,367$\frac{1}{2}$	4,024,040	1 4 11$\frac{7}{8}$
1898	3,232,057$\frac{1}{8}$	4,124,170	1 5 7$\frac{19}{20}$
1899	2,736,928$\frac{1}{2}$	4,135,583	1 10 2·64
1900	1,882,730$\frac{1}{2}$	3,434,822	1 16 5·85
1901	2,541,597	4,922,830	1 18 8·85
1902	2,650,110$\frac{1}{4}$	5,386,970	2 0 7·85
1903	2,516,684$\frac{3}{64}$	4,988,069$\frac{1}{2}$	1 19 7·73
1904	2,570,388$\frac{5}{8}$	5,336,982	2 1 6
1905	2,335,646$\frac{1}{8}$	4,850,660	2 1 6
1906	2,679,572$\frac{7}{16}$	6,834,369	2 11 0
1907	2,481,494$\frac{3}{4}$	5,978,531	2 8 2
1908	1,645,400$\frac{3}{4}$	3,191,582	1 18 9
1909	2,384,938$\frac{1}{2}$	4,510,642	1 17 9

Note.—Diamonds to the value of £643,047 were imported into Kimberley Division in 1909 from Barkly West Division, Hay, Herbert, and Bechuanaland. The actual output of the mines in the Kimberley Division alone, for 1909, was: Weight, 2,410,846 carats; value, £4,224,904; average price per carat, £1 15s.

[From the Statistical Register, etc.]

The Diamond Production of the Free State Province

Apart from the Wet, or River, Diggings, the earliest discovery of diamonds in South Africa was that made at Jagersfontein, in the Free State, in August, 1870; and, although this discovery was overshadowed by the magnitude of the Kimberley deposits, diamond mining has been carried on in this province of the Union since 1878. After the war, mining, like all other industries, developed rapidly under the Crown Colony Government, and this progress has been maintained up to the present. Taking the latest available returns, we find that in the year 1909–10 the eight diamond producing mines in the Free State, of which the Jagersfontein Mine is the most important, had a collective output of 787,613$\frac{5}{8}$ carats in weight, and £1,525,706 in value. In the same year they employed 1,015 Europeans and 6,943 natives and coloured persons.

The Diamond Production of the Transvaal Province

The Transvaal diamond output is practically the yield of a single mine, the Premier, which was discovered by Sir Thomas Cullinan, at Elandsfontein farm, 25 miles east of Pretoria, in November, 1902. The area of the pipe is stated to be 78 acres, and it has been ascertained by boreholes, sunk within this area, that diamondiferous ground is found at depths varying from 300 to 1,001 ft. Since mining operations were commenced in 1903, the directors of the Premier (Transvaal) Diamond Mining Company, Ltd., have spent very large sums of money in providing the machinery and plant necessary for the successful working of the mine. In addition to this equipment, which includes the largest washing plant of its kind in the world, a railway connecting the mine with the Pretoria-Delagoa Bay line at Rayton Junction has been constructed;[1] a water supply, fed by a reservoir of 400 acres in area, has been provided; and a little town of residences for the European employés and quarters for the native labourers has been built. In July, 1910, this latter held a population of 1,887 Europeans and 12,200 natives; and out of these totals 786 of the Europeans and 11,632 of the natives were men actually employed in the mine. From a financial point

[1] This brings the Premier within (about) an hour and a quarter's run of Pretoria.

of view, it is interesting to note that the whole of this vast expenditure—amounting at October 31st, 1910, to £1,603,201 19s. 2d.—has been met out of the annual earnings of the mine, and that the capital of the Company, therefore, remains at its original modest figure of £80,000.

The table on p. 302 shows the magnitude and value of the Premier output of diamonds.

And to this may be added the record of the Transvaal diamond output, as presented by the Chamber of Mines Report for 1910:

Year. (Every third only to '09.)	Weight (Carats).	Value.	Estimated Value per Carat.
		£	s.
1903	174,976	239,752	27·40
1906	1,069,391	1,563,141	29·23
1909	1,877,486	1,176,680	12·54
1910	2,090,068	1,317,715	12·61

The Premier Mine has been made familiar to the British world by the great Cullinan diamond, which was found on January 25th, 1905, and so named after Sir Thomas Cullinan, the Chairman of the Company. This, the largest diamond in the world, weighed in the rough 3,025½ carats; and to the writer, who was permitted to see and handle it before it left the Transvaal, it appeared to be of the size of a man's clenched hand, while in colour and general appearance it resembled a mass of coarse, whitish glass. By arrangement with the Company, it was presented, in 1907, by the Transvaal Government to the late King, Edward VII; and, after it had been cut and polished at Amsterdam, it was added to the Crown jewels of Great Britain. The clippings of the

Premier (Transvaal) Diamond Mining Company, Limited

Year Ending.	No. of Loads Washed.	No. of Carats found.	Value of Diamonds.			Yield per Load Carats.	Value per Carat.			Value per Load.			Cost of Production Per Load Washed.	
			£	s.	d.		£	s.	d.	£	s.	d.	s.	d.
Oct. 31, 1903	76,931	99,208¼	137,435	2	9	1·29	1	7	8·5	1	15	6·7	2	7·2
,, 1904	939,265	749,653½	866,030	0	5	·798	1	3	1·2	0	18	5·3	2	7·62
,, 1905	1,338,071	845,652	[1]994,687	5	7	·609	1	3	6·29	1	14	3·98	3	3·44
,, 1906	2,988,471	899,746	[1]1,277,739	6	4	·301	1	8	4·82	0	8	6·61	3	5·71
,, 1907	6,538,669	1,889,986¼	[2]1,702,630	19	8	·290	[3]0	18	0·20	[3]0	5	2·49	2	4·14
,, 1908	8,058,844	2,078,825¼	1,536,719	19	0	·258	[3]0	14	9·40	[3]0	3	9·75	1	10·24
,, 1909	7,517,793	1,872,136½	1,172,378	7	2	·249	0	12	6·29	0	3	1·43	1	11·42
,, 1910	9,331,882	2,145,832¾	1,496,641	5	5	·230	0	13	11·39	0	3	2·49	2	0·56
,, 1911	8,325,272	1,774,206	1,433,970	6	9	·213	0	16	1·97	0	3	5·34	2	2·02

[1] Includes Cullinan Diamond at nominal valuation. [2] Proceeds of Cullinan Diamond not included.

[3] Subject to the loss on Contract with Diamond Syndicate.

[From the Directors' Report for the year ending October 31st, 1911.]

great stone yielded six fine diamonds, the largest of which weighed 90 carats; and these lesser diamonds were presented by the Union Government in June, 1910, to Queen Mary, in commemoration of the Union of South Africa, which had come into being on the preceding May 31st.

The Danger of Over-Production

It is obvious that the Premier Mine constitutes a formidable rival to the De Beers Company, and that the danger of cheapening the diamond by over-production, which in the case of the Kimberley mines was avoided by the amalgamation of the competing interests brought about by Rhodes in 1889, now threatens the diamond industry of South Africa as a whole. It will be observed, however, from the tables given above, that the average value per carat of the De Beers' production is considerably in advance of that of the Premier. In 1909, to take a single year, the value of the Kimberley output per carat was returned at £1 17s. 9d., while that of the Premier is given as 12s. 6.29d. It is claimed by the De Beers Company that this difference in quality has enabled it to maintain its traditional policy of regulating its output in accordance with the demands of the market; and that its action, while exercising a beneficial influence upon the industry as a whole, has not injuriously affected the financial interests of its shareholders. The De Beers attitude is well stated in the article in *The Times* of November 5th, 1910, to which I have referred before.

Diamonds are luxuries, and the demand for them depends very much upon the general prosperity of trade, in America in particular. The refusal to put diamonds on the market when the demand is not there is in the interests of all concerned in the diamond trade. It has a steadying influence upon the industry, and protects the Company's customers who hold stocks of diamonds from any sudden and discouraging fall in the value of such stocks. There appears to be on the average something like 5,000,000 sterling per annum available for the purchase of the higher class of rough diamonds, and it is, therefore, simply courting disaster to attempt to put out a greater quantity of the stones than can be absorbed by the general public. De Beers do not believe that the production of a greater quantity of diamonds and the putting of them on the market would result, on the whole, in any larger amount of money being obtained from the consumer. . . .

But strong as is the financial position of De Beers, it is difficult to understand how so formidable and prolific a competitor as the Premier Mine can be disregarded permanently. It would, therefore, seem to be desirable, in the interests of both companies alike, that advantage should be taken of any suitable opportunity of arriving at an agreement on a question which so vitally affects the prosperity of the industry as the mutual limitation of output by competing producers.[1]

The Gold Mines

The importance of the mining industries as the dominant factor in the economic system of the Union has been exhibited in the table with which this Chapter commenced; the importance of the South African gold industry as a factor in the economic life of the great commercial nations, may be gathered from the circumstance that its contribution now constitutes more than a third of the world's total annual supply. Speaking roughly, nine-tenths of the South African gold production is won in the Transvaal, and of the Transvaal output 95 per cent. comes from the Witwatersrand District of the Rand. Before, however, we examine the origin and present conditions of gold mining in this predominant centre of the industry, it will be convenient to give a moment's consideration to the lesser goldfields of South Africa.

Apart from the Transvaal, then, gold is found and

[1] The entire South African diamond output for the year 1911 is reported to be 4,891,998 carats, of the value of £8,746,724.

worked in the provinces of the Cape and Natal, in the Bechuanaland Protectorate, in Swaziland, and in Southern Rhodesia. The three proclaimed goldfields of the Cape, lying respectively in the divisions of the Knysna, Prince Albert, and Mafeking, have a very small output—how small may be seen from the fact that their total registered production up to August 31st, 1910, amounted only to 5,565 ozs. 17 dwt. 7 qrs. In Natal (*i.e.*, in Zululand) the production of gold is somewhat greater; and the annual output in this province for 1909 was returned as being 1,595 ozs. in weight and £6,697 in value. The annual export from the Bechuanaland Protectorate for the same year amounted to 17,015 ozs., of the value of £55,619; and in Swaziland the five producing mines exported gold to the value of £44,499 in the year ended June 30th, 1910. There remains Southern Rhodesia, where alone among the countries here mentioned the gold production has attained considerable proportions. Here, as in the Transvaal, gold mining is the chief industry, and its progressive advance and present importance may be seen in the subjoined table:

RHODESIAN (SOUTHERN) GOLD PRODUCTION since 1900 (at three year periods to 1909):

Year.	Bullion ozs.	Estimated Value.
	ozs.	£
1900	85,367	308,249
1903	231,872	827,729
1906	551,894	1,985,099
	Fine Gold. ozs.	
1909	623,389	2,623,708
1910	609,956	2,568,198

Note.—In 1911 the value of the gold produced was £2,674,896.

Gold Mining in the Transvaal

Apart from the Rand, gold is won in considerable quantities from the Eastern (or outside) mines of the Transvaal, which formed the earliest seat of the industry in South Africa.[1] As early as 1873 the Landdrost of Lydenburg reported to the Boer Government of that day that gold had been discovered in his district; ten years later the diggers withdrew from the Lydenburg fields to the De Kaap valley, 50 miles southward, in the district of Barterton. Here, in 1886, owing to the discovery of the rich Sheba mine, no less than 10,000 diggers had assembled. In September of the same year, the Rand was proclaimed a public goldfield; but in spite of the continuous expansion of the great goldfield, mining has been carried on in the Eastern fields since 1884. In that year the entire Transvaal output of gold amounted to £10,000 in value; while, in 1910, these "Outside Districts" of the Transvaal alone produced 304,011 ozs. of fine gold, of the value of £1,291,354—an annual gold production which in any other country but the Transvaal would attract attention.

The Discovery of the Witwatersrand Goldfields

In the manner of its discovery, the Rand contrasts sharply with the Kimberley diamond mines. It was revealed not by accident, but as the crowning reward of a long-continued, persistent, and arduous search for gold, in which men's lives were lost, and deeds of heroism or of shame were witnessed only by the silent Bush, or the crags and torrents of the lonely mountain heights. It was known that gold was to be found in the Transvaal as early as 1854; but the emigrant farmers, whom the Sand River Convention of 1852 had made masters of the

[1] That is, *modern* South Africa: Rhodesia was, of course, the seat of the prehistoric, or Phœnician, gold mining.

land, closed the country to prospectors. In these circumstances, the pioneers, Hartley, Baines, and the rest, turned their footsteps towards the scene of the ancient gold workings beyond the Limpopo; and here, in 1865, the German explorer, Karl Mauch, discovered the Tati goldfield. Two years later, the Boers ceased to prohibit prospecting; and in 1872 the Volksraad, influenced by the more progressive spirit which led to the choice of Burgers as President, formally recognised the industry by passing the first gold law. Under this enactment "the right to mine for minerals" was vested in the State, but provision was made for the distribution of the prospective mineral wealth of the country as between the Government, the owner of private land, and the discoverer. At the same time the Government undertook to administer lands declared to be "public diggings," and rewards were offered to prospectors for the discovery of payable goldfields. During the period of the annexation (1877–80) an Australian mining expert, Armfield by name, was employed by the first administrator, Sir Theophilus Shepstone, to prospect for gold; and in the course of examining the more likely districts, Armfield was at one time at work close to the Rand basin. The discovery of the great goldfield was delayed, however, for some years longer, and was due to the skill and perseverance of two brothers, Mr. W. H. and Mr. F. P. J. Struben, who came to the neighbourhood in 1884. The former secured the mineral rights on the Wilge Spruit farm, erected a small battery, and found that the ore from a reef which he discovered, and named appropriately enough the "Confidence Reef," yielded a good return of gold. In the meantime, Mr. F. Struben, when engaged in prospecting at the western end of the Rand basin, came upon the actual conglomerate, or banket, beds, which he found to contain deposits of gold. In spite of the discouraging estimate of the value of the deposits which was formed at first by some authorities, Mr. Struben persisted in having test crushings of the ore carried out at his brother's mill. The results of these crushings were so satisfactory, that, when they had been reported to the Boer Government, on September 8th, 1886, the Witwatersrand was declared a public goldfield.

The Growth of Johannesburg

This event produced a change, the like of which for suddenness and world-wide effect has been equalled rarely, if ever, in the history of Anglo-Saxon colonisation. The scene of the Strubens' discovery, called the " Ridge of the White Waters " by the Boers from the pure streams which flowed down its northern slope, was the crown of the long stretch of rolling uplands, which, running almost due east and west for 300 miles, separates the valley of the Vaal from the head waters of the Limpopo, and constitutes the " High Veld " of the Transvaal. It was here, in this wind-swept desolate region, lying nearly 6,000 ft. above sea-level, that six years before the Vierkleur, raised by the Triumvirate on Dingaan's Day, December 16th, 1880, had signalled the outbreak of the first Boer war. No place could have seemed less likely to become the site of a great town than the remote and inaccessible region in which lay the newly proclaimed public diggings. None the less, by the end of the same year, Ferreira's Camp, with its fifty miners, was raised to the rank of a township, graced by President Kruger with the name of Johannesburg. At first, everything that the miners required, building materials, mining gear, tools, food, and drink, had been dragged for a hundred miles or more up and over the veld by processions of ox-wagons, travelling laboriously at an average speed of 1½ miles an hour. For there were no railways and very few roads at this time in the Transvaal or the Free State ; even in the Cape Colony railway communication had been

By permission of *The Consolidated Goldfields of S.A.*

TURNING THE FIRST SOD FOR THE FOUNDATION OF SIMMER DEEP MILLS

established between Kimberley and the coast for little more than a year, while in Natal the railway from Durban, the port, extended to only a few stations northward of Maritzburg. It was at this early stage that Kimberley capital and Kimberley experience were of supreme assistance. At the end of the first year Johannesburg had a population of 3,000, and the first annual return of the gold mines of the Rand, which showed an output of 19,080 ozs., of the value of £81,045, had been published. Five years later, the pithead gears, batteries, and surface works of the mines extended for 30 miles along the ridge; and Johannesburg itself, now recognised as the industrial centre of South Africa, had been brought by the extension of the Cape railways through the Free State within forty-eight hours of Capetown and three weeks of London. To Lord Randolph Churchill, who saw it at this date and described it as resembling an "English manufacturing town minus its noise, smoke, and dirt," its growth seemed "almost magical." In 1896, ten years after the proclamation of the goldfield, the first official census returned the population living within a three-mile radius of the Market Square as 102,078; and of this total 50,907 were Europeans.

From the first, better and more permanent materials were used for the construction of houses and buildings at Johannesburg than those with which the founders of Kimberley had been content. But it was not until after the war that Johannesburg shook off the sordid remains of the mining camp, and began seriously to furnish itself with a civic equipment worthy of its large population and commercial importance. Before peace was declared, Lord Milner had established a nominated town council, and this body was succeeded in 1903 by an elected council which contained some of the ablest members of a singularly able industrial community. From this time onwards rapid progress was made in the work of providing all things necessary for the health and convenience of the inhabitants of Johannesburg. The roads were reconstructed or improved, and this, together with the provision of a permanent and abundant water supply, enabled the municipality to lessen materially the evil of dust and to lay down a system of water-borne sewerage; whole streets of sordid and insanitary houses were demolished, and the native and coloured inhabitants removed to locations placed well beyond the municipal boundary; the old horse-drawn tramways were superseded by an electric system; new buildings, public and private, were planned under proper architectural supervision, and constructed on a scale commensurate with the wealth and importance of the commercial capital of South Africa; the existing parks were improved, and new parks and public spaces were laid out and suitably embellished. In the meantime, the population of the town has grown with the continued expansion of the gold industry; and in the census of May 7th, 1911, the total number of the inhabitants of Johannesburg is returned as 237,220, and of this total 120,411 are Europeans.

The Gold Industry of the Witwatersrand

The story of the gradual expansion of the gold industry is told in the subjoined table, which is taken from the Twenty-first Annual Report of the Transvaal Chamber of Mines for the year 1910:

Witwatersrand Gold Production from May, 1887, to December, 1910:

Year.	Tons Milled.	Fine Gold.	Value.
		ozs.	£
1887..	—	19,080	81,045
1888..	—	171,789	729,715
1889..	—	306,167	1,300,514
1890..	—	408,569	1,735,491
1891..	—	601,810	2,556,328

By permission of *The Consolidated Goldfields of S. A.*

SURFACE WORKS, SIMMER DEEP GOLD MINE

TRANSVAAL GOLD PRODUCTION

	Year.	Tons Milled.	Fine Gold.	Value.
	1892..	—	1,011,743	4,297,610
	1893..	2,215,413	1,221,171	5,187,206
	1894..	2,830,885	1,639,252	6,963,100
	1895..	3,456,575	1,845,875	7,840,779
	1896..	4,011,697	1,851,422	7,864,341
	1897..	5,325,355	2,491,593	10,583,616
	1898..	7,331,446	3,564,581	15,141,376
The war, Oct. 11th, 1899, to May 31st, 1902 ..	1899..	6,639,355	3,317,857	14,093,363
	1900..	—	—	—[1]
	1901..	412,006	238,877	1,014,687
	1902..	3,416,813	1,690,096	7,179,074
	1903..	6,105,016	2,859,482	12,142,307
Chinese Labour, June, 1904, to March, 1910; but reduced in 1908. Period of full operation, 1905-6-7 ..	1904..	8,058,295	3,653,794	15,520,329
	1905..	11,160,422	4,706,433	19,991,658
	1906..	13,571,554	5,559,534	23,615,400
	1907..	15,523,229	6,220,227	26,421,837
	1908..	18,196,589	6,782,538	28,810,393
	1909..	20,543,759	7,039,136	29,900,359
	1910..	21,432,541	7,228,311	30,703,912
	Totals		61,219,337	273,678,440
	Add	Estimated unrecorded output for 1887-9	34,607	147,000
		[1] Undeclared output for Oct., 1899, to May, 1900 ..	584,841	2,484,247
		Amount won in 1904, but undeclared ..	4,447	18,890
	Grand Totals ..	—	65,053,232	276,328,577

And to this may be added the following statement, which shows how the total gold output of the Transvaal has risen (in ten-year periods) from the commencement of gold mining in the Eastern fields:

Year.	Ozs.	Value.
		£
1884	2,377	10,096
1894	1,805,000	7,667,152
1904	3,773,519	16,028,883
1910	7,532,322	31,995,266
Totals for period 1884–1910 ..	68,291,928	290,085,590[1]

[1] The output for 1911 is reported to be £31,976,121 in value, showing a slight decline.

[From the same Report.]

The next two tables, taken together, will exhibit the present importance of the Transvaal gold production as a factor in the commercial intercourse of nations, and a chief source of the supply of the gold which is needed by the world both for currency and the arts.

Estimated value of the annual gold production of the world since 1887, at £4.24773 per fine ounce. The figures to 1907 are those given in the Report of the Director of the United States Mint:

Year.	Value.
	£
1887	21,735,000
1897	48,509,000
1907	84,769,000
1910	94,135,635

[From the same Report.]

Estimated gold output of the world during 1910:

Country.	Value.	Percentage of Total.
	£	
Transvaal	32,001,735	34·0
United States	19,737,838	21·0
Australasia	13,519,153	14·4
Russia	7,397,000	7·9
Mexico	4,991,000	5·3
Rhodesia..	2,590,924	2·7
Canada	2,142,000	2·2
West Africa	755,985	·8
Other Countries	11,000,000	11·7
Total ..	94,135,635	100·0

[From the same Report.]

By permission of *The Consolidated Goldfields of S. A.*

FOUNDATION BLOCKS (CONCRETE UNDER STAMPS)

The importance of the gold industry to South Africa itself is shown by the fact that the export of raw gold constituted 63 per cent. of the total value of the South African produce exported in the year 1910. Of this gold export, as we have seen, nearly 90 per cent. is provided by the Rand.

The Character of the Deposits

Before examining the financial results obtained by the gold mines of the Rand, in respect both of the profits earned for the shareholders and the benefits derived from the money spent in wages and in the purchase of materials, it is necessary to realise certain special features of the industry, which make gold-mining on the Rand different from gold-mining in other parts of the world. These may be grouped under the heads of: (1) The nature of the gold deposits; (2) the methods of recovery; and (3) the narrowness of the margin of profit.

Geologists have not yet determined at what stage gold was deposited in the layers of water-worn quartz pebbles, cemented together by their own detritus, and placed between layers of sandstone, which the Dutch called "banket" ore, because of its resemblance to almond toffee. But it is generally agreed that the original materials of these conglomerate beds of the Rand Basin were gradually gnawed by the waves of an inland sea from its coasts of sandstone hills with quartzite veins, and laid smoothly beneath its waters. In the course of time incalculable, the fragments of quartz and the grains of sand, now piled into a mass 4 or 5 miles, or more, in depth, were once again welded into horizontal bands of rock. Then—who shall say "when" or "how"—a great convulsion of Nature upheaved the floor of this primæval sea, drove the granite foundations of the continent and streams of molten rock through it, and, working with unequal pressure, tilted the northern edges of its interstratified bands of conglomerate and sandstone rocks to the surface. Then another interval of time, again incalculable, passed, and the face of the earth assumed the appearance that it bore when the first human footsteps pressed the veld.

Whether the gold entered the conglomerate beds when they and the sandstones were laid down upon the floor of the sea, or was filtered into them from the streams of molten matter that burst upwards and through them at the time of the great convulsion, remains uncertain ; but there is no question as to the wide extent and remarkable constancy of the deposits themselves. Speaking roughly, the upturned edges of the Rand Basin can be traced round an oval 100 miles long and 30 miles broad, and, although the conglomerate and sandstone bands were broken and dislocated by the intrusion of igneous rocks, they dip southwards from the northern outcrop with a certain regularity of curvature, falling at an average angle of 30 degrees. Assays from the conglomerate beds, when followed by boreholes along the dip down to a depth of 4,000 ft., have shown that the gold deposits remain constant down to this point. Mining, therefore, is no longer confined to the 30 miles of outcrop, but the ore is being extracted at increasing depths from the beds as they dip southward. And as on the Rand, mainly owing to its high elevation, the temperature rises only 1 degree Fahr. to every 208 ft. of vertical depth, as against a normal rise of 1 degree to every 65 ft., it follows that, other things being equal, mining can be carried on here at a much greater depth than elsewhere. It seems likely, therefore, that the extension of deep-level mining will be restricted not by the exhaustion of the gold deposits, but by the increasing cost of the equipment and labour required to raise the ore.

The conglomerate beds from which the gold of the

Rand is won are distinguished, therefore, by their great extent and the general constancy of their yield. It is these qualities that make the Rand the greatest producer among the goldfields of the world. But, in addition to them, the conglomerates have other qualities that must be noticed; since they impose certain conditions and limitations upon the successful prosecution of the industry. In the first place, the character of the ore is such that, in extracting the gold deposits, mechanical methods have to be supplemented by elaborate chemical processes; and, in the second, the yield per ton of ore thus obtained is very small as compared with the average returns shown by gold mines in other parts of the world.

Methods of Recovery

It has been calculated that the conglomerate ore contains one particle of gold to 80,000 or 90,000 particles of waste material. If, therefore, the industry is to be carried on at a profit, every possible particle of gold must be recovered, and every operation, whether of mining or of metallurgy, must be conducted with the highest economy of labour and materials. These conditions have caused the Rand Mines to become, in the character and perfection of their equipment, not merely mines, but manufactories of gold; and, as such, the first and most necessary condition of their success is an economic, as well as adequate, supply of manual labour. After the ore has been raised to the surface by the ordinary methods of underground mining, it is broken up by rock crushers, and fed into the batteries of stamps, where it is beaten fine, mixed with water, and passed out over copper plates coated with quicksilver. On these plates all particles of free gold are retained, and the gold thus secured is from 60 to 70 per cent. of the total deposits. It then remains to recover the 30 to 40 per cent. of the gold deposits which consist of atoms so minute, that they are still enclosed in the fine particles of ore left by the stamps. To effect this purpose, the particles of ore are first ground to a fine powder in the recently introduced tube mills, and then treated chemically. The ore, now a fine powder, is saturated with cyanide of potassium or sodium, with the result that the cyanogen, having a special affinity for gold, releases the potassium or sodium, and combines with the gold in the form of cyanide of gold. The auriferous solution thus obtained is carried to the precipitating boxes, where the cyanogen, as it has an even greater affinity for zinc than for gold, releases the gold upon coming into contact with zinc scrap, and the gold, when precipitated, falls in a fine powder to the bottom of the boxes. It is then carried to the retorting and smelting furnaces. By the employment of these combined mechanical and chemical processes the mines of the Rand are able to recover from 93 to 96 per cent. of the gold originally deposited by Nature in the hard pyritic ore.

The Paucity of the Yield per Ton

But these perfected methods of recovery cannot, of course, alter the primary fact that the ore of the conglomerate beds of the Witwatersrand is essentially low grade. On the occasion of Mr. Joseph Chamberlain's visit to Johannesburg, in January, 1903, the following figures were laid before him by the Mining Engineers of the Rand to illustrate this aspect of the gold industry.

On the Rand, 39,364,671 tons of ore, crushed by sixty-three companies from the date of their inception to the year 1903–4, produced gold of the average value of 42s. 11d., at a working cost, including depreciation of machinery and plant, of 30s. 11d., and a net profit of 12s. per ton.

By permission of *The Consolidated Goldfields of S.A.*

SLIMES TANKS

In other gold-bearing regions the value of the yield per ton was stated to be as follows in the under-mentioned mines:

	Per Ton. s.	d.
In New Zealand (Waihi Gold Mines)	55	4
In Australia—		
Queensland (Mount Morgan)	109	10
,, (Charters Towers)	103	7
,, (Gympie)	102	9
,, (Croydon)	68	11
,, (Ravenswood)	60	10
,, (Elberidge)	75	3
West Australia (Great Boulder)	102	0
,, ,, (Kalgoorlie)	140	0
,, ,, (Lake View)	120	0
Tasmania (Tasmania Gold Mines)	82	0
,, (New Golden Gate)	70	0
In India (Mysore)	108	7
,, (Champion)	107	3
,, (Ooregum)	83	5
,, (Nundydroog)	97	5
In the United States—		
Cripple Creek (Portland)	200	6
Nevada (Comstock)	205	4
Colorado (Camp Bird)	127	9
In Venezuela (El Callao)	152	2
,, Mexico (El Oro)	55	9
,, Canada (Le Roi)	49	6[1]

It will be observed that this statement of the Witwatersrand average yield per ton as 42s. 11d. is based upon the returns obtained during a period, which includes the early years of the industry, when the mines, working on the outcrop of the conglomerate beds, confined themselves to crushing the higher grade ore. With the introduction of deep-level mining, and the gradual improvement of the plant and equipment of the companies, however, an increasingly large percentage of lower grade ore has been brought to the mill. The

[1] The figures are taken from *Transvaal Problems*, by Sir Lionel Phillips (p. 79).

natural result of this has been that in recent years the average yield per ton has materially decreased. On a basis of the returns obtained in the year 1909, for example, it was 28s. 11d., as compared with an average yield per ton in Australia and New Zealand of 55s. to 140s.; in India of 83s. 5d. to 108s.; in the United States of 127s. 9d. to 200s.; and in Canada of 49s. 6d.

The improvement in equipment and methods of recovery, which has made it possible to extract the gold from the lower grade ores at a profit, has been accompanied by other efforts to reduce the cost of production, and, in particular, in view of the scarcity of native labour, labour-saving machinery has been invented and introduced. In this connection it must be remembered that the Rand gold industry, owing to the magnitude of its operations, has been able to secure the services of the most skilful mining engineers and metallurgists that the world has produced. The combination of professional talent with the practical experience and administrative ability of the leaders of the industry has been so far successful, that while, as we have seen, there has been an appreciable reduction of the average yield of the ore now brought to the mill—a reduction due in part to a slight deterioration in the yield of the conglomerate worked by the deep-level mines, but mainly to the fact that it has become possible to obtain a profitable recovery from ores of lower grade which were originally neglected—the net profit per ton has scarcely fallen at all. In fact, as the subjoined table (on the next page) shows, the gain in economy of production for a time exceeded the loss due to the fall in the grade of the ore milled.

The arrest of the fall in working costs in 1910 is attributed to a slight increase in the cost of labour, caused by the continued deficiency of native African labourers. The circumstance serves to emphasise the fact that, with ore of so low a grade, an abundant and economic

By permission of The Central Mining and Investment Corporation

STAMP MILL, VILLAGE DEEP

supply of manual labour is no less necessary for the Rand gold industry, than an abundant and economic supply of coal is for the manufactures of Yorkshire, Lancashire, and the Midlands.

Year.	Yield in shillings per ton.	Working Costs (do.).	Working Profit (do.).
1905	35·82	23·167	12·653
1906	24·51	22·167	12·343
1907	33·04	20·83	13·11
1908	31·39	18·00	13·39
1909	28·917	19·083	11·974
1910	28·50	17·583	10·971
1911 (Jan. to Sept. only) ..	27·917	17·917	9·66 [1]

The Life of the Rand

Assuming that an adequate supply of manual labour is obtained, and in the absence of any revolutionary changes in the economic life of the world, the duration of the gold industry will depend upon the extent of the workable conglomerate beds. Mr. Hamilton Smith, as the result of investigations on the possibility of mining at deep levels made on behalf of Messrs. Rothschild in 1893, estimated the value of the Rand deposits to be £325,000,000, or nearly £100,000,000 greater than the total yield of the Californian mines between the years 1849 and 1892 (£230,000,000). And a later estimate, that of Mr. Theodore Reunert, put forward the higher figure of £450,000,000. In 1895 Mr. Hamilton Smith wrote that "the chances were far greater then, than they were in 1892, of his conjectures of that date being

[1] Published in an article in *The Times* of January 2nd, 1912. A cable published in February in the same journal stated that the returns for the Witwatersrand alone for January, 1912, showed a total of 2,067,161 tons crushed, a total profit of £999,557, an average yield per ton of 27s. 6d., average working costs of 18s. 10d., and an average profit per ton of 8s. 11d.

realised" ; and he predicted that the maximum product of the Rand would be reached about the end of the nineteenth century, when it would probably exceed £12,500,000 per annum.[1] To-day, gold to the value of the whole of Mr. Smith's estimate has been produced, and the mines are still extracting further gold at a rate of production which is two and a half times as great as that indicated by his suggested maximum. These figures are interesting in themselves, and they show that at all events this early estimate of the extent of the gold-bearing conglomerates has fallen short of the reality. Writing ten years later (1905), Sir Lionel Phillips, whose right to be heard on such matters is unquestioned, has stated the broad data which enable us to form some conception of the extent of the payable beds. "The Main Reef Series of the conglomerates," he writes,[2] which has been "proved to persist in a continuous line, with but trifling interruptions," for 61½ miles, "is held under various forms of title, but principally under claim licences, for a horizontal distance from the outcrop that would involve sinking to a vertical depth of 8,000 to 10,000 ft. before its contents of precious metal could be secured." Assuming that the 61½ miles will be worked to an average depth of 4,000 ft. vertical—an assumption which is generally accepted as reasonable by most engineers—"we obtain an area of 40,200 claims."[3] On June 30th, 1905, when gold to the value of (roughly) £125,000,000 had been extracted since the commencement of mining in 1887, it was calculated that "4·279 per cent. of this area had been exhausted." Sir Lionel, in putting forward these figures, is careful to point out that "it is impossible to give any definite estimate,

[1] Article in *The Times* of February 19th, 1895.

[2] *Transvaal Problems* (Murray, 1905), p. 76.

[3] Under the Transvaal Gold Laws the extent of a claim (for quartz mining) was 150 (Dutch) ft. on the strike × 400 on the dip. 30 Dutch feet = 31 English feet.

either as to the length of life or the ultimate productive capacity of the Witwatersrand." Assuming, however, the same basis of calculation to hold good, it would follow that, as by the end of 1911 gold to the value of over £300,000,000 had been extracted; approximately, 12 per cent. of what may be called the "ascertained payable area" of the Rand has been exhausted up to the present time. And Sir Lionel Phillips, when writing at a still more recent date (1910), sums up the question of the permanence of the industry in the following passage:

> To-day no one has reason to fear either the loss of the beds or their serious impoverishment within the limits of practical mining. The economic question of the future resolves itself into one of ample development and equipment of mines for their exploitation at great depths. An adequate supply of unskilled labour is another essential factor. . . . Given an active and progressive policy on the part of the Government, a bright future for South Africa is assured. The opinion that used to be fashionable, especially in circles not too friendly to the gold mining industry, that its life would be a limited one, and that in a few years there would only be holes in the veldt to denote where it had once thrived, is no longer held by any one acquainted with the real conditions. The 60 miles in length of the sedimentary deposits now being worked will not be exhausted in the present century, and at present our knowledge is confined to the northern shore of the sea of bygone ages. We know of bends to the south at Modderfontein on the east and at Randfontein on the west, and we know of one little patch of the main reef series at Heidelberg, 35 miles to the south, where the formation dips to the north, and where we therefore know that we have arrived at the opposite shore; but, for all we know, there may yet be discovered areas of the sea in other parts of the Transvaal, especially towards Potchefstroom and Klerksdorp, which may add to the life and production of the gold mining industry of the country; in fact, for all practical purposes, gold mining here may be looked upon as a permanent industry.[1]

The Present Position of the Transvaal Gold Industry

We are now in a position to consider the present operations of the gold industry. Taking for our basis

[1] Article in *The Times* of November 5th, 1910.

the year 1910, as being the latest year in respect of which complete returns are available at the time of writing, we find from the Report of the Chamber of Mines[1] that the total annual output of gold amounted to £32,001,735 in value, and that of this amount £30,703,912 came from the Witwatersrand.

The profits earned by the industry are disclosed in the returns made for the purposes of collecting the 10 per cent. Profits Tax, which was imposed on the gold mines by the Transvaal Government in 1902. In the period July 1st, 1909, to May 30th, 1910—the financial year was shortened to ten months by the establishment of the Union on May 31st, 1910—ninety-four companies, of which seventy-three belonged to the Witwatersrand and twenty-one to the Outside Districts, were assessed; and of this total, seventy-seven made taxable profits, eight made taxable profits which, however, were less in amount than the amortisation allowed; and nine were worked at a loss.

On the seventy-seven mines making taxable profits—

The aggregate Profits made (as assessed by the Treasury) were	£11,280,988
The Allowance for amortisation was	1,862,879
The Taxable Profit was	9,418,109
The Tax charged (10 per cent. on Taxable Profit) was	941,925
The Tax paid during the period was	892,831
The Percentage of the Tax assessed to the aggregate profits made was	8·35
The Percentage of the Tax assessed to the dividends declared was..	9·59
And the Tax assessed per ton of ore milled was ..	11·27d.

The total sum paid by the companies in dividends for the year 1910 was £9,130,958; and of this total, the Rand furnished £8,887,185, while £243,773 came from the Outside Districts. The total amount of the dividends declared during the period 1887–1910 is £75,105,429, of which £72,436,579 was drawn from the Rand, and £2,668,850 from the outside districts. The average

[1] Published in 1911.

number of persons at work on the gold mines during the year, including coloured convicts, was: Whites, 24,520; coloured, 178,313; Chinese, 191. Of these totals, 23,421 whites, 167,415 coloured persons, and 191 Chinese were employed in the Rand, and 1,099 whites and 10,898 coloured persons in the outside districts. The amount earned in salaries and wages was £13,402,232; and of this total, £1,203,736 was paid in salaries, £6,858,035 in wages to the white skilled workmen, £5,333,204 to the coloured, (mainly native) unskilled workmen, and £7,257 to the small number of Chinese coolies who were not repatriated until March in this year (1910).

The sum of £11,886,931 was paid by the gold mines for the stores purchased for consumption during the year ended December 31st, 1910. These stores included, among other articles, foods for the Chinese coolies, coal, coke, electrical machinery, explosives, hand tools, iron, lead, lime, lubricants, machinery and machine tools, foods for the native labourers, paint, pipes, ropes (wire), steel, timber, Tube Mill requisites, etc., etc.

The Labour Employed

The average numbers of the respective classes of labour employed in the Witwatersrand and Transvaal gold mines in each year from 1904 to 1910 are shown in the subjoined table:

Year.	Witwatersrand.			Transvaal (as a whole).		
	Whites.	Coloured.	Chinese.	Whites.	Coloured.	Chinese.
1904 ..	13,027	68,438	9,668[1]	13,712	77,425	as before.
1905 ..	16,227	91,084	39,952	17,044	100,037	
1906 ..	17,210	84,897	51,427	17,980	93,628	
1907 ..	16,755	105,915	49,302	17,468	115,273	
1908 ..	17,593	139,893	21,027	18,434	150,317	
1909 ..	20,625	161,795	6,516	21,589	172,382	
1910 ..	23,651	183,613	305	24,757	195,246[2]	

[From the *Chamber of Mines Report*, 1911.]

[1] Seven months.

[2] For 1911 Native labour returns, see Part IV, Chap. I, p. 245.

The European workmen of the Rand gold industry, as a whole, constitute one of the most costly bodies of skilled labour in the world. The average wages per shift, usually of ten hours, of the principal occupations, as given in the Circular published by the Emigrants' Information Office on October 1st, 1911, are as under:

	s.	d.
Mine:		
Shift bosses	24	7
Timbermen	20	8
Pipemen	18	11
Platelayers (underground)	18	4
Banksmen and Onsetters	14	8
Gangers and Trammers	13	9
Pumpmen	18	11
Miners: Machinemen—		
Stoping	21	3
Developing	24	2
Miners—Handmen—		
Stoping	18	9
Developing	19	1
Engine-drivers—		
Underground	20	6
Skipmen	15	9
Mill:		
Firemen	22	10
Amalgamators	18	4
Millwrights	20	4
Cyanide Works:		
Firemen	23	3
Cyaniders	16	8
Smelters	18	2
Sundry Surfacemen:		
Firemen	15	2
Unskilled labourers	5	9

The actual value of these wages (*i.e.*, their power to purchase the necessaries and conveniences of life) depends very largely upon the character and circumstances of the individual wage-earner. As, however, the cost of living to the European in South Africa is a subject which is elsewhere discussed,[1] it will be sufficient to add here that it is generally recognised that a married artisan, with a family, cannot live in comfort on the Rand unless he earns an income of at least £300 a year. The fact is one of the highest significance, since it demonstrates the extreme difficulty of substituting unskilled European labour for unskilled native labour in the gold industry. The white man, whether a skilled or unskilled labourer, if he is going to marry and settle down in South Africa

[1] See Part V, Chap. V, p. 493.

—in other words, if he is to become an effective citizen —must be paid at a rate which will enable him to maintain himself and his family according to the standard of the white inhabitants. A glance at the relative numbers of the white and coloured employés, and the amounts respectively paid in wages to them, as set out above, will be sufficient to show that, under the existing economic and political conditions of the Union, the employment of white unskilled labour is absolutely prohibited by its costliness. It is generally calculated that a native labourer costs the companies £50 a year—£30 in wages, and £20 in food, quarters, and other expenses. If, therefore, the (in round numbers) 200,000 native labourers now employed at £50 a year per head were replaced by European labourers at £300 a year, the annual bill for unskilled labour would suddenly rise from £10,000,000 to £60,000,000—a sum approximately twice as great as the value of the output, and six times as great as the net profits of the industry.

Native Labour

The Witwatersrand Native Labour Association is the chief recruiting agency of the Transvaal mining industry. The two following tables, taken from its report for the year 1910, will exhibit the magnitude of its operations, and the number and variety of the sources from which the gold mines draw their supply of native African labour.

Table showing the number of natives employed and distributed in 1909 and 1910:

Year.	Allotted by W.N.L.A.	Recruited by Mines.	Average No. of Natives employed during the yr.
1909	106,201	52,071	159,084
1910	123,591	80,029	177,795

Table showing the territorial analysis of the natives employed by members of the W.N.L.A. at December 31st, 1910:

1. *British* :

				Per Cent.
Transvaal	13,404		= 7·49	,,
Swaziland	2,586		= 1·45	,,
Bechuanaland ..	697		= ·39	,,
Cape	46,599		= 26,06	,,
Basutoland	5,674		= 3·17	,,
Free State	715		= ·40	,,
Natal	11,952		= 6·67	,,
Rhodesia	627		= ·35	,,
British Nyasaland ..	3,370		= 1·88	,,
North-East Rhodesia .	60		= ·03	,,
		85,684		47·85

2. *Foreign* :

(*a*) Portuguese : Province of Mozambique—Lorenzo Marques, Gaza, and Inhambane Districts :	74,395		= 41·54	Per Cent.
Quilimane, Tete, & Barue	5,924		= 3·31	,,
Mozambique District .	9,435		= 5·26	,,
Portuguese Nyasa ..	3,067		= 1·71	,,
Beira and Chinde ..	248		= ·14	,,
		93,069		51·96
(*b*) German S.W. Africa	301		= ·17	
		301		·17
3. *Other Sources*	29		= ·02	
		29		·02
Total	179,083			100·00

Under the republic, the natives employed on the gold mines were subjected to petty exactions on the part of the Government officials, the arrangements for their accommodation were wholly inadequate, and the system of recruitment by private and unsupervised labour agents, or "touts," was productive of much evil alike to the employer and the employed. Upon the establishment of Crown Colony government in the new colonies after the war, Lord Milner impressed upon the leaders of the gold industry the urgent necessity of improving the condition of native labour on the mines, both on humanitarian grounds and as being the first and most obvious step in the direction of increasing the supply; and the supervision of the labourers and the inspection of the premises in which they lived were included in the regular administrative duties of the Transvaal Native Affairs Department. In the pursuance of this policy, the Witwatersrand Native Labour Association was formed to serve as a responsible agency for recruiting native labourers, and the sanitary and living accommodation provided for them in the mines, and the general conditions of their employment, were rapidly improved. The process of reform, thus inaugurated, has been continued in subsequent years; and to-day no reasonable measure is omitted which is calculated to secure not merely the just treatment, but the comfort and convenience, of the native labourer from the moment when he is recruited, or leaves his kraal independently, to the time—some six or twelve months later, as the case may be—that he again reaches the country, or district, in which his home is to be found. The one serious evil which remains to be removed is the high rate of mortality. The chief sufferers are natives who come from the warm, subtropical districts of the East Coast, and from Central Africa. Strenuous efforts, however, have been made to protect the native labourers—or, rather, to induce them to protect themselves—from the injurious effects of exposure to the comparative cold of the high veld in winter, after they have emerged heated from their work underground; and, although the rate of mortality is still too high, it has been, and is still being, steadily

22—(2139)

reduced. The subjoined table will show the progress which has been made since 1902, when the Native Affairs Department first began to combat the evil.

Table showing the highest death rate per 1,000 per annum, in any month in each year, from 1903, among natives employed on mines and industrial works in the proclaimed labour districts of the Transvaal:

Year.	Month.	Rate per 1,000.
1903	July	112·5
1904	November	56·5
1905	,,	63·1
1906	January	47·1
1907	May	38·3
1908	December	38·0
1909	March	41·1
1910	December	36·23

Based upon the Union Native Affairs Report for 1910.

If the mining companies desired to ill-treat the natives in their employment, they would find it difficult to do so in view of the system of Government inspection. The inspectors, who frequently visit the compounds, make it their business to ensure that every native who has a ground of complaint, real or imaginary, shall have an opportunity of laying it before them. The hours and conditions of employment under and above ground are carefully supervised; the dormitories are inspected to see that the air-space per head, required by the medical officers' regulations, is provided; the quality and quantity of the food is examined; and the sanitary arrangements and hospital accommodation are the subject of special vigilance. So far, however, from desiring to ill-treat the native labourers, the mining authorities realise that, in view of the scarcity of the supply, it is greatly to their

By permission of *The Central Mining and Investment Corporation*

COMPOUND, LANGLAAGTE DEEP

advantage that every labourer should be made as comfortable as possible, in order that he may return himself for further periods of service—as is done especially in the case of the Mozambique natives—and serve to attract his neighbours to the mines by giving favourable accounts of his own experiences.

The life of the natives in the compounds of the gold mines is very much the same as that which has been sketched in the case of the diamond mines at Kimberley. In one respect, however, the conditions of employment are more favourable. The Rand native labourer is not rigorously confined within a compound covered with a network of wires; he has the run of the mine premises, and leave of absence to visit friends, and for other reasonable purposes, is readily granted. His wages are not so high, but, on the other hand, he is provided with all his food and with medical attendance, in addition to his quarters. Here, too, as at Kimberley, arrangements have been made for the banking of his wages, and for the easy transmission of sums of money to relatives or creditors; only on the Transvaal mines this is done through the agency of the Native Affairs Department, and not through the company. Moreover, the mining companies have undertaken recently—and voluntarily—to pay suitable compensation to natives who may be injured by accident, or to their dependants in the case of death.

Taxation of Gold, Diamond, and Other Mines

Prior to the establishment of the Union, the gold mines in the Transvaal paid a tax of 10 per cent. on the net profits earned, which was imposed by the Crown Colony Administration in 1892, as being the equivalent of the charges indirectly levied by the Republican Government[1]

[1] Apart from a 5 per cent. direct tax (which, however, had not been levied), the gold industry was indirectly taxed through the Dynamite Monopoly and by excessive railway rates.

upon the gold industry before the war. The diamond mines were subjected to a tax of 60 per cent. under the provisions of the Diamond and Precious Stones Ordinance of 1903. This latter taxation was based upon the exceptional results obtained by the one great diamond mine, the Premier, in the first year that it was in operation.[1] In the Cape Colony there was no special tax on gold or diamond mines, but the De Beers Company was made to contribute to the Revenue through the income tax, which was so graduated as to fall most heavily upon the largest incomes.

Upon the establishment of the Union, the income tax in the Cape Province was abolished, and an Act was passed to consolidate the various mining revenue laws of the four provinces. This latter—the Mining Taxation Act, 1910—introduces (1) a uniform profits' tax of 10 per cent. in respect of precious metals and precious stones; and (2) a graduated tax in respect of other minerals ranging from 2½ per cent. upwards. So far as gold mining is concerned, the measure adopts the general principles of the Transvaal 10 per Cent. Profits Tax.

The following passage, which appears in the Report of Transvaal Chamber of Mines for the year 1910, gives some further particulars of the Act:

" . . . the new law re-enacts the [Transvaal 10 per cent. Profits Tax] but includes Diamond Mines in addition. Under

[1] Under the Republican law of 1898, the owners of diamond mines were entitled to retain one-eighth only of the diamondiferous area, the rest passing to the State. It was, therefore, considered to be a considerable concession to allow the private owner to retain four-tenths of the areas, leaving only six-tenths for the State. The proprietors of the Premier might have taken the four-tenths to which they were thus entitled, and worked this part of the mine without any further liability to taxation. In view, however, of the difficulty of ascertaining in which part of the diamondiferous area the pipe lay, they elected to work the whole area of the mine, thus, virtually, entering into a partnership with the Government, to whom they were required to pay six-tenths of the profits earned.

> the old law the calculation was based on the annual net produce, but this is now based upon profits, and the present Act defines the method by which the meaning of the term "profits" is to be arrived at. . . . . Under the new Act it is provided that where the Government are entitled to a share of profits under certain laws mentioned in the second schedule, or under any grant, lease, or agreement, to an amount not less than that leviable under the new law, such profits will be exempt from the tax. In respect of minerals other than gold and diamonds, the Act introduces the principle of the sliding scale of taxation, being from $2\frac{1}{2}$ per cent. on profits where those profits do not exceed 5 per cent. of the gross revenue; and thereafter for every additional 1 per cent. of profit an addition of $\frac{1}{10}$th per cent. to the rate of taxation. If in any year the profits of any Company affected by the Act do not exceed £1,000, those profits are entirely exempt from the tax for that year.

It will be observed that the Act does not affect the Premier Diamond Mine; since in the case of this mine the Government is entitled to a share of the profits six times as great as the 10 per cent. leviable under the Act.

CHAPTER IV

AGRICULTURE AND STOCK-RAISING

AGRICULTURE is the weak point in the economic system of the Union. All other new Anglo-Saxon countries have cultivated their virgin soils so successfully, that, from almost their first beginnings, they have been able to export large quantities of agricultural produce to Great Britain and other thickly populated countries. South Africa, so far from exporting wheat or cattle, has not yet succeeded in producing food sufficient for the requirements of her own population. The one considerable agricultural export of the Union is wool; but the South African wool export of £3,728,000 in value for 1909 makes a poor appearance in comparison with the £25,483,000 and £6,305,888 of wool sent respectively from Australia and New Zealand in the same year. Still less unsatisfactory is the contrast presented between the £681,575 worth of wheaten flour and meal imported into South Africa, and the £16,176,842 of the wheat export of Canada. Other examples scarcely less striking might be added; but the columns of the subjoined table will sufficiently reveal the undeveloped character of South African agriculture, and the paucity of its production as compared with that of the other dominions.

Value of the respective exports from the four dominions of the undermentioned articles of agricultural produce in 1909:

	Meat.	Wheat.	Wool.
Canada	£2,559,000 (Live Stock)	£3,054,630 (Wheaten Flour) 13,122,212 (Wheat)	£113,689
Australia.. ..	2,809,000	6,628,683	25,483,000
New Zealand ..	3,733,000	305,902	6,305,888
South Africa ..	—	—	3,728,000

[Compiled from the Statistical Abstract.]

On the other hand, in spite of its late start—for, as will appear, a start has been made in the course of the last few years—all who know the country are agreed that the effective development of the agricultural resources of the Union is now assured. In the first place, South African agriculture, owing to the variety of soil and climate, admits of a wider range of industries than is to be found, perhaps, in the agriculture of any other single State. Apart from corn-growing and stock-raising, as ordinarily understood, it includes the raising of Angora goats and ostriches; the cultivation of the vine, of fruits of all kinds, and of the Black Wattle; and the growing of tropical and sub-tropical produce such as sugar, tea, coffee, tobacco, and cotton. In the second place, the causes mainly responsible for its backward condition are now known; and it is recognised both by the State and by private enterprise that their injurious effects can be diminished largely, if not altogether removed. These causes are the prevalence of stock diseases and insect pests, and drought. Scientific knowledge, backed by wise regulations firmly enforced, has done much to eradicate the two former evils, and the extension of works for irrigation and water storage, already fruitful of good results, is gradually overcoming the latter.

After the war, the power of the State was employed actively in rendering assistance to the farmers of the New Colonies; and during the ten years which have passed since the Peace of Vereeniging, South African agriculture, especially in the Transvaal and Free State provinces of the Union, has made an appreciable advance. Prior to the war the standard of agriculture in the (then) republics was markedly inferior to that attained by the farmers in the British Colonies of the Cape and Natal. The extent of this inferiority may be inferred from the fact that, even in 1909, after five years of steady progress in the New Colonies (1904–09), the value of the agricultural exports of the Cape Colony constituted two-thirds of the total value of the agricultural produce exported from South Africa. But in this, as in other respects, Lord Milner made it possible for the New colonies to come into line with the rest of South Africa. He made full use of the opportunity presented by the destruction of horses, cattle, and sheep during the war, to introduce breeds of stock, which were both better in quality, and better able to thrive under the climatic conditions of the various districts, than those which they replaced. The Departments of Agriculture, established under his direct supervision, not only gave the farmers a practical exhibition of the methods of scientific agriculture by means of their experimental and stud farms, but they waged unceasing and successful war against animal diseases, insect pests, and noxious herbs; and gave advice and assistance in matters of irrigation and water storage. Moreover, in the great extension of the joint railway system of the two colonies, which was accomplished under the Crown Colony administration, Lord Milner vigilantly safeguarded the interests of agriculture, with the result that the main food-producing districts in both colonies were brought into direct railway communication with the Rand and other profitable markets.

The advance in agriculture since the war is, therefore, more marked, as, indeed, it was more needed, in the Transvaal and Free State provinces. The progress achieved may be seen from the table on p. 335, in which the year 1904, as being the first normal year after the repatriation, is compared with 1909, the last year before the Union was established.

At the same time, the Transvaal Agricultural Department, advancing upon the generous lines laid down by Lord Milner, had become the most completely equipped and scientifically organised institution of its kind within the Empire. And in this connection it is interesting to notice that Mr. F. B. Smith, who was chosen by Lord Milner to be the first Director of Agriculture for the Transvaal in 1902, retained his position under Responsible Government in 1906, and has since become, as Acting

	Wheat.		Maize.		Live Stock.		
	Acreage.	Bushels.	Acreage	Bushels.	Horses.	Horned Cattle.	Sheep.
The Transvaal—							
In 1904	—	255,906	—	4,738,948	52,159	553,388	846,939
" 1909	—	520,280	600,000	8,020,650	125,951[1]	899,673	3,011,906
The Free State—							
In 1904	140,477	211,947	281,536	1,291,810	76,251	363,204	2,999,547
" 1909	218,342	701,519	433,414	6,781,040	132,574	721,258	7,481,251

Secretary for Agriculture, the permanent head of the Union Department of Agriculture. As in the Transvaal, so now in the Union Government, General Louis Botha, the Prime Minister, is Minister of Agriculture; and it may be expected, therefore, that the methods and machinery so successfully employed by Mr. Smith under his official chief in the Transvaal, will now be employed no less effectively in the service of the Union. In short, what South Africa has always wanted—the development of her agricultural resources as a whole under scientific guidance—is well within sight.

Cereals

The national grain crop of South Africa is maize (mealie), which can be grown from the Cape to the Zambezi. In 1909, in spite of the large demands for local consumption, over 300,000,000 lbs. of this useful cereal, of the value of £665,597, were exported. Among the provinces of the Union, the Transvaal was the largest producer with 8,020,650 bushels; the Free State came

[1] Including mules and asses.

second with 6,781,040 bushels; while Natal and the Cape produced respectively 2,393,460 and 1,428,987 bushels. What is said to be the largest maize farm in South Africa is to be found in the neighbourhood of Vereeniging. It extends for 22½ miles diagonally, and in 1909 it had an area of 6,000 acres under maize.

In the production of wheat, barley, and oats, the Union, as we have noticed, is very deficient. These crops are grown, however, in all four provinces, although in Natal the quantities raised are very small. The Cape Province is the main producer, and here in the season 1909–10 there were harvested 2,345,223 bushels of wheat on 335,294 acres of land; 660,336 bushels of barley on 63,165 acres; and 2,395,401 bushels of oats on 331,766 acres. In the same year, the Free State produced 701,519 bushels of wheat on 218,342 acres, and 1,566,993 bushels of oats on 133,180 acres; and the Transvaal 520,280 bushels of wheat on an unreturned acreage.[1] For purposes of comparison, the subjoined table is added:

Wheat production of Canada, Australia, and New Zealand in the year 1909:

	Acres.	Bushels.
Canada	8,063,000	163,202,000
Australia.. ..	6,586,236	90,413,597
New Zealand ..	311,000	8,661,100

In the Cape Province the chief wheat-growing areas are to be found in the south-west districts, lying between the barrier ranges and the sea to the north and south-east of Capetown; but in recent years the utilisation of the flood waters of the Zak River for irrigation has brought promise of wheat production on a large scale

[1] The figures are taken from the "Statistical Abstract." There are no returns given of the Transvaal production of oats.

TYPICAL CORN FIELD

in the dry and sparsely populated north-western districts of the province. Outside the Cape there are areas specially suited for wheat both in the Free State and the Transvaal: such, for example, as the tract of land on the Basuto border, between Bethlehem and Wepener, known as the "Conquered Territory," in the east of the former province, and the country between Zeerust and Lichtenburg in the west of the latter.

In view of the backwardness of South Africa in the production of corn, it is worthy of notice that the levelness of much of the area of the great plateaux, joined with the general dryness of the atmosphere, make many districts of the Union specially suitable for cultivation by steam-drawn machines. On the high country in the Transvaal, steam ploughing by the double-engine system can be continued without interruption from the weather through the whole of the year, while at the same time the absence of rain and damp reduces the wear and tear of the machinery and gear. In the 160,000,000 acres of the Dry-land Zone of South Africa, therefore, there would seem to be an ample and remunerative field for the application of steam power and motor traction to farming operations on a great scale. The prevalence of animal diseases is another circumstance which makes the introduction, where possible, of the direct and cable system of mechanical ploughing, and the use of steam and petrol tractors in general, a matter of special importance to South African agriculture. To replace animal by mechanical traction, wholly or in part, would render the farmer independent of cattle, horses, mules, and asses for ploughing operations and transport in times of animal epidemics, and thereby relieve him of one of his greatest anxieties. It would, moreover, make it possible to cultivate some low-lying, but eminently fertile, districts, which are now closed to agriculture as being especially subject to such visitations.

Stock-raising

There are within the Union abundant areas eminently suitable for the breeding and pasturage of horses, horned cattle, and sheep; and in wool, of which there is a considerable export, the South African farmer finds his chief and most general source of revenue. In the pastoral industries, as in wheat production, the Cape is notably superior to the other provinces, and she claims that her live stock is more numerous than the collective stock of the three remaining provinces. Although the Free State is now a large producer of wool, most of the total South African export, amounting in 1909 to nearly four millions sterling in value, still comes from the Cape Province, which in the same year possessed 18,807,168 sheep, as against 7,481,251 in the Free State; 3,011,906 in the Transvaal; and 1,068,996 in Natal. The foundation of this—the staple agricultural industry of the Union—was laid early in the period of British rule by the introduction into the Cape of the Merino sheep, the producer of fine wool, by Reitz and Breda in 1812; by Lord Charles Somerset (1814–26); and by the Albany settlers in 1820. Forty years later, 23,172,785 lbs. of wool was exported from the Cape; in 1871 the export had risen to 48,822,562 lbs., while in 1909 it was 102,346,692 lbs., of the value of £2,854,835.

In connection with the subject of stock-raising, a brief reference must be made to the special liability of the South African farmer to visitations of animal diseases. These outbreaks, which have inflicted enormous losses upon the pastoral industries in the past, have been checked and controlled, since the war, more and more successfully by the gradual fencing of the farms, and the framing and enforcement of effective regulations for the prompt destruction or isolation of infected animals. That they have still to be reckoned with, however, may be seen from the record of the destruction recently worked in Natal by the rinderpest, or East Coast cattle fever. The disease was afforded an opportunity of passing through Zululand into Natal by the disorganisation of the local administration consequent upon the native insurrection of 1906. The effect of its presence is registered in the yearly returns of the live stock of the province. In 1905 Natal possessed 783,887 head of horned cattle. The number fell to 634,547 in 1906, and to 416,527 in 1907; while in 1909 it had risen only to 502,212. It is satisfactory to be able to add that provision has been made in the Union Budget of 1911 for the advance by Government of a considerable sum on loan to the Natal farmers, to enable them to erect fences; and it may be hoped, therefore, that in the future Natal will be able to offer a more effectual resistance to the progress of the disease.

The Angora Goat

Important as is the production of wool, in the raising of the Angora goat and the ostrich we have two industries of special interest as being more characteristically South African. The Angora, which furnishes mohair, is chiefly raised in the midland and upper districts of the Cape Province, but it is also bred in considerable numbers on the uplands in the south-east of the Free State. The value of the hair of this animal, which is a native of the central plateaux of Asia Minor, and is found also in the highlands of Persia and Kashmir, was known to antiquity; and it provided the material out of which the hangings of the Tabernacle of the Jews, and the robes of the ancient kings of Persia, were woven. The Dutch East India Company endeavoured to introduce the Angora into the Cape Colony as early as 1725, bringing the goats from Kashmir to Ceylon, then in its possession, and thence to the Cape. This endeavour, however, failed; and subsequent attempts, made early in the last century, met with no more than a partial success. The founder of the industry was Adolph Mosenthal, who, with the active assistance of Lord Stratford de Redclyffe, the British Ambassador at Constantinople, himself secured a number of these fine-haired goats from Asia Minor, out of which he succeeded in landing thirty rams and ewes at the Cape in 1856. These formed the parent stock from which the pure-bred Angoras of the Midland and Eastern districts have descended. To these names should be added that of Sir Titus Salt, the Yorkshire manufacturer, who did much to develop the use of the Cape mohair in England. In 1860 the export of Angora hair was 385 lbs.; in 1870 it had risen to 403,153 lbs.; and in 1909 it was returned at 19,649,053 lbs., of the value of £861,639. The Cape mohair is not quite so fine in quality as the Asiatic product, but the clip is said to be more ample. The value of the hair fluctuates somewhat in response to changes of fashion, and the Cape farmers are learning, therefore, to maintain a steady market by studying the varying requirements of the importers.

The Angora goat is a beautiful as well as a valuable animal. It has a small and shapely head, surmounted in both sexes by long, flat, corrugated horns extending from 18 to 24 in.; and its fleece, which is white and lustrous, falls almost to the ground in natural ringlets. The herds give little trouble to the farmer, and are left to roam unrestrained over the grass-clad uplands. The breeders of the Angora in the Cape Province are making it their object to produce a strain, which, without being inferior to the original Eastern stock in the quality of the hair, will be more completely adapted to the new environment; and with a view of protecting the industry against the endeavours which have been made to introduce and acclimatise the Cape Angora in the United States, the export of these animals has been forbidden by law.

AN ANGORA GOAT

Ostrich Farming

The feather industry originated in the Cape, and, although ostriches are now raised in all four provinces of the Union, the bulk of the large annual export of ostrich feathers is still produced in this province. The actual birth of the industry is to be attributed to Arthur Douglass, of Albany, who, in 1869, perfected an artificial incubator, and by this means brought about the domestication of the bird. Prior to Douglass' invention, the wild birds were hunted, killed, and stripped of their feathers. Then, at a later date, young birds up to seven months old were caught and kept with other farm stock. This plan, however, was attended by very slight success, since the birds, when full-grown, proved fierce and unmanageable; and in 1865 the number of ostriches included in the returns of the live stock of the colony was only eighty. The value of the incubator lay in the fact that it enabled successive generations of birds to be hatched and brought up by hand; and these birds, being accustomed to man from birth, became increasingly tame, until in course of time the domestication of the ostrich was accomplished.

The chief seat of the industry is in the Oudtshoorn district of the Cape, but it is also carried on largely in the eastern districts of the same province. The best pasture for the ostrich is lucerne; and it is calculated that an acre of lucerne will carry three birds, each of which will furnish in two years three crops of feathers, yielding an average profit of from £5 to £7 per plucking. Rape is another favourite pasturage, and the birds thrive on the bushes of the karroo in conjunction with other pasture. They are also fed by hand, and in this case maize is the principal article of diet. Since the problem of domestication was solved the ostrich farmer has made a further advance. By carefully selecting the parent birds, and observing the variations in the plumage of the respective offspring, he has gradually evolved strains which are capable of producing feathers far superior to those of the primitive bird in form, texture, and colour. It is by the production of these high grade feathers, and incidentally by the sale of stock for breeding purposes, that the ostrich farmer makes his largest profits.

The progress of the industry is exhibited in the increasing weight and value of the feathers annually exported. With the introduction of the incubator, the export rose from 17,000 lbs., in 1869, to 28,768 lbs. in 1870; in 1895 it was 353,651 lbs. in weight and £527,782 in value; and in 1909 it was returned as being 788,262 lbs., of the value of £2,091,280—a return which shows that a considerable appreciation in the value per lb. has taken place in recent years. It should be added that, in order to maintain the virtual monopoly of the ostrich feather industry possessed by the Cape, an export tax of £100 per bird and £5 per egg was imposed in 1883 by the (then) Government.

The following account, which is taken from an article contributed by Mr. Arthur Douglass to the *Cape Official Handbook for* 1893, gives an excellent picture of an ostrich farm:

We can imagine nothing more delightful and interesting to a traveller than a visit to a large ostrich farm. Let us try and describe what may be seen on one we know well. The size of the farm is 13,000 acres, situated in the Eastern Province of the Cape Colony. The herbage is a mixture of grass and succulent karroo bushes. The rainfall in this part of the Eastern Province is too uncertain to allow of cultivation without irrigation, so the cultivation is confined to a few acres of lucerne irrigated by pumps; some soft green food being indispensable for rearing the little ostrich chicks during droughts. On the farm are kept 600 ostriches and 400 breeding cattle. The whole property is enclosed by strong wire fences 5 ft. high, and subdivided into numerous camps, with similar fences. Near the homestead the camps are of about 100 acres each, being appropriated to the rearing of the young birds. Beyond these, again, are camps of about 25 acres each, these being given up to a superior pair of old birds in each camp for breeding, whilst beyond these again are large camps of

A PAIR OF OSTRICHES

about 2,600 acres in extent, with 150 birds in each. But let us take a stroll in these camps and see what is going on. Here, in the first, we find an old Hottentot with about 30 little ostriches only a few days old around him. These have all been hatched in the incubator, and he is acting as nurse to them, cutting up lucerne for them to eat, supplying them with fine gravel to fill their gizzards with to grind their food, breaking up bones to let them get a supply of phosphates, and giving them wheat and water; and at sundown he will bring them back to the incubator for warmth, or should the weather change and rain come on, he will be seen hurrying home with his thirty little children following him to a warm, well-lighted room with a clean sanded floor. In the next camp we have a pair of birds and about fifteen chicks, accompanied by a Kafir, who has been with them every day from the time they hatched to get them tame and accustomed to man. These have been hatched by the parent birds, who will brood them at night in the camp.

Wine Production

From a historical point of view, viticulture is the most interesting branch of South African agriculture. The foundation of the industry takes us back to the very birth of the Dutch settlement, to the days of the Van der Stells, and to the Huguenot immigration of 1688; and the country in which the wine farms are found is not only the earliest settled, but in many respects the most picturesque in South Africa. For, although the vine is cultivated successfully in the Free State and the Transvaal, the production of wine is almost entirely confined to the Cape Province. Here, on the mountain slopes and in the valleys of the south-western corner of "the old colony," are to be found the coal-black roofs of thatch, the quaint gables, the gleaming whitewashed walls and the tree-shaded courtyards of the old Dutch wine farms.

From the time of Simon Van der Stell onwards, a certain quantity of the wine produced at the Cape was regularly sent home to Holland for the use of the directors of the Dutch East India Company; and after the British occupation the English sovereigns, as the successors of the Dutch Company and Government, continued to profit by the right up to the middle of the last century, when, upon the petition of the Cape wine farmers of that day, this tax upon the industry—as they regarded it—was abolished. The sequel is significant. Constantia and other Cape wines, being no longer seen upon the Royal table, lost the vogue which they had hitherto enjoyed; and when, as the result of the commercial treaty negotiated with France by Cobden, the duties on French, and afterwards Spanish, wines were greatly reduced, the Cape vintages practically disappeared from the English market.

The districts in which the wine grape is chiefly cultivated are those of the Cape, Stellenbosch, the Paarl, Worcester, and Robertson. In the first three of these the rainfall suffices for the vines, but in the two latter the vineyards are watered by irrigation. The climate and soil of the Cape is especially suitable for the vine; and the degree in which the vineyards within this area exceed in productiveness those of any other country may be seen from the subjoined table.

Table showing the respective quantities of wine produced from 10,000 vines (or 1 hectare) in the undermentioned countries:

Hungary, 1863–72	24 hectolitres per hectare.
Germany	24 ,, ,,
Austria, 1874–80	18½ ,, ,,
Switzerland	42 ,, ,,
France, 1873–83	18½ ,, ,,
Italy	14½ ,, ,,
Spain	17 ,, ,,
Greece	17¾ ,, ,,
Algeria, 1882	25¼ ,, ,,
United States	14½ ,, ,,
Australia	14½ ,, ,,
Cape Colony—Coast districts	86½ ,, ,,
,, ,, —Inland ,,	173 ,, ,,

[From Baron von Babo's standard work on "Viticulture," as quoted in the *Cape Official Handbook*, 1893.]

The comparative failure of the Cape industry, and the virtual exclusion of its production from the English market, cannot be attributed, therefore, to any deficiency of soil or climate. The inferior quality of the Cape wines, according to a report furnished to the Cape Government in 1894, was due to the fact that none of the Cape farmers, with the exception of those in the Constantia district, understood the treatment of the grape, or the great cleanliness necessary for the production of good wine and brandy. Since the period of this report, however, great efforts have been made by the agricultural department of the (former) colony both to improve the character of the Cape vineyards and to reform the methods of wine production pursued by the farmers. These efforts have been facilitated by the destruction wrought by the phylloxera in 1886 and subsequent years, which has led to the complete reconstruction of the vineyards of the more important and progressive growers.

In view of the continued prevalence of the insect and the increase of the area attacked by it, the Government obtained supplies of the phylloxera resisting vines of America, and established nurseries in the various wine-producing districts, in which young plants were grown from the imported vines. It was thus learnt by gradually acquired experience that the Cape vines grafted on to the American stocks produced a yield of grapes superior both in quality and quantity to that of vines raised directly from the cuttings. The desirability of forming the reconstituted vineyards out of Cape vines grafted on to American stocks having once been established, it still remained to ascertain, again by experiment and observation, which stocks were most suitable for the respective growths of vines. The process was a long one, and it is stated that the question of appropriate stocks has not been decided even yet in all cases. In the course of the last twenty years, therefore, the greater part of the vineyards have been dug up, and replanted with the best varieties of the Cape vines grafted on to the American stocks ascertained to be most suitable in each case.

The reconstruction of the Cape vineyards has been accompanied by a gradual improvement in the methods of wine production and distillation of spirits. In this work a leading part has been played by the Government wine farm at Constantia, "Groot Constantia," the vineyard laid out by Governor Simon Van der Stell at the end of the seventeenth century, where examples of the best methods and appliances are exhibited for the benefit of the wine producers of the provinces. And, in addition to this model wine farm, an Experiment Station and Oenological Institute has been established recently at the Paarl by the Agricultural Department.

The earnings of the wine farmer are not confined, however, to the profits realised by the sale of wine and brandy. In addition to these, he obtains an appreciable return from the growing of fresh grapes for the market, and by the production of vinegar and raisins. Indeed, it is calculated that while the industry, as a whole, maintains a population of some 100,000 persons, 10 acres under vines will support a family in comfort. The industry, therefore, is a valuable one from an economic point of view, since it permits of close settlement in a conspicuous degree.

The table on p. 347 exhibits in detail the areas and localities of the Cape vineyards, and the classes and quantities of wine and brandy respectively produced in the season of 1909.

Nearly the whole of this production is consumed within the Union. The export of Cape wine, which has fallen in recent years, amounted in 1909 to 48,152 gallons, of the value of £9,229. The Australian export for the same year was 976,647 gallons, of the value of £125,443.

Wine, Brandy, Vinegar, and Raisins Produced in the Season 1909 in the Cape Province

Districts.	Wine.									Total Production of Wine.	Brandy:			Total Production of Brandy.	Area under Vines.
	Pontac.	Hermitage.	Cab. Sauvig.	Stein Grape.	Sauv. Blanc	Green Grape.	White French.	Muscatel: Wht.	Muscatel: Red.		Dop.	Grape.	Wine.		
	Leaguers = 128 gallons (nearly).														Morgen
Caledon..	8	53	—	115	3	328	20	23	—	550	68	14	—	83	225
Cape	15	843	117	553	4	70	9	7	6	1,626	60	70	23	154	714
Ceres	—	13	—	—	—	133	—	—	—	147	1	58	—	60	95
Clanwilliam	—	32	—	15	—	38	23	15	—	123	41	—	3	44	63
Ladysmith	—	—	—	2	—	—	70	22	—	95	3	51	—	55	47
Malmesbury	215	215	8	39	—	1,507	68	1	30	1,869	131	—	1	133	900
Montagu	13	—	—	—	—	35	84	88	1,195	1,416	92	645	2	739	331
Oudtshoorn	—	—	—	—	—	—	—	—	—	—	—	118	4	122	102
Paarl	213	1,933	51	927	16	4,236	222	89	37	7,726	670	—	29	700	3,028
Piquetberg	—	11	—	267	—	231	17	—	6	533	32	—	—	33	329
Prince Albert ..	—	—	—	—	—	6	—	—	17	24	8	27	—	36	26
Riversdale	—	—	—	—	—	1	8	—	9	18	83	17	—	101	25
Robertson	6	112	—	1,087	—	146	618	65	1,024	3,059	477	1,008	46	1,532	684
Stellenbosch ..	113	892	69	2,322	251	1,101	9	1	2	4,762	200	8	—	209	1,621
Swellendam	—	—	—	—	—	12	27	—	8	48	—	188	—	189	145
Tulbagh	1	128	—	67	—	774	73	3	52	1,100	65	—	—	67	364
Worcester	55	567	12	91	—	2,150	1,013	111	206	4,206	244	764	8	1,017	1,478
Grand Totals ..	425	4,801	257	5,488	275	10,773	2,262	426	2,594	27,312	2,182	2,974	118	5,274	10,183[1]

In 1908, 117,024 gallons of Vinegar and 2,198,089 lbs. of Raisins were produced.

[1] Morgen = rather more than 2 acres.

[From the Statistical Register of the Cape Province, 1910.]

It may be hoped, however, that when the reconstruction of the vineyards is completed, and the quality of the production has been still further improved, the oldest of South African industries will regain its footing in the English market. It should be added that in 1909, besides the 21,552 acres in the Cape Province, 1,075 acres were returned in the Free State as being under vines.

Sugar, Tea, and Wattle Bark

These three industries are grouped together as being peculiar to the Province of Natal. The sugar plantations and tea gardens are to be found in the warm, semi-tropical coast belt; while wattle planting is carried on mainly in the higher country of the second terrace, or midland belt, which is from 1,000 to 3,000 ft. above sea-level.

The sugar industry possesses a special economic interest, as affording an example of an industry which, up to the present time, has been worked regularly by indentured Asiatic labour. A commencement of sugar planting was made at the time of the Byrne immigration (1848–52), when, in 1849, several acres on the Compensation Flats, about 35 miles north of Durban, were planted with cane. The earliest mills were driven by oxen, the machinery and methods of cultivation were extremely primitive, and the planters were dependent upon native African labour. Some progress was made, but it was found that the refusal of the natives to undertake regular and continuous employment prevented them from acquiring even the minimum training necessary to produce efficient workmen; and in 1859 the importation of indentured labour from India was sanctioned by the Government. The new labour supply produced an immediate expansion of the industry, and added greatly to the general prosperity of the little colony. In 1866, when some 5,600 coolies had been introduced, the importation ceased. It was resumed in 1874, when a fresh expansion of the industry took place, which was accompanied, as before, by a quickening of the development of the colony as a whole, and an increase of its revenue. From this date onwards the importation of indentured Indian labour was continued, until in 1911, in view of the exclusion of Asiatic settlers from South Africa, the Indian Government refused to allow any more Indians to be employed as indentured labourers in Natal. At the present time, therefore, the sugar planters are confronted with the difficulty of having to find a fresh labour supply capable of replacing the indentured Indian coolies, whose services have been regarded hitherto as essential to the prosperity of the industry.

The growth of the industry may be traced by the annual returns. In 1858, before the introduction of Indian labour, there were 1,490 acres under cane. In 1869, 15,892 acres had been planted, of which 5,757 were reaped; 7,823 tons of sugar were produced; and 34,778 gallons of rum were distilled. Ten years later, expert planters and trained factory hands were brought over from Mauritius, the Natal Central Sugar Company's factory at Mount Edgecombe was erected, and improved methods of cultivation and sugar extraction were introduced. The expansion consequent upon these measures continued until the progress of the industry was checked by the general fall in the market price of sugar, which, beginning in 1883, showed a decline in 1889 of 50 per cent. as against the level of 1861, and has continued until quite recent years, when at last a slight recovery has taken place. The Natal planters, however, with few exceptions, survived the fall in price, and the returns from 1895 onwards record an almost continuous increase in both the acreage of the plantations and the production of sugar. In that year (1895) 14,732 acres were under cane, and 411,620 cwts. of sugar were produced. In

1905 the acreage had risen to 45,865 and the production to 532,067 cwts. In 1909, when a sudden increase of production had taken place, there were 46,567 acres under cane, and 1,549,821 cwts. of sugar were produced.

At present the sugar cane requires two years' growth to bring it to maturity, and only one-half of the plantations are reaped in each year. In 1909 there were thirty-four mills in full operation, and more than 8,000 persons were employed in the industry. The wages of the indentured Indians are from 10s. to 15s. a month for men, and from 7s. 6d. to 5s. for women, with quarters and rations; while "free" Indians (*i.e.*, Indians who, after completing the term of their contract, settled in the province prior to the prohibition of Asiatic immigration) are paid from £1 to £2 a month. The Natal sugar production, which is of the approximate value of £500,000, is absorbed almost entirely by South Africa. In 1909 only 2,705 cwts. were exported to the United Kingdom. As, however, South Africa spends over a million sterling annually on sugar, and at present imports largely from oversea countries, the Natal planters have every reason to increase their output. It is believed that this object can be secured not merely by extending the area of the plantations, but by increasing their productiveness. Efforts are being made, therefore, by experimental investigation and by the careful selection of the plants, to reduce the period required for bringing the cane to maturity, and to improve the yield both in quantity and quality. At the same time, it is hoped to increase the percentage of sugar extracted by an extended use of large central factories, in which the latest and most successful chemical and other methods will be fully employed.

The warmth and moisture which the tea garden requires are provided in a high degree by the coast belt of Natal, but its higher altitudes (roughly, 1,000 ft. above

SUGAR CANE

sea-level) are most favourable to the growth of the plant. In 1895 the acreage under tea was 2,297, and the production 737,000 lbs.; in 1909 the returns showed that the acreage had advanced to 5,909, and the production to 1,773,202 lbs. There is an ample supply of land suitable for the tea plant in the Province, and it is calculated that the Lower Tugela Division of the Victoria county alone contains an area of such land sufficient to supply all South Africa with tea. As a pursuit, the industry has much to recommend it. Unlike coffee and cinchona, the plants, when once established, are permanent. Drought is practically the only evil to be feared by the planters, since the tea garden cannot be burnt and is not subject to the attacks of locusts. Natal tea is less full-flavoured than the produce of India and Ceylon; but it can claim to be more wholesome, as containing 7½ per cent. less tannic acid, while it is richer in the stimulating caffein. As in the case of the sugar industry, Indian labour has been employed hitherto in the tea gardens and factories for the cultivation of the plants, and the picking and preparation of the leaf.

The cultivation of the black wattle, the famous producer of tan bark, and a native of South-eastern Australia, is one of the most lucrative of the agricultural industries of South Africa. It is carried on chiefly in the Midland Belt of the Natal Province, where it has added a new and pleasing feature to the countryside, and in certain districts raised the value of the land, since the wattle can be grown on soils unsuitable for ordinary cultures. It is the only timber industry which gives a full return within a few years, and henceforth maintains this return in perpetuity. It yields ample profits; and the capital expenditure is small, as little or no machinery is necessary. Moreover, the culture of the tree and the preparation of the bark are both so simple, that no expert knowledge is required, and they can be carried on successfully by a farmer of average experience and intelligence. The industry began in 1886 with an export of bark valued at £11. Ten years later, 67,580 cwts., of the value of £16,450, were exported; and in 1909, with some 150,000 acres planted with black wattle, the Province exported 705,849 cwts., of the value of £192,950.

The European demand for bark for tanning purposes is very large, and almost the entire production of Natal is shipped to the United Kingdom; as, however, black bark does not suit the requirements of the English market, the Natal export is largely sold to Continental buyers.

Tobacco and Cotton

Certain areas in the Transvaal and Rhodesia are as eminently suited, alike in climate and soil, for the production of tobacco and cotton, as are the south-western districts of the Cape for the cultivation of the wine grape. Prior to the war, tobacco was grown by the farmers in the Transvaal, and the leaf cured by very primitive methods; and this "Boer tobacco" was used largely in South Africa and not unknown in England. After the war, the question of the cultivation of tobacco, cotton, and other tropical plants was taken up by the newly-established Department of Agriculture, and by private enterprise. Tzaneen, a derelict tobacco estate in the Zoutpansberg district, was purchased by the Administration; expert direction, and the latest appliances were obtained from the United States, and an efficient plant for the preparation of the leaf was installed in the factory. The estate developed into an experiment station, to which was added subsequently a School for Tropical Agriculture. The two combined formed an institution in which the future planters of the Transvaal can learn, from competent instructors, the best methods of tropical and semi-tropical culture, and see these methods applied in the cultivation of the Government plantations. The work thus begun under Lord Milner's administration has been continued and expanded under Responsible Government, and to-day the Union Department of Agriculture maintains, in addition to Tzaneen, experiment stations at Rustenburg, Barberton, and Pict Retief. In these stations the experimental cultivation of tobacco and cotton is carried on under the direction of the thirteen officers of the Department, who are charged with the promotion of this branch of agriculture. The provision made for the annual maintenance of the four stations in the estimates for the financial year 1911–12 was £4,045, while a sum of £8,000 was appropriated out of loan funds to capital expenditure in connection with Tzaneen. In addition to thus collecting the data necessary for the production of tobacco of fine quality on a scale which will permit the Transvaal grower to compete successfully in the European market, the Government has rendered direct assistance to the farmers growing tobacco in the Rustenburg District by erecting a large warehouse there to which the tobacco crops can be brought uncured, and by itself carrying out the preparation of the leaf, under expert direction, in the fermenting rooms at Pretoria. In this way the proper treatment of the leaf is secured, before the crop is offered to the manufacturers in the open market. During the eight months ended August 31st, 1911, the Union export of tobacco was 51,509 lbs., of the value of £5,097.

In Southern Rhodesia the tobacco crop is only second in importance to maize. In order to assist the planters, the Chartered Company established a central warehouse for the Virginia leaf at Salisbury, and a second warehouse for the Turkish leaf at Buluwayo. These warehouses have been taken over now by the Tobacco Company of Rhodesia and South Africa, Ltd., and the system under which the industry is worked is this: The leaf is received by the Tobacco Company from the planters at the warehouses, graded, placed in keeping condition, re-packed, and finally sold by auction at a small charge calculated to cover the cost of these services. Each crop is valued by the Company's Manager upon delivery at the warehouse. The planter can at once obtain payment of a sum equal to 50 per cent. of the value of the crop as thus ascertained, and a further advance of 25 per cent. of the estimated value, when the leaf is ready for sale. The Tobacco Company also undertakes to provide expert advice in respect of methods of cultivation, and to make cash advances for the erection of curing barns and packing sheds to planters of approved standing. During the year 1911 the export of tobacco from Southern Rhodesia was 413,852 lbs., of the value of £34,810.

There is every prospect of South Africa becoming an effective contributor to the raw cotton supply of the United Kingdom. In Northern Rhodesia the production of cotton of high quality is being actively developed, and it is believed that the crop will be grown successfully in many districts of Southern Rhodesia. In the Transvaal, cotton has been raised as an occasional crop by the farmers in the Zoutpansberg for some years past, but it is only quite recently that the Agricultural Department and the Railway Administration have been able to open the way for the development of the cotton areas of the Province on an adequate scale. Now, however, experiments in planting, carried out at the Tzaneen, Rustenburg, and Barberton stations, have shown that suitable varieties of American cotton can be grown in the Transvaal to produce crops of equal quality and superior yield per acre as compared with those obtained from the same plants in America. It has been ascertained, also, that the area of the province suited for cotton growing is larger than has been supposed hitherto; and, in particular, that the Zoutpansberg district alone contains a cotton belt which equals any other cotton area within the Empire. As the new railway running from Komati Poort to Louis Trichardt and Messina on the northern border of the province, now under construction, will traverse this belt of country, the future cotton planters will have the advantage of direct railway communication with Delagoa Bay. As this new railway is an essential factor in the development of the industry, it may be added that the first section of the line, covering the 74 miles from Komati Poort to Newington (Selati), was opened in May, 1910; and that the second section, the 140½ miles from Newington to Tzaneen, is expected to be handed over by the contractors at the close of the present year (1912). There is one further consideration which favours the prospect of the establishment of the industry. The period during which rain falls in the Transvaal in normal years, namely, October to April, is sufficiently long to permit of the crop reaching maturity within it; while, on the other hand, the practically rainless period of May to September will afford ample time and opportunity for gathering the cotton in excellent condition.

In the returns of the Statistical Abstract for 1909, the South African export of raw cotton is shown as being 84,782 lbs., of the value of £1,686. Tobacco does not appear in the list of the "principal articles" exported, but the production within the Union is stated to be: In Natal, 2,527,012 lbs.; in the Free State, 646,260 lbs.; and in the Transvaal, 2,891,000 lbs. The latest available figures of the Union and Rhodesian exports, obtained from the South African returns, have been given above.

Fruit-Growing

A few words on the subject of fruit-growing will serve to complete this brief review of the more distinctive features of South African agriculture. In all provinces of the Union and in Southern Rhodesia the cultivation of fruit as a separate industry is being carried on in an increasing degree. The extension of railways, by bringing the producer into cheap and rapid communication with the larger centres of population, has made the business of providing fruit for the local market both more important and more lucrative ; and the export trade is being developed with success, especially where, as in the Cape and Transvaal provinces, the attention of the grower has been directed to the production of the Citrus family (oranges, lemons, etc.). In the matter of supplying fruit to the European markets, it must be remembered that South Africa, like Australia, has the advantage of the reversal of the seasons, and that its summer produce can, therefore, reach Europe in the winter and spring months, when, naturally, it is most welcome.

In Natal the distinctive tropical fruits, such as mangoes, plantains, guavas, pineapples, and bananas, are raised in the Coast Belt, and the citrus fruits, in the country inland, up to an altitude of 3,000 ft. above sea-level. Almost every description of fruit can be grown in the Cape Province ; but the industry flourishes most in the south-western districts, and it is here that the fruit farms established by Rhodes are to be found. In the Transvaal fruit-growing has made great advances since the war. Apples, pears, peaches, apricots, and nectarines are grown universally, and orchards of from 500 to 20,000 trees are no longer an uncommon sight. But it is by the production of citrus fruit—the orange, lemon, mandarin, tangerine, and grape—that the Transvaal fruit farmer expects to win a place in the European market. Indeed, as a producer of oranges, the Transvaal promises to rival California. The soil and climate of the province are, in general, well suited for the fruit, but Rustenburg, lying between the mountain ranges, with its

FRUIT DRYING

soil of a deep chocolate loam, and its slopes at once sunlit and protected, seems destined by Nature to be the especial home of this new and delightful industry.

State Aid to Agriculture

It remains to consider the important question of State aid. As we have seen, the attention paid to agriculture by the administration created upon the close of the war by Lord Milner in the New Colonies, gave an impulse to the development of the agricultural resources of all the four colonies now forming collectively the Union of South Africa. In the New Colonies, owing mainly to the success of Lord Milner's financial arrangements, the progress was most marked; and it could be said of the Transvaal, on the eve of the Union, that there was no other country in the world in which agriculture received a more complete and generous assistance from the State. As the former Transvaal Agricultural Department has been adopted virtually as the basis of the Union Department of Agriculture, the same may now be said of the Union as a whole.

The relation of the State to the agricultural and pastoral industries of the Union will be exhibited most clearly by considering (1): What the Government proposes to do, and (2) what at the moment it is actually doing:

(1) The duty of the Union Government in this respect is defined by the Commission[1] for the reorganisation of the Departments of the Public Service in its Fourth Report, which deals with the Department of Agriculture. In reviewing the assistance which South African agriculture received from the State prior to the Union, the Commissioners point out that three methods were pursued: (1) The exemption of farmers from taxation, and the

[1] Appointed under the provisions of the Union Act (see p. 157).

imposition of protective duties upon agricultural produce. (2) The method of "doles," which, prior to the late war, "had attained considerable dimensions"; since in its expenditure the State devoted the greater portion of the money "to the immediate relief of the farmers rather than the advancement of the agricultural art." (3) State aid on scientific lines, which was the result of a great outburst of activity after the war on the part of the Colonial Governments in the direction of the development of agriculture. In the case of the Cape and Natal, however, this activity was checked by commercial depression, and was followed by a period of reduction and retrenchment.

At the time of the Union there were Agricultural Departments in all four colonies, and the respective departmental votes for the ten months—July 1st, 1909, to April 30th, 1910—were as follows:

	Out of Revenue.	Out of Loan Funds.	Total.	Estimated to be recoverable as revenue.
	£	£	£	£
The Cape	215,000	96,000	274,000	27,000
Natal	152,000	105,000	195,000	42,000
Transvaal	292,000	60,000	314,000	38,000
Orange River Colony (Free State)	76,000	26,000	92,000	9,000

In considering the form which State aid to agriculture should assume under the Union, the Commissioners distinguish between services, the performance of which is a recognised duty of a Central Government, and those which may be delegated to local authorities, or left to private enterprise, as the circumstances of a particular State may seem to require. The first class comprises "matters upon which uniformity of legislation and action is essential for the security and well-being of the community as a whole, such as measures relating to diseases of animals and plants, including locust destruction, where these pests obtain; noxious weeds; the adulteration and sale of artificial manures and foodstuffs for live stock, and other articles required by farmers; the purity and viability of seeds; brands; the transport of live stock and agricultural produce; the grading and marketing of produce; the obtaining of markets abroad and the supervision of exports; and the preparation of statistics."

In respect of these duties, the Commissioners are of opinion that it is essential that they should be retained, without exception, by the Union Government.

"The second class," they write, "includes subjects such as the application of science to agriculture; agricultural education; the maintenance of experimental and stud farms and stations; assistance to agricultural societies; the organisation of the farmers' land banks, land settlements, forestry, game, and fisheries; and the conservation of natural resources."

And they record the opinion that, in the existing circumstances of the Union, the functions mentioned in this second class should also be undertaken by the Central Government; and they add that both classes of duties should be performed through the agency of a single Union Department, and not by a central department with sub-departments in the several provinces.

In the formation of this single Department of Agriculture out of the four Departments of the former Colonial Governments, the Commissioners recommended certain changes of *personnel*. As the table in which these changes

24—(2139)

are exhibited conveys an excellent idea of the scope of the new Union Department, it is reproduced below:

Divisions of the Department of Agriculture.	No. of Officers up to May 30/10	No. now proposed.
Administration (Head Office)	93	69
Veterinary (*a*) Control	65	69
,, (*b*) Research	45	39
Entomology	24	20
Plant Pathology	5	7
Botany	30	9
Horticulture	7	5
Chemistry	17	7
Tobacco and Cotton	11	13
Co-operation	5	5
Inspection of Grain	13	6
Sheep and Wool	4	6
Dairying	4	7
Publications and Library	13	4
Brands and Fencing	42	40
Guano Islands	4	4
Viticulture	4	5
Cold Storage	9	—
Transport	18	—
Field Cornets	131	—
Colleges and Experimental Farms (excluding stockmen, storemen, grooms, and similar subordinate officers) ..	72	95
Total	616	417

With reference to this table, it should be stated that the reduction of officers from 616 to 417 is more nominal than real. It was proposed to attach the 131 field cornets, and some of the 18 Natal transport employés, to another department; and the actual net reduction was confined to members of the clerical and non-professional staff, and to certain other officers whom the consolidation of the four services had rendered redundant. And in this connection it may be noticed that the Report, which was published on August 3rd, 1911, bears the signature of Mr. F. B. Smith, the Acting Secretary for Agriculture and permanent head of the new department, in addition to those of the three Commissioners.[1]

The Union Department of Agriculture

(2) In order to learn what the Union Government is actually doing for agriculture, we cannot do better than ascertain the provision which was made on its behalf in the Estimates for the first complete year of the Union (1911–12). From the statements issued by the Union Treasurer, we find that provision was then made for a large expenditure upon agriculture, and the agricultural interests, out of loan funds. The character and purposes of this capital expenditure will be seen from the following table.

Capital expenditure upon agriculture and allied interests to be provided out of loan funds :

Votes.	Head.	In period 1910–12.	After March 31st, 1912	Total.	Estimated Amounts required to complete the works and services comprised in the votes.[2]
		£	£	£	£
D	Agriculture .	153,835	100,149	253,984	370,000
E	Lands.. ..	188,680	51,904	240,584	660,256
F	Irrigation ..	493,579	1,000	494,579	590,742
H	Land Banks.	568,000	260,438	828,438	1,000,000
			Grand	Total ..	2,620,998

The expenditure under Vote D is concerned with the

[1] Messrs. H. C. Campbell, A. Browne, and A. B. Hofmeyr.

[2] These amounts comprise expenditure by Colonial Governments prior to the Union, as well as that from May 31st, 1910, onwards.

erection of vermin-proof fencing, and of cattle-dipping tanks in the Cape Province, and the provision of fencing loans, etc., for the prevention of East Coast (cattle) fever in Natal. There is also a provision of £20,000 for loans to creameries in the Free State Province.

The expenditure under Vote E is mainly devoted to agricultural development in Natal (£277,656) and land settlement in the Transvaal (£337,300).

The chief provisions under Vote F are:

	£
Irrigation loans to farmers and public bodies (under £25,000) in the Cape Province	287,290
Kopjes irrigation works, including purchase of land, in the Free State	80,000
Potchefstroom dam in the Transvaal	24,000
Pelitzi project, Tzaneen (including distributories and surveys)	8,000
New irrigation works in the Transvaal—	
(*a*) Works	15,000
(*b*) Purchase of land	120,000

Under Vote H, £500,000 is provided for the Natal Land and Agricultural Loan Fund, and the same amount for a similar fund in the Free State Province. In connection with this vote, it should be remembered that it is to the advantage of the community that every encouragement should be given to the farmers to fence their farms, since the absence of fences largely increases the difficulty of enforcing the regulations necessary to check the East Coast cattle fever. Prior to the Union, in all four colonies fencing was made compulsory by law upon the outbreak of the disease; since, the respective governments had power to fence farms or areas infested, or suspected of being infested, with the disease, in the event of the owner failing to do so. When this was done, half the expenditure in the case of the Transvaal, and the whole of it in the case of Natal, was recovered from the owners, or adjacent owners. But, in order to promote voluntary fencing, funds were supplied on loan by the

Land Banks to the farmers, and in both the Transvaal and Natal the necessary materials were supplied at cost price by the Government.

The sums to be expended out of the revenue for the year 1911–12 in the interests of agriculture were not confined to the vote for the Ministry so-named, but included items in the votes for the Ministries of Land and Education.

The total sum provided for the Department of Agriculture was £751,284, of which £8,983 was allocated to the " Prime Minister and Minister of Agriculture," and £742,299 to the Department. Among the sub-heads of the vote the more significant, as indicating the character of the services which the Department is called upon to perform, are the following: £365,310, nearly one-half of the whole vote, is assigned to the Veterinary Division; and out of this sum £67,000 is provided for giving compensation to the owners of stock slaughtered compulsorily to prevent the spread of the East Coast fever. For bacteriological research and the investigation of disease, a sum of £30,458 is provided; and for botany and agronomy, £23,213, of which £5,780 is allocated to the destruction of noxious plants and weeds. To " tobacco and cotton " is assigned £12,618; to the wool industry, £4,937; to the dairy industry, £9,010, of which £4,000 is for the equipment of model dairies; to viticulture, £9,089, the Groot Constantia Farm costing £1,575, and the Paarl Oenological Institute, £800, in maintenance. The provision for " Entomology " is £14,837, out of which is £3,000 is for locust destruction, and £860 for the suppression of fruit and plant pests. " Chemistry " receives £7,514; " Publications " (which includes the most useful agricultural journals, distributed gratuitously), £4,296; " Co-operation," £5,036; and " Brands and Fencing," £77,901. The " Dry Land Experiment Stations " at Lichtenburg, Warmbaths, and Bloemhof cost £4,551; cold stores and abattoirs, £9,419; and for the "Inspection of Grain" £5,169 is provided. Lastly, we notice that the net increase of this vote for agriculture, as against that of 1910–11, is £49,893; that the departmental receipts are estimated to bring in £126,900; and that the services rendered by the Department to other departments are valued at £21,700.

Irrigation and Water Supply

Large and comprehensive as is this provision for the Ministry of Agriculture, it does not by any means represent the whole of the annual expenditure of the Union Government by which agriculture is especially benefited. The business of the Ministry of Lands is to a large extent concerned with matters calculated to promote directly the development of the agricultural and pastoral resources of the country. Of the £279,753, which forms the total vote for this department, nearly one-half (£122,860) is allocated to irrigation. Of this sub-vote, £28,301 is required for the maintenance of the Irrigation Director and his staff of seventy-two persons; £9,000 is provided for reconaissance surveys and the preparation of irrigation projects; £9,140 for the year's work upon the Hydrographic Survey, which is to record the water capacity of the Transvaal; and a further sum of £1,500 for the securing of the effective administration of the Irrigation Acts of the Cape and Transvaal Provinces. But the Department assists the farmers to obtain adequate water supplies by boring for Nature's underground reservoirs, as well as by the storage and distribution of the rainfall. The provision for "Boring" is £54,384, of which the greater part is assigned to finding water for farmers under the regulations of the Department, and the balance to similar work for other Government departments.

Stock Diseases, Locusts, Etc.

The reader will have noticed the largeness of the staff of the Veterinary Division of the Department of Agriculture—sixty-nine officers for veterinary "Control" and thirty-nine for "Research"; apart from forty officers for "Brands and Fencing," and some hundreds of field cornets and other subordinate officials employed locally to see that the Government regulations are carried out—and the ample provision made under this head in the estimates for the year 1911–12. He will have noticed, also, the largeness of the sums assigned out of loan funds to Land Banks for the purpose of enabling landowners and farmers to fence their holdings, and how the prompt destruction of infected cattle is facilitated by the payment of compensation to the owners; and the regular annual expenditure upon the scientific investigation of cattle diseases. These facts and figures will be sufficient to show that the Union Government is both taking effective measures for the prevention of the spread of stock diseases, and endeavouring at the same time to discover remedies which will enable the South African farmer ultimately to eradicate them.

The significance of the vote of £5,780 for the destruction of noxious plants and weeds, and of £14,857 for entomology (out of which £3,000 is allocated to locust destruction, and £860 to the suppression of fruit and plant pests) is not, however, quite so obvious; since the operations thus provided for are unfamiliar. Indeed, as we have had occasion to notice before, few countries occupied by Europeans are subject to insect pests in the same degree as South Africa. The assistance which the Union Government renders to agriculture in these respects is, therefore, of great importance, and an illustration of the methods of the Department will not be superfluous. A recent example of the practical manner in which the farmers are aided in their warfare against the worst of these insect plagues—the locust—is afforded by the records of the Agricultural Department of the (former) Orange River Colony. In the early months of the South African summer of 1909–10 vast numbers of locusts were hatched, from the eggs deposited earlier in the year, both in the south-west of this colony and in the neighbouring districts of the Cape. The laying of the eggs in unusual profusion was, however, duly reported to the Department; a large amount of poisonous material was prepared and then distributed to the farmers for use when, with the earliest rains, the hatchings were expected to begin. At the critical period the Agricultural Inspector arrived upon the scene to direct the operations. These were so successful, that the young locusts were destroyed wholesale directly they came out of the ground, and only a few swarms were able to develop their wings and escape. As the result of this prompt and effective action, a visitation of locusts which would have caused widespread devastation to the crops throughout the colony, was averted, and averted, too, at a cost of less than £2,000.

Agricultural Education

Provision for institutions offering a special training in agricultural pursuits is made in the vote for the Ministry of Education, and a sum of £95,629 was allocated to the sub-head of Agricultural Education in the estimates for 1911–12. As the subject will be discussed as part of the educational system of the Union as a whole, it will be sufficient to state here that six Agricultural Colleges, or Schools, are maintained; that the sum of £4,015 was provided for Agricultural Scholarships; and that the Colleges are so equipped as to be able (1) to provide a complete course of instruction, extending over two or three years, in the science and art of agriculture to persons intending to become farmers; and (2) to give instruction to persons already engaged in farming. A recommendation made by the Reorganisation Commissioners in dealing with the Department of Agriculture may be mentioned also in this connection. It is that these Agricultural Colleges should be made to serve as centres for the collection of information concerning the localities in which they are established. While novel phenomena and matters of national importance would be referred to the head-quarters of the Department at Pretoria, questions of local requirements and inquiries from agriculturists in the neighbourhood of the Colleges could, in the opinion of the Commissioners, be dealt with satisfactorily and promptly by the College professors and instructors. The recommendation is an excellent one, and if put into effect, it would help to prevent the Union Department of Agriculture from overlooking the special conditions and needs of this or that particular district, and thereby remove the one valid objection to the course adopted in the creation of a single department at Pretoria in preference to a central office with sub-departments in the several provinces.

Stud Farms

Apart from the Colleges, a number of institutions of an eminently practical nature are included, somewhat incongruously, in the education vote. In particular, it is here that provision is made for the maintenance of the Government stud farms, which enable farmers of moderate means to obtain the best and most suitable progenitors in the breeding of horses, cattle, sheep, and donkeys. For the Standerton Stud Farm, established in an upland district of the South-east Transvaal especially suited for horse-breeding, the sum of £5,468 was provided in the estimates of 1911–12; and from the (mainly) thoroughbred stud maintained here, the farmers can secure horse sires at moderate charges to run with their mares for the season. To the Ermelo Sheep Stud Farm £3,628 was allocated; to the Besterspruit Experiment Station and Donkey Stud Farm, £1,960; and to the "Purchase of Pedigree Stock," £5,000. The fact that both the Standerton and Ermelo Stud Farms now produce a considerable revenue—estimated in the case of the former to be £2,100, and in the case of the latter, £1,000—shows that the utility of these institutions has begun to be appreciated.

Forestry

With the exception of the forests of the Knysna and the Amatola Mountains (north of King William's Town), in the Cape Province, and the few isolated forest areas in the eastern border of the Transvaal, there is no indigenous growth of forest trees in South Africa west of the Drakenberg. The two forest areas of the Cape Province once formed part of a belt of evergreen forest, which ran continuously along the seaward slopes of the southern ranges, and thus united with the existing forest belts on the mountain slopes which fall to the eastern coast in the Transkei, Natal, and Zululand. Elsewhere, the central plateau is furnished naturally with no other woodland covering than the scanty scrub of thorny bush, or the grotesque Euphorbia. The most useful of the indigenous timber trees provided by these evergreen forests are the yellow-wood (*Podocarpus elongata* and *latifolia*); the black stinkwood or laurelwood (*Ocotea bullata*); the sneezewood (*Ptæroxylon utile*); the wagon woods (assegai, *Curtisia faginea*, and white pear, *Apodytes dimidiata*); and the black ironwood (*Olea laurifolia*). Although these woods are hard, durable, and serviceable for furniture-making and other purposes, the timber produced by the indigenous forests is quite insufficient for the needs of South Africa; and large quantities of wood—of the value of over £1,000,000 in 1911—are imported from oversea annually for the building and other industries of the country. Moreover, such trees are very slow-growing; and the fact is generally recognised that much better results are to be obtained from artificial afforestation than from the conservation of the existing indigenous forests. Experience has shown that the most suitable exotic trees, if planted in favourable localities, will yield in twenty years as much timber per acre as the indigenous forest yields in two hundred. And it is said, in particular, that the plantations established little more than thirty years ago in the Cape now produce more revenue than all the natural forests in the province. The rapid growth of the black wattle in Natal, and the lucrative results obtained from this industry, afford another case in point.

In these circumstances it becomes the duty of the State not so much to preserve the existing forests, as to provide the country with new forest areas, planted with exotic trees capable of producing the kinds of wood especially required by the various industries. In the Cape Peninsula the process has been accomplished. Here the indigenous forest has been replaced long ago by the pine, the white poplar, and the oak, all of which are naturalised and thrive excellently. The Cape oak, in particular, has twice the density of foliage of the English oak, and its acorn is twice as large. The blue-gum (*Eucalyptus globulus*), which was introduced in 1828, has been planted here and throughout South Africa. The pine woods of Parktown, the residential suburb of Johannesburg, and the eucalyptus woods of the Sachenswald and Frankenwald show how rapidly these exotic trees will grow even on the High Veld of the Transvaal.

The conservation of the forest areas of the Crown lands by the State was begun in the Cape and Natal about 1880; and it is stated that there are now 1,200,000 acres of such areas (of which 400,000 only are actual forest) in the former province, and some 30,000 acres in the latter. The area of the Crown forest in the Transvaal is said to be 30,000 acres; and in this province and the Free State afforestation dates from the beginning of the Crown Colony administration. In this, as in all other agencies for the development of the material resources of the New Colonies, Lord Milner took an active personal interest, and tree-planting was begun even before the war had reached its close.

THE FORESTRY DEPARTMENT

For the accomplishment of the twofold purpose of preserving the existing forests, and providing the country with a new and more ample supply of timber and ornamental trees, the Union is divided into forest conservancies, in each of which nurseries and plantations are established. The staff of the Forestry Department consists (in 1911) of a chief conservator, 9 conservators and assistant conservators, 27 district forest officers and their assistants, 112 foresters and assistant foresters, 15 rangers and forest guards, and 38 native forest guards—241 persons in all.

The conservancies, with the number of the nurseries and plantations in each, are as under:

Conservancy.	No. of Nurseries, etc.
Cape Province—	
Western	14
Midland	3
Eastern	5
Transkeian	3
Transvaal	11
Free State	7
Natal	3
Cape Extension	5

FORESTRY DEPARTMENT

In addition to the regular work of forest protection and development thus carried on, the Department makes grants to public bodies and private owners of land engaged in tree-planting, assists them with expert advice and supplies young trees from the various nurseries at practically cost price. The Government also maintains schools of forestry, and makes provision for the scientific training of the forest officers employed by the department.

CHAPTER V

TRADE AND COMMERCE

In the two preceding chapters we have seen what South Africa produces, and the industries in which the European population are chiefly engaged. The sea-borne trade of the Union, which amounts in round numbers to £90,000,000 in value, is divided between exports and imports in the proportion of 5 to 3, with an increasing balance, however, in favour of the former. Of the 50,000,000 of exports, more than four-fifths are the produce of the mines, while less than one-fifth comes from stock-raising and agriculture. The imports are the complement of the exports. With practically no manufactures of her own, the Union purchases manufactured goods of all descriptions; being deficient in agriculture, she imports the articles of food necessary to supplement the local supplies; and being actively engaged in the development of her natural resources, she buys railway, mining and building materials, machinery, tools, and utensils, and a vast amount of miscellaneous "plant."

An attempt is made in the table on the next page to focus the main features of the Union oversea trade. It shows the principal exports and imports of the Union for the year 1911, with the approximate percentage of each to the totals; while for comparison are added the Union totals for 1910, and the all-South Africa totals for 1911.

Distribution of Sea-borne Trade

We have now to observe how this sea-borne trade is distributed between the United Kingdom, other British

Sea-borne Trade of the Union in 1911

Exports.			Percentage to Total.
	£	£	
Mines:			
Gold	35,064,344	—	
Diamonds	8,281,907		
Coal	1,079,007		
Copper	573,537		
Tin, etc. (approx.) ..	250,000		
		45,248,795	82
Principal other Industries			
Wool	3,899,828		
Ostrich Feathers ..	2,253,140		
Angora Hair	917,874		
Hides and Skins ..	1,211,527		
		8,282,369	15
Total of all Merchandise (South African Produce)		55,389,353	100
For comparison:			
Total Union Exports for 1910		52,228,174	
Total of all Merchandise (South African Produce) exported from British South Africa, as a whole, in the year 1911		57,784,875	—

Imports.			Percentage to Total.
	£	£	
Food and Drink		6,336,262	18
of which:			
Corn, Grain, & Flour ..	1,285,157		
Meats	485,969		
Spirits (portable) ..	354,862		
Wines	83,259		
Cheese	143,640		
Manufactures (principal):			
Cotton	2,992,720		
Iron and Steel (except machinery)	1,229,005		
Machinery (except locomotives) ..	2,958,955		
Hardware and Cutlery	1,778,407		
Leather Goods ..	1,549,346		
Woollen	891,769		
Furniture	694,956		
Vehicles	788,058		
		12,883,216	36
Wood and Timber		1,106,143	3
Stationery and Books. ..		740,055	2
Total of all Merchandise imported ..		34,945,447	100
For comparison:—			
Total Union Imports for 1910 ..		34,007,178	
Total of all Merchandise imported into British South Africa, as a whole, in the year 1911 ..		36,423,539	

[Compiled from the Union Trade Returns, 1912.]

possessions, and foreign countries. The exports are taken almost entirely by the United Kingdom, the percentages (from British South Africa as a whole in 1911) being 90·9 for the United Kingdom, 3 per cent. for the other British possessions, 6·8 for all foreign countries, and 2 per cent. for articles shipped as stores or going through the post. In fact, apart from the United Kingdom, the only appreciable customer is Germany, whose purchase is 3 per cent. of the whole. Of the imports, 58·3 per cent. come from the United Kingdom, 10·1 per cent. from elsewhere within the Empire, and 31·6 per cent. from foreign countries. Among these latter, Germany and the United States are prominent with percentages respectively of 9·6 and 8·0. It may be added, that while in the last five years (1906–11) the United Kingdom imports have gained slightly, foreign imports have also gained at the expense of other British possessions. Germany is the most strenuous competitor; and the import from this country of the value of three and a half millions sterling forms approximately one-third of South Africa's purchase from outside the Empire, and 9·6 of her total imports.

The following tables, which are taken from the Returns issued by the Union Department of Commerce and Industries, will exhibit the movement of South African sea-borne trade in greater detail.

Return showing the total value of imports of merchandise and of exports of South African produce into and from the various ports, and also the total value of the imports and exports of British South Africa during the month of August, 1911, and the eight months ended August 31st, 1911, as compared with corresponding periods of 1910:

TRADE OF SOUTH AFRICA AS A WHOLE—IMPORTS

Ports.	Month ended August 31st. 1911.	Month ended August 31st. 1910.	Eight Months ended August 31st. 1911.	Eight Months ended August 31st. 1910.
	£	£	£	£
Viâ Cape Town	564,368	572,807	3,941,847	3,604,507
,, Capetown, through Parcel Post ..	43,321	60,813	409,134	422,293
,, Port Elizabeth	814,379	709,642	5,561,104	4,818,596
,, East London	326,141	385,513	2,424,388	2,393,142
,, Mossel Bay	51,482	35,216	311,028	270,631
,, other Cape Ports..	3,450	3,693	47,876	47,308
,, Port Natal (Durban)	945,249	842,112	6,886,303	6,430,752
,, Delagoa Bay (Lorenzo Marques) ..	434,836	464,701	3,449,795	3,957,607
,, Beira	120,110	102,815	1,024,170	734,113
,, Feira (on the Zambezi) and overland ..	6,055	398	9,286	925
Total *Viâ* British Ports	2,754,455	2,550,464	19,590,966	17,988,154
,, Portuguese Ports ..	551,946	567,576	4,473,965	4,685,720
Total Merchandise: British South Africa ..	3,309,401	3,117,980	24,064,931	22,673,874
Gold, Raw, in transit	991	3,264	12,733	24,152
Articles for South African Governments ..	183,905	169,211	1,288,888	1,751,458
Specie	205,000	21,596	895,578	1,964,204
Grand Total: British South Africa	3,699,297	3,312,037	26,262,070	26,413,686

PROPORTION OF BRITISH TO FOREIGN MERCHANDISE IMPORTED

	Month ended August 31st.				Eight Months ended August 31st.			
	1911		1910		1911		1910	
	£	Per Cent.	£	Per Cent.	£	Per Cent.	£	Per Cent.
From United Kingdom	1,949,374	58·9	1,953,173	62·6	14,119,576	58·6	13,385,746	59·0
,, other British Possessions ..	287,058	8·7	244,379	7·8	2,392,094	10·0	2,475,701	10·9
,, Foreign Countries	1,072,969	32·4	920,428	29·6	7,553,261	31·4	6,812,427	30·1
Total ..	3,309,401	100	3,117,980	100	24,064,931	100	22,673,874	100

In connection with the above table, it may be observed, that the arrangement made by the Transvaal-Portuguese Railway Agreement of February 2nd, 1909, under which 50 per cent. of the "Competitive Zone" traffic, *i.e.*, the carrying trade to and from the Rand, was assigned to Delagoa Bay, 30 per cent. to Durban, and 15 to 20 per cent. to the Cape ports, has not proved effective in preventing the diversion of this traffic from the Cape Ports. And, according to the Union Railway Report, the process still continues, in spite of the fact that the Cape through rates have been lowered twice since the Agreement came into force.

The growth in the business of the Cape ports, as shown by the increases recorded in this and the next table, is to be attributed, therefore, solely to the improvement in the general trade of the Union as a whole.

EXPORTS

Ports.	Month ended August 31st.		Eight Months ended August 31st.	
	1911.	1910.	1911.	1910.
South African Produce:	£	£	£	£
Viâ Cape Town — Diamonds	900,645	626,321	5,600,017	5,843,204
Viâ Cape Town — Gold, Raw	3,580,911	3,094,262	23,977,557	23,035,870
Viâ Cape Town — Other S. A. P.	111,284	89,224	799,212	660,600
,, Port Elizabeth	185,812	273,610	2,070,713	2,103,939
,, East London	63,578	88,687	1,016,656	1,161,387
,, Mossel Bay	62,977	105,478	733,454	784,272
,, other Cape Ports	2,279	68	330,600	164,275
Natal (Durban) — Gold, Raw	185,672	73,253	727,376	681,774
Natal (Durban) — ,, Concentrates	—	—	1,831	621
Natal (Durban) — Other S. A. P.	192,907	225,807	1,829,867	2,027,995
Delagoa Bay: Lorenzo Marques	31,209	14,329	253,543	299,192
Beira — Gold, Raw	—	16,352	53,134	62,502
Beira — ,, Concentrates	366	5,665	35,445	46,177
Beira — Other S. A. P.	23,376	17,106	122,186	97,715
Feira (on the Zambezi) and overland	6,378	332	23,627	1,602
Total — *Viâ* British Ports	5,295,443	4,577,081	37,110,910	35,465,539
Total — ,, Portuguese Ports	54,771	53,452	461,308	505,586
Total: South African Produce	5,350,214	4,630,533	37,575,218	35,971,125
Imported Goods Re-exported:				
Through the Parcel Post	3,431	4,885	33,690	36,577
Gold, Raw (in transit)	991	3,264	12,733	24,152
In Bond and Duty Paid	77,888	50,705	658,672	372,250
Specie	12,600	8,341	78,279	97,720
Grand Total	5,445,124	4,697,728	38,358,592	36,501,824

Trade of Union of South Africa—Imports

Articles.	Month ended August 31st.				Eight Months ended August 31st.			
	1911.		1910.		1911.		1910.	
	Quantity.	Value.	Quantity.	Value.	Quantity.	Value.	Quantity.	Value.
		£		£		£		£
Cotton Manufactures ..	—	314,124	—	291,341	—	2,055,147	—	1,849,712
Corn, Grain, and Flour, lbs.	29,078,953	111,049	15,589,161	66,782	214,276,327	825,690	289,678,398	1,213,284
Meats .. .. .. ,,	1,962,249	39,893	739,623	24,689	14,525,159	305,572	7,574,175	234,107
Total Articles of Food and Drink (including Corn, etc., and Meats)	—	530,738	—	443,016	—	4,032,742	—	3,920,823
Iron and Steel Manufacture (except Machinery) ..	..	109,052	—	111,744	—	827,361	—	1,018,049
Machinery (except locomotives & parts thereof	—	260,911	—	256,085	—	2,037,397	—	2,453,702
Total (including Imports) for the Government, and Specie)	—	3,561,585	—	3,204,243	—	25,243,773	—	25,683,087

EXPORTS (SOUTH AFRICAN PRODUCE)

Articles.	Month ended August 31st.				Eight Months ended August 31st			
	1911		1910		1911		1910	
	Quantity.	Value.	Quantity.	Value.	Quantity.	Value.	Quantity.	Value.
		£		£		£		£
Maize .. .. lbs.	20,874,200	39,975	59,991,391	113,859	66,894,764	116,953	228,676,625	463,636
Coal .. .. tons	146,622	105,289	101,892	68,120	1,013,477	715,338	953,068	621,480
Copper: Tons of 2,240 lbs.	279	13,605	208	11,570	15,842	408,661	9,800	273,378
Ostrich Feathers .. lbs.	62,810	149,334	75,226	250,655	546,150	1,598,452	496,710	1,611,083
Diamonds .. ..carats	485,966	900,645	431,107	626,361	3,150,547	5,600,017	3,838,194	5,842,407
Coffee .. .. lbs.	1,362	53	1,693	78	43,913	1,311	8,568	347
Sugar .. .. ,,	27,346	260	1,274,864	7,875	156,620	1,313	4,156,241	25,190
Sugar Products (Syrup, Molasses, etc.) .. lbs.	954,739	1,577	798,175	1,710	3,760,053	6,259	4,735,066	10,215
Tea .. .. .. lbs.	1,034	30	850	23	69,724	2,222	75,086	2,515
Wines .. .. galls.	6,245	1204	5,005	701	42,397	8,263	37,007	7,366
Gold, Concentrates ,,	—	512	—	342	—	11,610	—	21,892
,, Raw .. .. ozs.	939,013	3,480,267	780,410	2,900,628	6,216,649	23,092,284	5,681,311	21,074,099
Angora Hair .. lbs.	1,281,194	61,371	1,034,314	60,897	9,977,637	469,268	9,449,997	480,704
Tin .. .. .. tons.	212	23,167	17	1,397	4,241	163,861	1,835	133,817
Tobacco .. .. lbs.	7,138	808	22,504	954	51,509	5,097	144,532	10,377
Wool .. .. ,,	2,046,855	69,136	3,257,282	113,279	73,756,571	2,189,195	71,279,181	2,243,121
Total Merchandise ..	—	5,077,767	—	4,389,391	—	36,022,591	—	34,435,103

Southern Rhodesia (Summary)

	Month ended August 31st.		Eight Months ended August 31st.	
	1911.	1910.	1911.	1910.
Imports—	£	£	£	£
(including Specie and Government) ..	254,523	258,487	1,968,920	1,648,008
Exports—				
Diamonds	—	210	—	1,382
Gold	286,316	283,239	1,665,783	1,706,047
Other S. A. Produce .	18,999	33,626	217,808	187,125
Total ..	327,853	357,028	2,040,298	2,047,511

Fiscal Policy

That England should have been able to hold her own in the South African market, in spite of the keen competition of her two great industrial rivals—Germany and the United States—is due very materially to the preferential treatment accorded to her goods in the South African Customs Tariff.

The foundation of the present fiscal system of the Union was laid at the Customs Conference held at Bloemfontein, in March, 1903, under the presidency of Lord Milner. The main principles embodied in the Customs Convention of that year were based upon a frank recognition of the economic conditions of South Africa as a whole ; and, as such, they are not likely to be abandoned, in any material respect, in the immediate future. They were :

1. The free admission of the necessaries of life, including foodstuffs, except in cases where free admission was found to be inconsistent with the due encouragement of South African industries.

2. The moderate taxation, or free admission, of building materials, agricultural implements, machinery, mining plant, and generally of all articles necessary either for

the development of the industrial resources of the country, or for providing the various communities with the full equipment of civilisation.

3. The maintenance of the revenual efficiency of the tariff by the imposition of heavy duties on luxuries, intoxicants, and narcotics, and on articles ministering mainly or exclusively to the convenience or pleasure of the wealthier classes.

4. The grant of preferential treatment to articles produced within the British Empire. In the case of articles being "the growth, produce, or manufacture of the United Kingdom," where such articles were classified under an *ad valorem* rate, they were accorded a rebate of 25 per cent. of such *ad valorem* duty; and where they were subject only to a 2½ per cent. *ad valorem* duty, the whole duty was remitted. And similar rebates were accorded to the goods of any other country within the Empire, provided, however, that the country so favoured granted equivalent reciprocal privileges to South Africa.

Upon the eve of the establishment of the Union, the four Colonial Governments gave notice to the other parties to the Customs Convention—Rhodesia, Basutoland, and the Bechuanaland Protectorate—of their intention to retire from the Convention on June 30th, 1910. The Union Government thereupon entered into fresh Customs agreements—cast, however, on the same lines as the expiring Convention—with Rhodesia and with the High Commissioner on behalf of the Native Territories not included in the Union; and thus the essential features of the Customs Convention of 1903 have been maintained. For trade purposes, therefore, the Union, Rhodesia, and the Native Territories constitute a single system.

The principles upon which this single system, comprising British South Africa as a whole, has based its common Custom tariff have been set out above; but the fiscal relations established by these new agreements also claim consideration. All domestic goods and produce, with the exception of spirits and beers, pass duty free to and from the Union, Rhodesia, and the Native Territories; and customs levied upon external goods in transit are rebated. In the case of goods imported from outside British South Africa into the Union, but in transit for Rhodesia, the Union Government collects the duties leviable under the common tariff and pays them over, *less* 5 per cent. for cost of collection, to the Rhodesian Government; and, in the converse case, the Rhodesian Government collects and pays over the duties to the Union Government. Also, in the case of domestic (South African) manufactures containing imported materials to an appreciable extent, the duties collected on the imported materials are paid over similarly by the government in whose territory the manufactures are made to the government of the territory into which the manufactures are imported. The same arrangements obtain as between the Union Government and the governments of the Native Territories with one exception. In the collection and payment of Customs duties upon imports and imported materials destined for the Native Territories, the method laid down in the schedule[1] to the South Africa Act is adopted. The Section runs:

> There shall be paid into the Treasury of the Union all duties of Customs levied on dutiable articles imported into and consumed in the Territories, and there shall be paid out of the Treasury annually, towards the cost of administration of each territory, a sum in respect of such duties which shall bear to the total Customs revenue of the Union in respect of each financial year the same proportion as the average amount of the Customs revenue of such territory, for the three completed financial years last preceding the taking effect of this Act, bore to the average amount of the whole Customs revenue for all the colonies and territories included in the Union received during the same period.

[1] This schedule sets out the terms upon which the Territories may be admitted to the Union.

The proportions of the annual Customs revenue of the Union respectively due on this basis are paid quarterly to the several administrations of the three Native Territories.

One other point is worthy of note, as indicating the partial admission of Portuguese East Africa to the commercial system of British South Africa. The arrangement made by Lord Milner in the *modus vivendi* of 1901, and modified in 1904, under which the Transvaal admitted the products and genuine manufactures of Mozambique, with the exception of spirits, duty free, was maintained in the Transvaal-Mozambique Agreement of February 2nd, 1909; and, since this latter agreement is binding upon the Union, these Mozambique products continue to enter the Transvaal Province duty free. Such products, however, upon removal from the Transvaal to another province of the Union, or to Rhodesia or the Native Territories, become subject to the duties leviable upon external imports of the like description under the Customs tariff.

The Ministry of Commerce and Industries

The Union Department of Commerce and Industries was formed out of the Customs and Excise services of the four colonies, together with the staff of the Statistical Bureau created in August, 1905, for the common benefit of the States constituting the former Customs Union of South Africa. In the Estimates for 1911–12, provision is made for 490 Customs officers under the Commissioner of Customs and Excise, and for 72 Excise officers under the Controller of Excise. As the number of the Customs officers employed by the four Colonial Governments on May 30th, 1910, was 523, the Union staff shows an appreciable reduction in this service; but practically no change, except a slight rearrangement of the staff, was made in the case of the Excise officers, since it was only in the Cape province that any considerable establishment for Excise purposes had been maintained.

In this connection it may be mentioned that in this province there is a considerable production both of spirits and beer. In 1909, 1,205,326 proof gallons of spirits was produced from the produce of the grape, and 44 gallons from materials other than produce of the grape; and the revenue collected thereon amounted to £185,252 and £7 respectively. In the same year, 2,606,647 gallons of Colonial beer were produced, in respect of which the revenue was £23,949. It may be added that the number of pot stills in the Cape Colony was estimated in the same year to be 3,550. There were thirty-five distillers who used stills other than pot stills, only one of whom distilled from materials other than the produce of the grape. In respect of the manufacture of beer, there were eight firms of brewers in the Cape Colony: three at Capetown, and one each at Port Elizabeth, Queenstown, East London, Oudtshoorn, and Kimberley.

Apart from the collection of Customs and Excise, the work of the Department would seem for the present to be confined to the preparation and issue of the trade returns and other statistics of the Union. Here, again, the establishment of the Union made little difference to the staff of the Statistical Bureau, since these officials were either transferred to the "Statistical Branch" of the Union Department, or otherwise absorbed by the head office.

Banks and Banking Facilities

The Union and British South Africa as a whole are well supplied with banking facilities. Apart from the great joint-stock banks by which the traders, merchants, and large industrial undertakings are served, the system of land banks, established, as we have seen, in the interests of farmers and landowners, is being extended by the

Union Government, and the number of Post Office savings banks is being rapidly augmented in response to the increasing requirements of the small depositor. In respect of the joint-stock banks, the maintenance of due reserves in cash and securities, and the frequent publication of statements showing the financial position of each bank and the character of its business, are required by law; and in these and other respects the interests of the depositors and of the general public are protected by enactments regulating the conditions under which the transactions of banks and banking firms must be conducted. The largest of these institutions, the Standard Bank of South Africa, Ltd., has branches in every considerable town in the Union and in South Africa; and at Maseru, in Basutoland; Blantyre, in Nyassaland; Delagoa Bay and Beira, in Mozambique; and at Mombasa, Nairobi, and Zanzibar, in British East Africa.

The table on the next page, which is taken from the Statistical Register of the Cape Province, will exhibit sufficiently the general character of the South African banks. It shows, as regards the several joint-stock banks transacting business in the Cape Colony on December 31st, 1909, the amount of authorised, subscribed, and paid-up capital and reserve fund; the number, nominal value, and amount paid up per share; together with the rate per cent. of the last dividend declared and the number of branches.

THE POST OFFICE SAVINGS BANK

The character and objects of the Government land banks have been indicated in the account of the assistance rendered by the State to agriculture, which has been laid before the reader in the preceding chapter. It will not be necessary, therefore, to refer now to these

BANKS ON DECEMBER 31ST, 1909

Banks.	Authorised Capital.	Subscribed Capital.	Paid-up Capital.	Reserve Fund.	Shares Number.	Nominal Value per Share.	Amount paid per Share.	Rate per cent. of last Dividend.	No. of Branches: In Cape Colony.	No. of Branches: In Natal, Orange Free State, and Transvaal.
	£	£	£	£		£	£	£		
Standard Bank of South Africa, Ltd. ..	6,250,000	6,194,100	1,548,525	1,900,000	61,941	100	25	10	97	34
Bank of Africa, Ltd.	3,000,000	3,000,000	1,000,000	500,000	160,000	18¾	6½	5	22	30
African Banking Corporation, Ltd. ..	2,000,000	800,000	400,000	140,000	80,000	10	5	5	24	8
Stellenbosch District, Ltd.	20,000	16,800	8,400	10,000	840	20	10	7½	1	—
National Bank of South Africa, Ltd. ..	1,100,000[1]	1,100,000	1,100,000	50,000	110,000	10	10	3	5	59
Netherlands Bank of South Africa ..	750,000	400,000	400,000	45,722	4,000	60	60	3	1	6
Totals	13,120,000	11,510,900	4,456,925	2,645,722	416,781	—	—	—	150	137

[1] With power to increase to £4,000,000.

All the above collectively had, on December 31st, 1909:

	£	s.	d.
***Liabilities*—**			
Deposits payable after notice or on a fixed day	17,211,160	17	2
Deposits on Current Account	18,589,569	10	2
Total Liabilities (including above)	61,569,210	8	3
***Assets*—**			
Coin in Bank Coffers	6,381,868	4	11
Loans and Advances other than Bills, secured	15,878,549	12	2
Total Assets (including the above)	61,569,210	8	3

useful institutions, but a few figures exhibiting the progress of the Post Office Savings Bank will not be superfluous. The Report of the Union Postmaster-General for the year 1910 records a great expansion in the business of the savings bank; and the fact is worthy of notice, since a rise or fall in the savings bank deposits is recognised as affording an excellent measure of the prosperity or the reverse of the general mass of any European community. According to the report, then, the number of accounts open in the Post Office Savings Bank of the Union on December 31st, 1910, had increased during the preceding twelve months by 20,399, or at the rate of 19·36 per cent.; and the amount standing to the credit of depositors in ordinary accounts had risen by £724,921, or 14·73 per cent. During the same period the percentage of the accounts withdrawn was only 2·66, representing 4·09 per cent. of the total value of the sums deposited. On June 30th, 1910, there were 644 savings bank offices open in the Union.

The progress of the bank, and the volume of the work which it transacts, is shown in the subjoined table:

PROGRESS OF THE POST OFFICE SAVINGS BANK in 1908–10:

	1910.	1909.	1908.
Deposits: Number	433,395	363,086	353,751
,, Amount	£3,708,918	£3,162,525	£3,222,809
Withdrawals: Number ..	182,555	177,830	187,447
,, Amount ..	£2,983,996	£2,866,815	£3,390,162
Accounts remaining open ..	218,617	198,218	187,363
Certificates: Issued	£284,300	£236,600	£174,100
,, Paid	£145,500	£125,600	£150,400
Balance due to Depositors:			
In Ordinary Accounts, including Interest.. ..	£5,202,599	£4,477,678	£4,237,511
In Certificate Accounts ..	£851,300	£712,500	£601,500

DUTCH MEASURES

Currency and Weights and Measures

English money is used throughout British South Africa. Owing, however, to the lower purchasing power of the sovereign—a subject to be discussed subsequently in connection with the cost of living—copper coins are rarely used, except in the coastal districts. In the Transvaal, in particular, for shopping purposes, the threepenny bit will be found to be the ordinary equivalent of the penny piece in England. For the convenience of travellers and others, letters of credit and drafts are granted by the banks on their branches and agencies, and circular letters of credit can be obtained.

While weight is measured commonly by the English standards, for land areas and bulk the use of the Dutch measures is more general.

In the accompanying list of special weights and measures employed in South Africa, the English equivalents are given :

Weight	—Short Ton	= 2,000 lbs. Avoirdupois.
	—Long ,,	= 2,240 ,, ,,
Area, etc.	—Cape Foot	= 1·033 English feet.
	— ,, Rood	= 12·396 ,, ,,
	—Morgen	= 2·116 Acres.
	—Hectare	= 2·471 ,,
	—302·38 Morgen	= 1 sq. Mile (English).
Bulk (Liquid)	—Leaguer	= 152 Dutch Gallons = 128 Imperial Gallons.
	—Pipe	= 110 Dutch Gallons
	—Aum	= 28 ,, ,,
	—Anker	= 9½ ,, ,,
	—Dutch Gallon	= ·7895 of an English gallon.
,, (Dry)	—Muid	= 3 bushels = 12 pecks = 24 gallons = 96 qrts.
Diamond Measure	—Carat	= 4 grains Troy.
,, ,,	—120 Carat	= 1 ounce ,,

Census of 1911—Union of South Africa

Table showing Movement of Population

	Total Population.			European Population.			Native Population.		
	No. on May 7th, 1911.	Numerical Increase since 1904.	Percentage Increase since 1904.	No. on May 7th, 1911.	Numerical Increase since 1904.	Percentage Increase since 1904.	No. on May 7th, 1911.	Numerical Increase, since 1904.	Percentage Increase, since 1904.
Cape..	2,563,024	153,226	6·36	583,177	3,436	0·59	1,545,308	120,521	8·46
Natal	1,191,958	83,204	7·50	98,582	1,473	1·52	951,808	47,767	5·28
Transvaal	1,676,611	406,660	32·02	420,831	123,554	41·56	1,224,155	287,028	30·52
Free State	526,906	139,591	36·04	175,435	32,756	22·96	339,811	110,662	48·29
Union of South Africa ..	5,958,499	782,675	15·12	1,278,025	161,219	14·44	4,061,082	565,978	16·19

	All other Coloured Races.			Total Population other than White.			Proportion per cent. of European to Total Population.		Number of European Females to every 100 males of European Population.	
	No. on May 7th, 1911.	Numerical Increase, since 1904.	Percentage Increase, since 1904.	No. on May 7th, 1911.	Numerical Increase, since 1904.	Percentage Increase, since 1904.	1911.	1904.	1911.	1904.
Cape	434,539	29,263	7·22	1,979,847	149,784	8·18	22·75	24·06	93·28	82·00
Natal	141,568	33,964	31·56	1,093,376	81,731	8·08	8·27	8·76	86·60	71·09
Transvaal ..	31,625	(Decrease) − 3,922	(Decrease) − 11·03	1,255,780	283,106	29·11	25·10	23·41	77·35	66·78
Free State ..	11,660	− 3,827	− 24·71	351,471	106,835	43·67	33·30	36·84	85·42	74·91
Union of South Africa ..	619,392	55,478	9·84	4,680,474	621,456	15·31	21·45	21·58	86·18	75·84

[Compiled from the Preliminary Returns of the Census, 1911]

PART V

POLITICAL AND SOCIAL CONDITIONS

CHAPTER I

FINANCE AND TAXATION

In discussing the financial position of the Union, and the taxation imposed by the Government to meet the cost of administration, the conditions which differentiate South Africa from the United Kingdom and the other oversea dominions must be recalled once more.

The Union, then, has a composite population, partly European and partly native or coloured; and the latter element outnumbers the former in the proportion of nearly four to one. In the census of May 7th, 1911, the European population was returned as being 1,278,025, and the non-European population as being 4,680,474, making together a total population of 5,958,499. While, moreover, the native or non-European population had increased since 1904 at the rate of 15·31 per cent., the European rate of increase for the same period was only 14·44. Both the composite nature of the Union population, and the relative smallness of its European element, are facts which must modify appreciably any conclusions based upon the mere figures of the financial returns of the Union; or on these figures as compared with the corresponding figures for the United Kingdom or for any one of the other three dominions. In Australia, New Zealand, and Canada, for all practical purposes, there are solid European populations; and, of course, this is the case in the United Kingdom. In any statement, therefore, made for the purpose of ascertaining how the Union stands in respect of indebtedness, taxation, or trade per head relatively to these other members of the Empire, neither the number of the European population alone nor that of the total population afford a valid basis of comparison. Indeed, as nearly all the purely manual labour of the Union is done by the natives and coloured people, the million and a quarter of European population, from the point of view of taxpaying, find a counterpart in the few millions of income tax-paying population in the United Kingdom, rather than in the 45,000,000 of its total population. Still less can the European population of the Union, taken alone, be placed on the same footing as the total populations of the other oversea dominions; since these latter, in all cases, comprise the numerically predominant classes of a community from which the manual labourers are drawn.

There is a second condition, and one of a different order, to be noticed, if it is desired to realise the actual, as against the nominal, value of the financial figures of the Union. This condition is the high cost of living to Europeans; or, to put it in another way, the fact that the sovereign has less purchasing power in South Africa than it has in England or in any of the other oversea dominions. The Union cost of administration, therefore, expressed in pounds sterling appears, when exhibited in comparison with that of other States, to be larger than it really is, since an expenditure of £100 in the Union is equivalent to an expenditure of (say) £75 in England or in the other dominions. A rough measure of the degree in which the Union figures are swollen by this circumstance is afforded by a comparison of the salaries received by the respective Prime Ministers of the four dominions. While the heads of the Governments in Canada, Australia, and New Zealand are paid respectively £2,500, £2,100, and £1,600 per annum, the salary of General Louis Botha, the first Prime Minister of the Union, was fixed at £4,000. The explanation of the difference is not, of course, to be sought in the assumption that the work of the Prime Minister of the Union is to be more arduous and responsible than that of, say, the Prime Minister of the Dominion of Canada, but is to be found in the fact that this and other official salaries in the Union, as expressed in pounds sterling, are higher than similar salaries in the other three dominions, where the purchasing power of the sovereign is greater.[1]

According to the statements issued by the Union Treasurer in 1911, the amount of the Public Debt for which the Union Government became responsible on May 31st, 1910, was £116,502,628 6s. 11d.; and for the financial year ending March 31st, 1912, the revenue was estimated to be £16,052,000, and the expenditure £16,890,281. At the same time, a further expenditure from loan funds, estimated to be £12,017,105, was required to carry out new works and services approved by the Government; and of this sum, £7,453,829 18s. 9d. was provided by loan funds in hand or authorised to be borrowed, while £4,563,275 1s. 3d. remained to be raised by a new loan.

Subject to the modifying conditions noticed above, the following statement will afford a rough measure of the significance of these figures.

Table showing the population, and total and per head revenue, expenditure, debt, and sea-borne trade of the United Kingdom and the dominions. The figures are approximate only, and are based on the latest available statistics in the case of each State:

[1] In England the Prime Minister receives £5,000 as First Lord of the Treasury, and an official residence. The Prime Minister of the Union also has Groote Schuur, bequeathed by Rhodes to be the official residence of the Prime Minister of South Africa, when union should be accomplished.

	Population.	Revenue.	Per Head of Population.	Expenditure.	Per Head.	Debt.	Per Head.	Sea-borne Trade.	Per Head.
		£	£	£	£	£	£	£	£
Dominion of Canada..	7,000,000	24,000,000	3½	18,000,000	2½	95,000,000	14	140,000,000	20
Commonwealth of Australia ..	4,500,000	44,000,000	9	(?) 44,000,000	(?) 9	257,000,000	59	134,000,000	30
Dominion of New Zealand	1,000,000	10,000,000	10	9,000,000	9	81,000,000	81	40,000,000	40
Union of South Africa { European..	1,250,000	16,000,000	13	16,000,000	13	116,000,000	93	90,000,000	70
Union of South Africa { Coloured ..	4,750,000	—	3½	—	3½	—	—	—	19
Union of South Africa { Total ..	6,000,000	—	2⅔	—	2⅔	—	—	—	15
United Kingdom	45,250,000	181,000,000	4	181,000,000	4	685,000,000	15	1,200,000,000	25

THE UNION DEBT

The particulars of the respective debts of the four colonies at the date of the establishment of the Union are contained in the statement on page 394, which was issued by the Union Treasurer in 1911.

And, in a second statement (shown on pp. 395–6), the allocation of the Public Debt assumed by the Union Government, as it existed on May 30th, 1910, is there set out.

Of the expenditure out of loan funds thus appropriated, the Union Treasurer, as the result of a further analysis, states that £80,211,786 9s. 6d. has been reproductive and £30,290,841 17s. 5d. non-reproductive. And of the reproductive expenditure, he assigns £75,234,694 0s. 2d. to "Railways and Harbours," and £4,977,092 9s. 4d. to "Other Services."

The purposes for which the further expenditure of some £12,000,000, already incurred by the Union Government, is required are shown in the subjoined statement.

ESTIMATED EXPENDITURE, to meet which Loan Funds are required:

Vote.	Estimated Expenditure in 1910–12. 6.	Estimated Expenditure after Mar. 31st, 1912. 7.	Total Estimated Expenditure. 8.
A. Railways and Harbours ..	£2,015,731	£5,217,868	£7,233,599
B. Public Works and Buildings ..	1,307,754	695,047	2,002,801
C. Telegraphs and Telephones ..	548,661	251,399	800,000
D. Agriculture	153,835	100,149	253,984
E. Lands	188,680	51,904	240,584
F. Irrigation	493,579	1,000	494,579
G. Local Works and School Loans	103,120	60,000	163,120
H. Land Banks	568,000	260,438	828,438
Totals	5,379,360	6,637,745	12,017,105

PUBLIC DEBT OF THE CAPE, NATAL, TRANSVAAL, AND ORANGE RIVER COLONY at May 30th, 1910, showing loans authorised and issued, the existing debt and interest thereon, and the unexercised borrowing powers.

Colony.	Loans Authorised.	Loans Issued.	Capital as affected by Conversions.	Paid off.	Existing Debt.	Interest for one Year.	Unexercised Borrowing Powers.
	£ s. d.	£ s. d.	£ s d.	£ s. d.	£ s. d.	£ s. d.	£ s. d.
Cape	61,018,425 2 7	57,745,689 16 1	57,904,820 16 11	5,338,786 11 10	52,566,034 5 1	1,921,763 17 9	3,624,265 12 0
Natal	23,321,862 9 10	22,665,378 16 3	22,940,094 1 10	253,500 0 0	22,686,594 1 10	797,557 13 4	656,483 13 7
Transvaal	32,317,591 11 10	32,317,591 11 10	32,317,591 11 10	—	32,317,591 11 10	969,527 14 11	—
O. R. Colony ..	9,182,408 8 2	8,932,408 8 2	8,932,408 8 2	—	8,932,408 8 2	274,222 5 1	250,000 0 0
Totals	125,840,287 12 5	121,661,068 12 4	122,094,914 18 9	5,592,286 11 10	116,502,628 6 11	3,963,071 11 1	4,530,749 5 7

ALLOCATION OF PUBLIC DEBT

PUBLIC DEBT OF THE CAPE, NATAL, THE TRANSVAAL, AND THE ORANGE RIVER COLONY, showing existing debt and interest thereon at May 30th, 1910, allocated to works and services :

Service.	Colony.	Existing Debt. £	s.	d.	Interest. £	s.	d.
A. Railways ..	Cape	30,884,252	12	5	1,132,932	13	5
,, ..	Natal	14,178,154	4	6	497,740	17	0
,, ..	Transvaal ..	15,618,447	16	0	468,553	8	8
,, ..	O. R. Colony	4,991,118	9	11	151,527	3	1
		65,671,973	2	10	2,250,754	2	2
B. Harbours ..	Cape	5,967,275	11	1	215,721	18	1
,, ..	Natal	3,595,445	6	3	122,039	7	5
		9,562,720	17	4	337,761	5	6
C. Posts and	Cape	489,740	18	0	18,294	7	1
Telegraphs	Natal	367,795	15	8	12,409	19	4
,,	Transvaal ..	319,671	14	3	9,590	3	0
,,	O. R. Colony	30,353	1	9	910	11	10
		1,207,561	9	8	41,205	1	3
D. Public Works	Cape	925,155	17	5	35,226	10	1
and Build-	Natal	1,289,313	18	9	43,431	11	7
ings	Transvaal ..	1,263,963	16	2	37,918	18	4
,,	O. R. Colony	516,974	17	1	15,716	17	8
		3,995,408	9	5	132,293	17	8
E. Agriculture ..	Cape	1,127,082	7	0	38,908	12	10
,, ..	Natal	368,061	10	10	13,184	19	7
,, ..	Transvaal ..	1,611,235	4	2	48,337	1	1
,, ..	O. R. Colony	1,327,816	1	3	40,150	0	1
		4,434,195	3	3	140,580	13	7
F. Local Works Loans ..	Cape	565,950	15	6	19,026	5	9
G. School Loans	Cape	818,377	17	8	28,023	16	0
H. Acquisition of	Cape	266,657	0	0	10,809	5	7
New Terri-	Natal	752,134	9	6	26,585	7	7
tory and Li-	Transvaal ..	2,529,911	0	0	75,897	6	7
abilities of Territories annexed ..		3,548,702	9	6	113,201	19	9
J. Immigration	Cape	172,510	5	0	7,153	16	11
,,	Natal	163,874	5	1	6,865	19	4
		336,384	10	1	14,019	16	3

Service.	Colony.	Existing Debt. £	s.	d.	Interest. £	s.	d.
K. War and	Cape	7,409,428	13	6	279,544	11	1
Defence ..	Natal	1,170,774	11	0	46,688	12	0
,, ..	Transvaal ..	1,579,174	13	0	47,375	4	9
		10,159,377	17	6	373,608	7	10
L. Deficiency	Cape	3,922,608	2	10	135,579	1	10
in Revenue	Transvaal ..	547,883	5	3	16,436	10	0
		4,470,491	8	1	152,015	11	10
M. Public	Cape	2,001	5	2	70	0	10
Stores	Natal	100,000	0	0	4,000	0	0
		102,001	5	2	4,070	0	10
N. Land and Agricultural	Natal	157,578	3	5	5,515	4	8
Loan Fund	Transvaal ..	2,112,692	9	6	63,380	15	6
	O. R. Colony	25,000	0	0	875	0	0
		2,295,270	12	11	69,771	0	2
O. Repatriation	Transvaal ..	5,305,100	14	0	159,153	0	5
,,	O. R. Colony	1,279,413	3	1	38,382	7	11
		6,584,513	17	1	197,535	8	4
P. Balances ..	Cape	14,992	19	6	472	18	3
,, ..	Natal	543,461	16	10	19,095	14	10
,, ..	Transvaal ..	1,429,510	19	6	42,885	6	7
,, ..	O. R. Colony	761,732	15	1	26,660	4	6
		2,749,698	10	11	89,114	4	2
	Totals ..	£116,502,628	6	11	£3,963,071	11	1

	£	s.	d.	
Of this debt,	£114,986,977	11	0	is in respect of Stock, Debentures, and Treasury Bills;
	1,050,000	0	0	was advanced by Banks pending the raising of Loans;
	465,650	15	11	is represented by advances made by the Public Debt Commissioners of Natal.
	£116,502,628	6	11[1]	

[1] *Union of South Africa.* Statement of Public Debt of the Cape, etc., at the date of entering the Union.

Of this total estimated expenditure of £12,017,105, loan funds in hand on May 30th, 1910 (£2,923,080 13s. 2d.) and authorised to be borrowed (£4,530,749 5s. 7d.) would provide £7,453,829 18s. 9d., leaving £4,563,275 1s. 3d. as the additional sum to be borrowed.

The most noteworthy items included under the head of "Public Works and Buildings" in the above statement may be mentioned. The new Union Government buildings at Pretoria are estimated to cost £1,130,000; and of this sum £43,148, provided out of revenue by the late Transvaal Government, was spent prior to May 30th, 1910; £535,000 was to be spent by March 31st, 1912; and £531,852 after this date. In addition to this great building, the capital is to be further enriched with a post office, and a museum and public library, at a respective cost of £112,000 and £90,000. At Capetown, the seat of the Legislature, a sum of £101,000 is to be expended on additions to the Houses of Parliament and £187,895 on law courts. Johannesburg and Durban are also to be provided with law courts, the former at a cost of £135,000 and the latter at a cost of £90,000.

The statements on pp. 398 and 399, issued by the Controller and Auditor-General of the Union, show respectively the cash balances of the constituent colonies at the date of the Union, and the position of the Union debt at March 31st, 1911.

Actual *v.* Nominal Indebtedness

With the assistance of these statements, it is possible to distinguish between the actual and nominal indebtedness of the Union. As the railways showed on the earnings for the year 1910 a net profit, after payment of interest on capital expenditure, renewals, and all working expenses, of £3,339,583, it is not very likely that the taxpayers of the Union will be called upon to meet the interest and other annual charges due upon the £80,000,000

STATEMENT OF CASH BALANCE (with certain realised assets) OF THE UNION IN REPORT OF CONTROLLER AND AUDITOR-GENERAL, dated Pretoria, December 11th, 1911

	Available Cash Balance as per page 13 of White Book [*i.e.*, at time of Union].	Corrections in Audit involving additions to Amount.	Corrections in Audit involving deductions from Amount.	Corrected Figure of Cash Balances.	Certain Assets realised in Cash in 1910–11 and 1911–12, and credited to Consolidated Revenue Fund.	Cash Balance, with certain realised Assets added thereto.
	£ s. d.	£ s. d	£ s. d.	£ s. d.	£ s. d.	£ s. d.
Cape	163,354 18 8	—	10,667 9 8	152,687 9 0	—	152,687 9 0
Natal	98,697 15 10	—	22,579 11 11	76,118 3 11	15,300 0 0	91,418 3 11
Transvaal	871,667 2 0	4,420 17 11	260 9 10	875,827 10 1	—	875,827 10 1
Orange River Colony	343,929 19 6	—	—	343,929 19 6	1,614 17 0	456,824 15 11
					110,000 0 0	
					1,029 19 5	
					250 0 0	
Totals	1,477,649 16 0	4,420 17 11	33,507 11 5	1,448,563 2 6	128,194 16 5	1,576,757 18 11

Union Debt existing at March 31st, 1911

	Loans Authorised.			Loans Issued.			Capital as affected by Conversions.			Paid Off.			Existing Debt.			Interest for One Year.		
	£	s.	d.	£	s.	d.	£	s.	d.	£	s.	d.	£	s.	d.	£	s.	d.
Cape	61,018,425	2	7	57,745,689	16	1	57,904,820	16	11	6,494,212	2	5	51,410,608	14	6	1,880,681	12	2
Natal	22,856,211	13	11	22,199,728	0	4	22,474,443	5	11	1,903,500	0	0	20,570,943	5	11	716,931	12	11
Transvaal ..	32,317,591	11	10	32,317,591	11	10	32,317,591	11	10	—			32,317,593	11	10	969,527	14	11
O. R. Colony ..	9,182,408	8	2	8,932,408	8	2	8,932,408	8	2	1,250,000	0	0	7,682,408	8	2	230,472	5	1
Union of S. Africa	—			[1]2,255,416	17	3	2,255,416	17	3	—			2,255,416	17	3	78,939	11	9
Totals ..	125,374,636	16	6	123,450,834	13	8	123,884,681	0	1	9,647,712	2	5	114,236,968	17	8	3,876,553	2	10

[1] Temporary Loan, being deposits by various funds to the Exchequer Account.

The accumulations of assets of the Sinking Funds at March 31st, 1911, amounted to £5,054,732 6*s.* 5*d.*

[From Report of the Controller and Auditor-General on Finance Accounts, etc., of the Union for period May 31st, 1910, to March 31st, 1911.—U.G. No. 12–'12.]

of reproductive expenditure, which constituted more than two-thirds of the original debt. Of the £12,000,000 further expenditure from loan funds already incurred, some part—*e.g.*, the votes for "Railways and Harbours" and "Telegraphs and Telephones"—will be reproductive. Assuming, however, the whole of it to be non-reproductive, and to be added, therefore, to the £30,000,000 of original non-reproductive expenditure, the resultant total of such expenditure is, in round numbers, £42,000,000. The actual, as distinguished from the nominal, Public Debt of the Union may be considered, therefore, to be approximately £40,000,000.

The statements of revenue and expenditure which now claim our attention will enable us to make the same distinction between the nominal and actual sum to be provided by the taxpayers of the Union for the annual service of the debt. As we have seen from the above statements, the annual interest on the original £116,500,000 of debt amounts to some £4,000,000. The actual sum to be met out of the revenue for the financial year 1911–12 in respect of the debt is to be found by deducting the sum of £2,806,000, appearing as interest due to the Union Government in the revenue account, from the sum of £4,560,326, which is shown in the expenditure account [1] as the interest, etc., payable on the Public Debt. That is to say, while the annual liability imposed by the Public Debt upon the people of the Union amounts nominally to £4,500,000 (in round numbers), its actual amount is only £1,750,000. Before leaving this aspect of the financial position of the Union, one other point may be noticed. The distinction between nominal and actual indebtedness can be recognised the more easily in the case of the Union, owing to the circumstance that

[1] See forward for Revenue and Expenditure Statements at pp. 402 and 410.

under the Constitution the railways and harbours, the works in respect of which the bulk of the reproductive loan expenditure has been incurred, are administered separately, and on the lines of an ordinary commercial undertaking, by the Railway and Harbour Board.

The Union Revenue

At the time of writing, the Financial Relations Commission, duly appointed under the Constitution Act to determine what sources of revenue should be assigned respectively to the Union Government on the one hand, and to the Provincial Administrations on the other, has presented its report; and it only remains for the Union Parliament to pass the necessary legislation to give effect to its recommendations. The nature of the proposals of the Majority Report, and the extent to which they were modified by the Union Government, have been stated before.[1]

The finance of the Union is, therefore, at the moment, in a transitional stage; and it is necessary to remind the reader that from the financial year 1912–13 onwards the sources of revenue now to be surrendered to the Provincial Administrations will disappear from the Union Estimates; while the Union expenditure will be lessened by (roughly) one-half of the £3,309,418 voted under the head of "Provincial Administrations" in 1911–12. It must also be remembered that the large contribution from the Railway and Harbour Fund, which the Union Treasurer now receives, is only a temporary source of revenue to be withdrawn, under the Constitution Act, in 1914.

The revenue which appears in the estimates of 1911–12 is, therefore, the combined revenue of the Central and Provincial Administrations; and the subjoined statement of this revenue, showing, as it does, the sources

[1] Part II, Chap. III, p. 159.

from which it is derived, presents a comprehensive and convenient account of the revenual resources and taxation of the Union as a whole.

Estimate of the Revenue for Year ending March 31st, 1912

£16,052,000

Statement showing Sources from which Revenue is Derivable

Head of Revenue.	Taxation.	Services rendered.	Income from Public Estate.	Fines, Forfeitures, Interest, etc.	Total.
	£	£	£	£	£
Customs	4,300,000	1,600	—	400	4,302,000
Excise	295,000	—	—	—	295,000
Posts, Telegraphs, & Telephones	—	1,385,000	—	—	1,385,000
Mining	1,568,000	—	675,000	—	2,243,000
Licences and Taxes on Trades and Vocations	483,000	—	—	—	483,000
Other Licences	58,000	—	—	—	58,000
Stamp Duties and Fees	380,000	150,000	—	—	530,000
Transfers of Property and Successions	506,000	—	—	—	506,000
Personal (Poll) Tax, Natal ..	100,000	—	—	—	100,000
Native Taxes	742,000	—	—	—	742,000
Pass Fees (Native)	355,000	—	—	—	355,000
Land Revenue (Quit Rents) ..	—	—	160,000	—	160,000
Rents of Government Property..	—	—	89,000	—	89,000
Sales of Government Property ..	—	—	100,000	—	100,000
Interest	—	—	—	2,806,000	2,806,000
Departmental Receipts	—	312,000	200,000	—	512,000
Fines and Forfeitures	—	—	—	147,000	147,000
Miscellaneous	1,000	—	79,000	—	80,000
	8,788,000	1,848,600	1,303,000	2,953,400	14,893,000
Contribution from Railway and Harbour Fund	—	—	—	—	1,159,000
Total	—	—	—	— £	16,052,000

As against the receipts for the preceding year (1910–11), the estimated returns from the various sources of revenue show slight normal increases in most cases; but the yield from mining taxation, as estimated here under the New Union Law, gives an increase of nearly £500,000. On the other hand, the yield from the income tax in the Cape Province (£218,000 in 1910–11) has disappeared, since this tax was abolished concurrently with the imposition of a profits tax on mines throughout the Union; and the contribution required from the Railways and Harbours Fund has decreased by £61,000.

The distinction between revenue and taxation, and the character of the several sources of revenue, are, in general, sufficiently indicated in the statement; but the particulars of certain heads of revenue, given below, will be found none the less useful.

Details of Heads and Items of Revenue

Posts, Telegraphs, and Telephones:

	£
Posts	917,000
Telegraphs	313,000
Telephones	155,000
	£ 1,385,000

Mining:

Government Ownership Revenue	
Diamond Mines[1]	202,500
Gold ,,	170,000
Other ,,	11,000
Licences and Mynpacht Dues	291,500
Taxes on Mining—	
Diamond Mines	445,000
Gold ,,	1,075,000
Other ,,	48,000
	£ 2,243,000

Licences and Taxes on Trades and Vocations:

Licences—General	250,000
Liquor Licences	177,000
Bank Note Duty	13,000
Companies' Capital Duty	20,000
Auction Dues	8,000
Taxes on Totalisators	15,000
	£ 483,000

[1] Includes the Government's six-tenths share in the profits of the Premier Mine.

Other Licences :	£
Arms, Game, Dog Licences, etc.	58,000
Stamp Duties and Fees :	
Stamps on Documents	455,000
Cigarette Stamps[1]	75,000
	£ 530,000
Sales of Government Property :	
Land	62,000
Other Assets	38,000
	£ 100,000
Interest :	
Railways Loan Capital	2,196,474
Harbours ,, ,, ...	357,929
Miscellaneous	251,597
	£ 2,806,000
Departmental Receipts :	
Agriculture	126,900
Public Health	5,770
Asylums	16,500
Printing and Stationery	47,500
Agricultural Education and Farms	57,600
Forestry	14,000
Irrigation	15,500
Etc., etc.	[to Balance]
Primary Education and Hospital Fees, etc. :	
Province of Cape	1,250
,, Natal	25,150
,, Transvaal	27,280
,, Orange Free State	19,450
	£ 512,000[2]

[1] This is really an excise levied on cigarettes produced within the Union or in Rhodesia, equivalent to the surtax on imported cigarettes.

[2] *Estimates of Revenue, etc.*, second and final print, Capetown, 1911.

Taxation

It will be seen from the admirably presented statement of the Union Treasurer that only a part of the revenue—in round numbers, £9,000,000 out of £16,000,000, or little more than one-half—is derived from taxation, the remainder being the yield of services rendered by the State and of various forms of State property and rights. But this separation of taxation from revenue does not by itself put us in possession of the amount of actual taxation which is paid respectively by the European population and the native or coloured populations of the Union; nor does it reveal the contribution of the non-South African shareholders in the mines.

Actual *v.* Nominal Taxation

In order to ascertain the amount of taxation borne by the European population, the railway taxation must be added, the native taxation must be withdrawn, and considerable reductions must be made under other heads.

According to the Railway Report for 1910 the net profits for the year were £3,339,583; but in view of the fact that certain reductions of rates had been made, and certain others were in contemplation, at the date of the report, the amount to be added on account of railway taxation may be put at £3,000,000. Of this sum, probably £2,750,000 would come out of the pockets of the European population, while £250,000 would be paid ultimately by the natives.

It is difficult to assign the precise amount of the native contribution to the taxation of the Union. The native taxation is shown in the statement under the two heads of native taxes, £742,000, and pass fees (native), £355,000, making a total of £1,097,000; but it is doubtful whether the pass fees fall ultimately upon the native or upon the European employer. On the other hand, the native contributes under other heads of direct taxation (*e.g.*, licences). On the whole, perhaps, £1,000,000 may be withdrawn as the native contribution to the direct taxation of the Union.

The heads of taxation from which deductions must be made are: Customs (£4,300,000), Excise (£295,000), and mining (£1,568,000). As a purchaser of certain classes of imported goods, the native makes a small but appreciable contribution to the Customs revenue; and he also helps to swell the Excise returns. A deduction of, say, £250,000 may be made, therefore, from the amounts shown under these two heads.

The deduction to be made from the mining taxation is one which affects beneficially the European and native populations both alike. As a large number of the shareholders in the gold, diamond, and other mines are persons resident in England or elsewhere outside the Union, and as the mining tax is merely a levy of 10 per cent. on the profits of the various undertakings, an amount of this tax proportionate to their holdings falls upon these external shareholders, and not upon the people of the Union. In the absence of any statements distinguishing the respective holdings of Union and non-Union shareholders, it is impossible to say what this amount may be; but assuming that one-half of the share capital of the mining companies is held by persons resident outside the Union, then £750,000 must be deducted under this head.

Summarising these adjustments, we have the following figures as a basis for computing the actual taxation paid by the European population of the Union under present conditions:

	£
Taxation Revenue, as shown in 1911–12 Estimates	8,788,000
Add for Railway Taxation, *less* Native proportion..	2,750,000
Total Nominal Taxation	£11,538,000

Deduct for—		
Native (Direct) Taxes	£1,000,000	
Native proportion of Customs and Excise	250,000	
Mining Taxation paid by non-Union Shareholders	750,000	
Total		2,000,000
Total actual Taxation of European Population..		£ 9,538,000

Putting the European population at 1,250,000 (in round numbers) and the actual taxation borne by it at £9,500,000, the Union taxation per head of European population will be £7 12s. It remains to notice that of this £9,500,000, more than two-thirds (the yield of the Customs, Excise, and Railway Taxation), is raised by indirect taxation, while of the direct taxation a considerable proportion (*e.g.*, the stamp duties, and transfers of property and successions) falls almost exclusively upon the property-owning section of the community. It must be remembered, moreover, that the Constitution provides for the entire abolition of railway taxation within four years from the date of the Union. We may conclude, therefore, that the European population of the Union, although the main burden of taxation necessarily falls on them and not on the native population, are none the less as well situated in this respect as their fellow-citizens in the other oversea dominions.

Native Taxation

On the same basis of calculation the native and coloured population pays £1,500,000 of actual taxation, as against the £9,500,000 borne by the European population; but in this case the greater portion is raised by direct taxation. The full and interesting statement of the total contribution of the natives to the State, so far as it is distinguished in the official returns, both central and local, published in the Union Blue Book on Native Affairs for 1910, is here reproduced.

27—(2139)

Summarised Return of Revenue Derived from Native Sources during 1910

Head of Revenue.	Cape.		Natal.	Transvaal.	Orange Free State.	Totals.
	Cape Prop-r.	Transkeian Territories.				
	£ only.	£ only.	£ only.	£ only.	£ only.	£ only.
Poll Tax (in Transvaal Consoli-dated Native Tax)	—	—	42,243	399,248	50,316	491,807
Hut Tax	24,991	89,846	195,663	—	—	310,500
Pass Fees	400	—	3,794	26,769	1,629	32,594
Labour Agents' Licences ..	1,335	1,830	616	5,479	180	9,440
Runners' Permit Fees	1,024	2,645	1 373	10	43	5,096
Marriage Licences	185	195	536	201	19	1,137
Medical Licences	5	15	3,255	—	—	3,275
Native Beer Licences	—	—	69	—	1,001	1,070
Private Location Fees	6,813	1,524	—	—	—	8,338
Registration Fees	4,892	—	—	—	—	4,892
Squatters' Rent	—	494	1,885	9,005	180	11,565
Quit Rent	12,721	10,890	48	48	125	23,790
Rents, etc., Urban Reserve Locations	5,246	7	—	—	—	5,253
Dog Tax	—	—	18,290	4,339	1,254	23,884
Judicial Fines and Fees of Court	11,093	12,671	31,033	3,949	7,787	66,535
Totals (true-showing £, s. & d.) .	68,707 7 11	120,120 10 2	298,763 5 11	449,051 15 9	62,538 6 3	999,181 3 3

Head of Revenue.	Cape.		Natal.	Transvaal.	Orange Free State.	Totals.
	Cape Proper.	Transkeian Territories.				
	£ only.	£ only.	£ only.	£ only.	£ only.	£ only.
Miscellaneous (local Government revenue not brought to account as public revenue)— Hospital Tax	9,633	—	—	—	4,828	14,462
Municipal Revenue from Natives	9,264	—	26,615	13,712	16,096	65,688
Revenue, Native Village Management Boards	285	49	—	—	—	334
Revenue, Divisional Council	5,096	—	—	—	—	5,096
Revenue, Glen Grey District Council (including Dog Tax).	3,902	—	—	—	—	3,920
Revenue, Transkeian Territories General Council ..	—	62,264	—	—	—	62,264
Dipping Tank Tax	954	6,286	—	—	—	7,241
Survey Fees	72	5,893	—	—	5,965	
Remittance Agency Receipts ..	—	—	—	15	15	
Compound Fees..	—	—	—	7	—	7
Rents, Mission Reserves ..	—	—	5,127	—	—	5,127
Scab Licences	—	—	6	—	—	6
Revenue, Natal and Zululand Native Trusts	—	—	322	—	—	322
Refunds, Purchase of Donkeys..	—	—	270	—	—	270
Railway Receipts from Natives (Johannesburg only).. ..	—	—	—	250,000	—	250,000
Totals (true-showing £, *s.* & *d.*).	29,227 12 6	74,493 16 8	32,341 1 7	263,735 17 0	20,924 12 3	420,723 0 0
Add Revenue as above.. ..	68,707 7 11	120,120 10 2	298,763 5 11	449,051 15 9	62,538 3 6	999,181 3 3
Grand Totals	97,935 0 5	194,614 6 10	331,104 7 6	712,786[1] 12 9	83,462 15 9	1,419,904 3 3

[1] The writer has not attempted to correct any slight errors, such as this, in the Union tables.

THE UNION OF SOUTH AFRICA

Expenditure

The General Abstract of the estimated expenditure of the Union Government for the year 1911–12, which is printed below, will serve a twofold purpose. It not only sets out the various services to which the revenue is appropriated, but it exhibits the entire framework of the Union administration in a convenient form.

Estimate of the Expenditure for the Year ending March 31st, 1912

£16,890,281

General Abstract

Vote.	Estimates 1911–12.	Total.
	£	£
The Governor-General and Parliament :		
H.E. the Governor-General	23,776	
Senate	21,365	
House of Assembly	54,415	
Joint Parliamentary Expenses ..	7,967	
		107,523
Prime Minister and Minister of Agriculture :		
Prime Minister	8,983	
Agriculture	742,299	
		751,284
Minister of the Interior, Minister of Defence, and Minister of Mines :		
Interior	229,141	
Public Health	109,292	
Asylums	263,492	
Printing and Stationery	187,239	
Defence	440,699	
Mines	217,906	
		1,447,769
Minister of Justice :		
Justice	61,417	
Superior Courts	184,358	
Magistrates	434,879	
Masters of the Supreme Court ..	26,079	

HEADS OF EXPENDITURE

Vote.	Estimates 1911–12.	Total.
Minister of Justice (continued):	£	£
Police	1,324,510	
Prisons and Reformatories	512,250	
		2,543,493
Minister of Education:		
Higher Education	109,520	
Agricultural Education	95,629	
Forestry	121,945	
		327,094
Minister of Finance:		
Finance	47,015	
Inland Revenue..	36,084	
Audit	60,779	
High Commissioner in London ..	28,228	
Public Debt	4,560,326	
Pensions	419,300	
Provincial Administrations	3,309,418	
Compensation to Colonial Capitals ..	40,682	
Miscellaneous	48,789	
		8,550,621
Minister of Lands:		
Lands	59,766	
Irrigation	122,860	
Deeds Offices	27,992	
Surveyors General	69,135	
		279,753
Minister of Native Affairs:		
Native Affairs	292,323	
		292,323
Minister of Commerce and Industries:		
Commerce and Industries	11,413	
Customs and Excise	157,299	
		168,712
Minister of Public Works, and Minister of Posts and Telegraphs:		
Public Work Department	463,558	
Buildings, Bridges, and Roads ..	472,846	
Post, Telegraphs, and Telephones ..	1,485,305	
		2,421,709
Total	16,890,281	

The total estimated expenditure of £16,890,281 for 1911–12 shows a net increase of £34,178, as against the estimated expenditure for the previous year (*i.e.*, the actual expenditure for the ten months—May 31st, 1910, to March 31st, 1911—*plus* one-fifth of this expenditure as being the equivalent of the two months omitted).

Among the heads of expenditure set out in the Abstract, two only demand attention. The provision for the annual service of the Public Debt, in the vote for the Ministry of Finance, is a nominal figure, since, as already noticed, the £2,806,000 brought into account as "interest" in the revenue statement, must be set against the £4,560,326 shown here, making the charge actually to be provided for (in round numbers) £1,750,000. And in this connection it may be observed that the Debt assumed by the Union Government has been borrowed, as a whole, on favourable terms. The largest single item, the £40,000,000 joint loan of the Transvaal and Orange River Colonies, owing to the Imperial guarantee, bears interest at 3 per cent.; some £27,000,000 have been borrowed at 3½ per cent., £22,522,000 at 4 per cent., and the balance at slightly higher rates. The fact that the provision of £3,309,418 for the Provincial Administrations, which appears in the same vote, will be largely reduced in subsequent years, has been noticed above.

It may be added, in conclusion, that the Union expenditure for the financial year 1912–13, as shown in the estimates introduced in the House of Assembly on January 29th, 1912, amounted to £16,782,343, being a decrease of some £100,000 as against the estimates for 1911-12. An additional £100,000 was provided under the head of "Defence," in view of the proposed legislation for the creation of a citizen army;[1] and the Education

[1] The net annual increase on this head was subsequently (May 6th, 1912) estimated at £500,000. See forward at 447.

Vote was swollen by £145,000 to permit of the extension of elementary education. On the other hand, mainly through the termination of the recent extraordinary expenditure for eradicating the East Coast cattle fever, the vote for the Ministry of Agriculture showed a reduction of some £250,000.

NOTE.—The resignation of Mr. Hull, the first Finance Minister of the Union, was announced by General Botha in the House of Assembly on May 21st, 1912.

CHAPTER II

LOCAL GOVERNMENT AND MUNICIPAL DEVELOPMENT

THE fact that under the Constitution Act the Provincial Councils are charged with the supervision and control of the machinery of local government has been noticed before.[1] But it will be convenient to recall here, that the sixth of the thirteen "classes of subjects" in respect of which the councils are empowered to make ordinances is "Municipal institutions, divisional councils, and other local institutions of a similar nature" (§ 85). And in respect of these "classes of subjects," the Executive Committees of the Provinces have the "powers, authorities, and functions" which pertained to the respective Governments of the four colonies prior to the date of the Union. There will be no break, therefore, in the continuity of the development of local government. Each of the varying systems established prior to the Union will be free to advance under the Union on its own lines, since the authority in whose control it is placed is identical in character with that by which it was created.

The provision is a wise one; for, while the urban centres of population throughout the Union may be said to have reached an equal plane of political and social development, there is a considerable disparity in these respects between the rural populations, both European and coloured, of the several provinces. In the Cape Province, for example, which is almost two

[1] Part II, Chap. II: "The Provincial Administrations," p. 150.

centuries older than any of her sisters, and where, naturally, the greatest progress had been made in the work of educating and civilising the coloured races, local government institutions had been introduced even among the native population. On the other hand, although in the Transvaal local government has not been extended so fully to the rural European population as it has been in the Cape, and has not, of course, been introduced at all among the natives, none the less the Transvaal system, which was an entirely original creation of the Crown Colony Administration under Lord Milner, is more highly developed and efficient than that of the older province. And the Orange River Colony, now the Free State province, was endowed also with the same system, differently applied, however, in accordance with the differing needs and conditions of its more settled but predominantly agricultural population. Subject to these and the like variations, the general framework of local government is practically the same in all four provinces: The municipality for the towns; the divisional council, or its equivalent, for country districts; and the village council, or sanitary board, for growing, but still rural, centres of population.

The excellence of the Transvaal system of local government is to be attributed to the circumstance that the legislation under which it was established was made by men actually engaged at the time, and on the spot, in the working of municipal machinery. Nominated town councils were set up by Lord Milner in Pretoria, Bloemfontein, and Johannesburg while the war was still in progress, and upon the cessation of hostilities other councils and sanitary boards were added. The nominated Town Council of Johannesburg (May 8th, 1901—December 9th, 1903), the most important of these transitional authorities, included not a few of the leading men of a singularly competent industrial community among its members, and some of their number applied themselves to the study of municipal institutions under the guidance and inspiration of the first Town Clerk, Mr. Lionel Curtis, who was himself endowed with a natural genius for administration. British legislation was, of course, examined and largely followed, but a close study was made also of the recorded experience of such municipal models as Birmingham and Glasgow, and full weight was given to local needs and conditions. Although the Johannesburg Town Council called the lesser local authorities to take counsel with it in a Municipal Conference held at Johannesburg in June, 1903, this body, under the direction and oversight of Mr. Curtis (who now held the post of Assistant Colonial Secretary for Local Government in the Transvaal Administration), became the authors, to all intents and purposes, of the Municipal Government Ordinances passed by the Legislative Council of the colony in 1903 and 1904, and of the system of local government thus established and afterwards perfected by subsequent legislation.[1]

In 1904–5 the Municipal Government Ordinances were applied throughout the colony, and before the Crown Colony Administration terminated, thirty-seven municipalities had been established. Further legislation provided for the introduction of Sanitary Committees in districts where the conditions, being semi-urban, made sanitary measures and regulations necessary for the comfort or safety of the inhabitants; and six of these committees were established.

In thus applying the system of municipal government

[1] Upon the establishment of the elective Town Council (in December, 1903), all the members of the nominated Council who came forward as candidates were elected by popular vote, with one exception. This exception, oddly enough, was a member of the nominated Council appointed by Lord Milner to represent the working men of the Rand.

to the towns of the Transvaal as a whole, it was found that whereas some old but comparatively unimportant towns were well endowed with town lands, no similar provision had been made by the late Government in the case of Johannesburg and other populous but recently-founded industrial centres. In these circumstances, the necessity for determining what proportion of the national estate and revenues could be properly assigned to the various localities in need of endowments was recognised on all hands. In 1904, therefore, a Commission was appointed by the Crown Colony Administration to inquire into, and report upon, the subject of the financial relations of the Central Government with the municipalities or other local government authorities. The question was eminently one in respect of which it was desirable that whatever was done should command the general approval of the inhabitants of the colony; and action was deferred accordingly until the advent of Responsible Government. Before, however, the Transvaal was merged into the Union, most of the recommendations of the Commission were put into effect by General Botha's ministry; and, in particular, substantial justice was done to Johannesburg by the grant to the municipality of so much of the public land lying within its area as was not required for any immediate purposes of the Central Government.

The qualifications of municipal electors are the ownership of rateable property assessed at £100; or the occupation of rateable property assessed at £300, or of premises of the gross annual rental of £24. But aliens, coloured persons, and natives are excluded from the municipal franchise. The councillors are elected by ballot; and in this and other respects the customary English procedure is generally followed, with, however, the notable exception that in the Municipalities of Pretoria and Johannesburg the principle of proportional representation is in operation. The system of the "single transferable vote," which is provided in the Constitution for the election of senators and members of the Executive Committees of the Provinces, has been adopted.

In the purely rural districts of the Transvaal the elementary functions of a local authority are still performed by the Resident Magistrate, an official who, in this province, and generally throughout South Africa, is charged with administrative as well as judicial duties. It was customary, however, under the Crown Colony Administration, for the officials of the various departments of the Government, especially those of Agriculture, Education, and Public Works, to seek the advice of persons resident in the various localities; and under Responsible Government formal recognition was given to the practice by the establishment of popularly elected local boards, charged with the duty of advising the department concerned upon matters of education, road maintenance and construction, water supplies, and the enforcement of regulations for the prevention of animal diseases. Under the Constitution, these Boards are maintained, but they now lend assistance either to the Union Government or to the Provincial Administration, according as the matters upon which they severally advise fall within the administrative sphere of the former or the latter. It should be noted also in this connection that the field cornets, the old Dutch local officials whose business is to aid the resident magistrate in securing the due administration of the law in country districts, have been re-established by General Botha's Ministry.

The control exercised over these local Government authorities by the Provincial Administration, as the successor of the Government of the Colony, consists of the usual powers of audit and surcharge, and the right to approve or disallow loan proposals, improvement schemes, and the alienation of municipal property.

Local Government in the Cape Province

In the Cape Province local government is fully developed among the European inhabitants, and, as we have noticed, is being extended tentatively to the native population. The institutions comprised in the system are municipalities, divisional councils, and village management boards. Municipalities were established as early as 1836, twenty years before the advent of Representative Government; but the law under which these institutions are regulated (with few exceptions) to-day—the Act of 1882, which consolidates and amends the earlier legislation—was passed ten years after the grant of Responsible Government to the (then) colony. Divisional councils were first established in 1855, two years after the grant of Representative Government. With the exception of the Cape Division, which returns a council of fourteen members, they consist of six to eight members, elected every three years by owners of fixed property of the value of £500 being also registered as Parliamentary voters, with the resident magistrate of the division as chairman. With the introduction of separate administrative machinery for land and education matters, the business of these councils has somewhat diminished; and to-day their chief functions are to maintain the roads, to decide questions of disputed land boundaries, to initiate and carry out works of public utility by means of loans secured on the rates, to nominate the field cornets, and to return three members annually to serve on the District Licensing Court. Provision for the establishment of village management boards was made in 1881. The adoption of the Act is voluntary, not compulsory; and its provisions are declared to be in force within a specified area upon the petition of the inhabitants concerned. The boards consist of three or five members, elected annually by the Parliamentary voters of the respective areas. They have no power to levy rates; but a village board can request the Divisional Council to levy rates up to 3d. in the £ within its area, and thus raise money for the prosecution of approved undertakings.

In 1909 there were in the Cape Province 120 municipalities, with aggregate receipts (for that year) of £1,164,712, and aggregate payments of £1,228,253; eighty divisional councils, with aggregate receipts of £365,987 (derived from road rates, pound sales, and toll fees), and aggregate payments of £348,427; and eighty-seven village Boards, with aggregate revenues (for 1908–9) of £12,753 2s.

On December 31st, 1909, the 120 municipalities had an aggregate assessed rateable value of fixed property of £54,866,101; a total indebtedness (after payment of sinking funds to date) of £6,478,874; total liabilities of over £7,000,000; and total assets of over £9,000,000, leaving £1,844,728 as the surplus of assets against liabilities. The assets (apart from bank balances, investments, arrears of rates, etc.) are the estimated present value of the property owned by the municipalities. Under this head are included the town halls or municipal offices, markets, houses, drainage works, waterworks, electric works, gas works, gardens and recreation grounds, the plant and stock of municipal undertakings, and street improvements.

The table on p. 421, which, like the foregoing figures, has been taken from the Statistical Register of the Cape Province, will exhibit the resources and financial position of each of the six chief municipalities.

The second table on p. 421 is interesting as showing the tramway service of the capital and other large towns of this province.

Local Government Among the Natives

The first establishment of local government institutions among the native population was the work of

Rhodes, who, in 1894, when Prime Minister and Secretary for Native Affairs, carried the Glen Grey Act through

TABLE SHOWING THE FINANCIAL POSITION OF THE SIX PRINCIPAL MUNICIPALITIES OF THE CAPE PROVINCE

Municipalities.	Receipts in 1909.	Payments in 1909.	Debt at Dec. 31st, 1909.	Assessed Rateable Value of Fixed Property.	Date of Valuation.	Balance Sheets (Surplus of Assets over Liabilities).
	£	£	£	£		£
Capetown	346,856	344,322	2,902,750	13,634,874	1907	706,374
Port Elizabeth ..	122,166	137,020	808,596	6,396,191	1909	175,087
Kimberley	79,649	76,893	43,074	2,161,905	1907	29,814
East London ..	96,883	93,756	382,882	3,039,655	1909	1,606
Grahamstown ..	20,060	23,522	79,308	851,990	1908	—
King William's Town[1]	32,563	112,564	186,675	844,991	1908	189,984

TABLE SHOWING THE MUNICIPAL AND PRIVATE TRAMWAYS IN THE CAPE PROVINCE IN 1910

Name of Company.	District Served.	Length of Line in miles.	Mode of Traction.	Approximate No. of Passengers carried during the 12 mos.	Car Mileage Run.
The Metropolitan Tramways Co., Ltd. The City Tramways Co., Ltd. .. .. The Southern Suburbs of Capetown Co., Ltd.	Capetown, Green & Sea Point, Woodstock, Mowbray, Randesbosch, Claremont, and Wynberg	28·9	Overhead Electric	11,905,561	1,865,481
The Camp's Bay Tramway Co., Ltd.	Camp's Bay, Sea Point, Green Point, portion of Capetown and *viâ* Kloof Road to Camp's Bay	7·8	,,	300,627	158,399
The Port Elizabeth Electric Tramways Co., Ltd.	Town of Port Elizabeth	8	,,	2,978,217	448,450
Victoria Tramway Co., Kimberley ..	Kimberley, Beaconsfield, Kenilworth	2·5 3	,, Horse	365,720	269,000
De Beers Consolidated Mines, Ltd. ..	Kimberley to Alexandersfontein	5·8	Overhead Electric	127,708	76,272
East London Municipal Tramways ..	East London Municipality	5·6	,,	1,322,023	249,855

the Cape Parliament. Under this measure a voluntary system of village and district councils was applied to the

[1] Raised in 1909 £170,771 loan, by issue of stock at 4 % interest, redeemable in 1909.

Glen Grey area and Fingoland, and subsequently extended to other districts within the Cape proper. In recommending the proposal to Parliament, Rhodes pointed out that, by the establishment of European control the natives had lost the occupation furnished by " War and Councils of War," and that as yet nothing had been substituted for it. Under his proposal, however, they would be able to occupy their minds with matters " like bridges, roads, education, plantation of trees, and various local questions." They were to be allowed " to tax themselves " for these local purposes, and under such a system " the Transkei would be able to pay the cost of its own development." The Glen Grey Act provided for the establishment of location Boards and district councils. The former, consisting of three holders of land appointed by the Government " after consideration of the wishes and recommendations of the resident holders of land in the location," could be invested with the powers of the village management boards. The latter were composed of twelve members, of whom six were to be appointed by the Government and six selected by the members of the Location Boards, with the Resident Magistrate of the district as *ex-officio* chairman ; and these councils were empowered to levy rates up to 2d. in the £ on the rateable property within the district, and a rate of not more than 5s. on every adult male native (with certain exceptions).

In the Transkei, at the present time, seventeen out of the twenty-six districts to which the system may be applied, possess district councils ; and a General Council of the Transkeian Territories, which had a revenue in 1910 of £62,264, has been established.[1] The present

[1] A general idea of the financial powers of these Native Local Government institutions in the Cape Province will be obtained by a glance at the table (second part) given in the preceding chapter, at page 408, which shows the native contribution to the local Government revenue of the Union in the year 1910.

condition and the future prospects of these institutions will appear in the following passage, written by the Under-Secretary for Native Affairs in the Cape Province, under date March 7th, 1911, and published in the recently-issued Union Blue Book on Native Affairs.

Local Government: The Native Councils.

More than usual activity characterised last session of the Transkeian Territories General Council, at which the several interests of the population, social, municipal, and agricultural, received a fair share of attention.

On the question of education, several proposals were submitted, which, however, were not likely to meet with the approval of the Government. . . .

The development and maintenance of roads and bridges, for example, should not wait upon the advocacy of a measure of compulsory attendance of children at Native Schools—a measure that is at present neither in accordance with the wishes of the people as a whole, nor consistent with their social requirements, besides being wholly impracticable from a financial point of view, involving, as it would do, an increase of schools and teachers to treble the present number.

But, subject to the moderation of wiser counsel when necessary, the Report continues:

The deliberations of Native Councils continue to serve a useful and important part in the work of local government. The system now embraces seventeen out of the twenty-six districts to which it may be applied in the Transkeian Territories. The later additions are Mount Aylif (1907), St. Mark's (1909), and Mount Frere (1910); while, at the present moment, steps are being taken to extend it also to the locations in the district of Matatiele. . . .

In extending the Council Proclamation to new areas, the policy has been not to thrust it upon the people, but to regard their favourable reception of the measure as a condition precedent to its application. Pondoland has been slow in attaining to the wisdom of its adoption. But it is significant of more enlightened views and the contentment of peaceable administration, that the chiefs of Western and Eastern Pondoland are at present conferring together as to the advisability of their coming voluntarily under its operation.

No definite advance has as yet been made in regard to the establishment of District Councils and a Ciskeian General Council.

28—(2139)

for the administration of local affairs in native areas of the Cape Province proper. The proposal is one that has received consistent support from local magistrates, and it was strongly advocated by the recent Native Affairs Commission, which made several important suggestions as to the steps by which it might most satisfactorily be introduced. Its adoption is only delayed by the opposition of the numerical majority of the inhabitants.

The necessity of placing local affairs in the districts concerned on a satisfactory footing is, however, being more and more keenly felt ; and the question presents itself . . . whether the Government should not assume the responsibility of initiating the measure —which is approved by the more enlightened minority—and undoubtedly for the common good.

The Free State and Natal

The local Government institutions possessed by the Free State and Natal present no features of special interest. The Free State system, subject to the modifications already indicated, is framed upon the same lines as that of the Transvaal, while the mainly British European population of Natal have followed naturally the English models in their local Government legislation. In this latter province there were, in 1910, three other Municipal Corporations besides Durban and Maritzburg, and local boards in seven lesser towns. The aggregate revenue and expenditure of these authorities in the preceding year amounted respectively to £530,686 and £518,154 ; and their aggregate indebtedness was £3,749,538.

The Chief Towns of the Union

The material progress achieved by South Africa since the termination of the War, remarkable as it is in all respects, has not as yet altered the economic conditions which, by making townsmen of the British and countrymen of the Dutch, have kept apart the two nationalities for nearly a century. State effort, such as the establishment of British agricultural settlers in the Transvaal and Free State, and the marked improvement which has taken place in both the profits and amenities of agricultural pursuits, have done something to break down the line of cleavage; but for the present the small towns and villages still remain frankly Dutch, while the large towns and industrial centres are scarcely less characteristically British. The fact that the presence of the native leaves little or no employment available in South Africa for the British emigrant of the unskilled labourer class, and the backwardness of agriculture, have limited most effectively the growth of the British element in the past; and the Union has only one centre of population at all comparable with the great cities of Australia and Canada. These Dominions have each alone four towns of over 100,000 inhabitants. In Australia, where the city growth is the more distinctive, Sydney and Melbourne have populations respectively of 637,102 and 591,830 inhabitants; Montreal, the largest city of Canada, has 466,197; and Toronto, 376,240 inhabitants. The one great centre of population in South Africa is, of course, the Rand; and Johannesburg, together with the lesser municipalities of this district, now form a European community of some 175,000 persons, while the composite coloured and European population numbers approximately 400,000.

The table on p. 426 gives the precise figures for Johannesburg, Germiston, Krugersdorp, Boksburg, and Roodepoort-Maraisburg—the chief municipalities of the Rand—and exhibits the movement of population in these and the other large towns of the Union.

Johannesburg

The municipality of Johannesburg has an area of 81½ square miles, all parts of which, whether built over or not, are connected together by an excellent system of streets and roads. It is furnished with a water supply[1]

[1] This is the joint undertaking of the Mines and the Municipalities of the Rand, whose several interests are equitably represented on the Administrative Authority—the Rand Water Board. The Board supplies water in bulk to the municipalities, and the latter distribute and retail the water to the consumers (other than the Mines).

TABLE SHOWING MOVEMENT OF POPULATION IN SIXTEEN PRINCIPAL URBAN CENTRES

Urban Centre.	All Races. Population: 1911.	All Races. Population: 1904.	Increase (+) or Decrease (−): Numerical.	Increase (+) or Decrease (−): Per Cent.	European or White. Population: 1911.	European or White. Population: 1904.	Increase (+) or Decrease (−): Numerical.	Increase (+) or Decrease (−): Per Cent.	Other than European or White. Population: 1911.	Other than European or White. Population: 1904.	Increase (+) or Decrease (−): Numerical.	Increase (+) or Decrease (−): Per Cent.
Johannesburg	237,220	155,642	+81,578	+ 52·41	120,411	83,363	+37,048	+ 44·44	116,809	72,279	+44,530	+ 61·61
Durban	69,187	67,842	+ 1,345	+ 1·98	31,896	31,302	+ 594	+ 1·9	37,291	36,540	+ 751	+ 2·06
Capetown	67,170	77,668	−10,498	− 13·52	29,933	44,203	−14,270	− 32·28	37,237	33,465	+ 3,772	+ 11·27
Germiston	54,327	29,477	+24,850	+ 84·30	15,589	9,123	+ 6,466	+ 70·88	38,738	20,324	+18,384	+ 90·32
Krugersdorp	53,881	20,073	+33,808	+168·43	13,187	6,946	+ 6,241	+ 89·85	40,694	13,127	+27,567	+210·00
Pretoria..	48,609	36,839	+11,770	+ 31·95	29,660	21,114	+ 8,546	+ 40·48	18,949	15,725	+ 3,224	+ 20·50
Boksburg	43,626	14,757	+28,869	+195·63	11,596	4,175	+ 7,421	+177·75	32,030	18,582	+21,448	+202·68
Benoni *	32,472	—	—	—	8,649	—	—	—	23,823	—	—	—
Roodepoort-Maraisburg	32,458	19,949	+12,509	+ 62·70	7,708	5,266	+ 2,242	+ 46·37	24,750	14,683	+10,067	+ 68·56
Port Elizabeth.. ..	30,676	32,959	− 2,283	− 6·93	18,216	21,987	− 3,771	− 17·15	12,460	10,972	+ 1,488	+ 13·56
Kimberley	29,519	34,331	− 4,812	− 14·02	13,656	13,556	+ 100	+ ·74	15,863	20,775	− 4,912	− 23·64
Pietermaritzburg ..	29,347	31,199	− 1,852	− 5·94	13,759	15,086	− 1,327	− 8·80	15,588	16,113	− 525	− 3·26
Woodstock	28,692	28,990	− 298	− 1·03	18,037	14,674	− 3,493	− 16·22	10,655	7,460	+ 3,195	+ 42·83
Bloemfontein	26,929	33,883	− 6,954	− 20·52	14,760	15,501	− 471	− 4·78	12,169	18,382	− 6,213	− 33·80
East London	24,277	25,220	− 3,943	− 15·63	12,552	14,674	− 2,122	− 14·46	8,725	10,546	− 1,821	− 17·27
Wynberg	16,017	18,477	− 2,462	− 13·31	7,386	9,335	− 1,949	− 20·88	8,632	9,142	− 511	− 5·59

[1] Declared a municipality in 1907. No figures for 1904.

[Compiled from the Preliminary Returns of the census of May 7th, 1911.]

By permission of *W. Basil Worsfold*

RISSIK STREET, JOHANNESBURG, IN 1905

and a system of water-borne drainage; with electric tramway, lighting, and railway services; with telephones, abattoirs, hospitals, parks, recreation grounds, and swimming baths; and its general equipment and its more recent public and private buildings are on the scale of an English town of the first rank. Its advance has been achieved in the face of unfavourable physical and political conditions; but both in the surmounting of such special difficulties, and in the administration of its finances and numerous municipal undertakings, Johannesburg has been well and generously served by individual citizens of public spirit and high ability.

The rateable value of the fixed property within the municipal area is £27,320,275; and the gross revenue and expenditure have risen respectively from £650,796 0s. 7d. and £648,299 16s. in the two years' period—May 16th, 1901, to June 30th, 1903—to £1,438,976 10s. 0d. and £1,504,042 6s. 4d. in the year June 30th, 1910, to June 30th, 1911. The expenditure of this last year, however, was swollen by a sum of £182,984 12s. 7d. appropriated out of revenue to capital expenditure and special payments. The revenue of the municipality comprises: (1) Rates; (2) fines; (3) licence moneys; and (4) charges for the supply of electricity, gas, water, and for sanitary and other services. Of these sources, the rates and charges for supplies and services are the most important. No rate of more than 3d. in the £ can be levied unless the consent of the Provincial Administration has been obtained, and up to the present time the maximum rate has been 2½d. This rate (2½d. in the £) levied on the above valuation produced £287,637 in the year 1910–11. The net profits derived from the various trading concerns of the municipality for the three years 1908–11 are shown on the next page.

The total indebtedness of the municipality is £5,750,000, which is made up of the issue of £5,500,000 Johannesburg

NET PROFITS OF MUNICIPAL UNDERTAKINGS

Year.	Gas & Electric Supply.			Water Supply.			Tramways.			Market.		
	£	s.	d.	£	s.	d.	£	s.	d.	£	s.	d.
1908–9 ..	76,196	4	5	1,476	19	11	44,296	9	5	6,222	19	1
1909–10 ..	88,737	8	11	20,000	15	9	50,953	0	5	7,149	15	9
1910–11 ..	35,508	5	5	27,702	18	1	46,186	19	11	5,913	12	0 [1]

Municipal Four Per Cent. Inscribed Stock and £250,000 Johannesburg Municipal Bills. Provision is made for the repayment of the three issues of Municipal Stock, by means of a redemption fund, on the dates in 1933 and 1934 upon which the principal sums respectively fall due; and a reserve fund has been established to make good the depreciation and obsolescence of the municipality's assets. As the redemption fund amounted to £880,838 8s. 10d. on June 30th, 1911, the net debt of the Municipality at the same date was only £4,869,161 11s. 2d. The cost of Discount on Loans and Flotation Expenses was £315,829 10s. 0d.; but to the £5,184,170 10s. 0d., being the net produce of the loans, and the £250,000 raised by the Municipal bills, a further sum of £403,028 2s. 7d. has been added from revenue and the like sources, making at June 30th, 1911, a total appropriation to expenditure on capital account of £5,837,198 12s. 7d. Of this total, a sum of £5,270,846 6s. 8d. had been spent up to June 30th, 1911, as shown on page 429.

Among the unfinished or newly projected works to which the unexpended balance, shown below, is appropriated, the most important is the Town Hall. The main contract for this building, which involves a total outlay of over £300,000, was accepted by the Municipal Council at a special meeting held on February 7th, 1912. [2]

[1] The net profit on the Market was actually £19,055 3s. 4d., but a sum of £6,473 3s. 7d. was taken therefrom for the establishment of an Art Gallery, and £6,668 7s. 9d. was held in suspense.

[2] Municipal Council of Johannesburg. "Town Treasurers' Report, etc., for the year ended June 30th, 1911." and "Minutes of Special Meeting of Feb. 7th, 1912."

By permission of *W. Basil Worsfold*

THE CORNER HOUSE, COMMISSIONER STREET, JOHANNESBURG, IN 1905

MUNICIPAL EXPENDITURE

EXPENDITURE ON CAPITAL ACCOUNT

	£	s.	d.	£	s.	d.
Public Health				896,153	7	5
Sanitary System	306,946	19	9			
Sewerage ,, ...	511,835	4	6			
Abattoir	51,617	17	7			
Road Construction and Improvements				701,559	5	5
Stormwater Drainage ..	342,036	7	3			
Sundry Properties				249,022	3	3
Klipspruit (Native) Location	48,072	12	7			
Town Lands, etc... ..	136,986	3	2			
Other Undertakings				1,323,622	9	2
Expropriated Area ..	1,145,046	11	5			
Fire Department	75,692	17	3			
Parks ,, ...	51,986	3	2			
Swimming Bath	11,947	6	3			
Trading Departments				2,100,469	1	5
Gas Department	35,776	4	11			
Electric ,, ...	894,448	2	1			
Tramways ,, ...	744,538	5	4			
Water ,, ...	329,792	18	3			
Market ,, ...	95,913	10	10			
				5,270,846	6	8
Balance—Capital Unexpended				566,352	5	11
				£5,837,198	12	7
Add Discount on Loans and Flotation Expenses				315,829	10	0
Total gross expenditure on Capital Account				£6,153,028	2	7

Johannesburg, unlike Capetown and Durban, had no original endowment of natural beauty. With the exception of its incomparable air, more translucent than the atmosphere of Athens and of Cairo,[1] it owes everything—even the woods of the Sachenswald—to the energy of its founders and their successors, the Municipal Council. Happily, the pioneers, captains of industry though they were, were not insensible to the æsthetic deficiencies of their new home. With what good effect they planted, and how quickly the genial nature of the place gave fruition to their efforts, is plain for all to see. Though it is barely twenty-five years since the Witwatersrand was a treeless waste of rolling uplands, to-day the

[1] The writer has sketched in water-colours at all three places (and in Australia).

traveller who looks southward from Hospital Hill upon Johannesburg sees islands of lofty buildings set in a sea of foliage; while northward the groves and gardens of Parktown rise only to sink again into the darker and more ample mantle of green spread by the woods of the Sachenswald. With no conspicuous marks of hill or mountain, no gleaming stretches of river or lake, it is a town that needs to be ennobled by monumental works of architecture, wide spaces arranged for decorative effect, avenues, terraces, and statuary grouped in gardens. With such works, the sun, shining through the dry, clear air of the High Veld, deals kindly, giving full value to their forms by sharp contrasts of light and shade, and striking rainbow hues from their surfaces by its horizontal rays in the morning and evening hours.[1]

Capetown

The site of Capetown is one of those scenes which have come from Nature's mint stamped for all time with the hall-mark of beauty. The traveller, who sees Table Mountain for the first time from the deck of the mail-boat in the early morning, and, therefore, sees it suddenly (since the ship has glided quietly to its moorings in the darkness), is not likely to forget the vision which meets his eyes. The square mass of granite, rising sheer for 3,500 ft. from ground to sky and showing a delicate blue through the morning air, is more like the wall of some gigantic fortress than any mountain that he has seen before. The thing is so straight and solid, so near—it so dominates the eye—that the green slopes of the Devil's

[1] The principal public buildings, erected or to be erected, are: The Town Hall, the Technical Institute, the Law Courts, the Stock Exchange, the Art Gallery, and the new St. Mary's Church. Among private buildings, the Rand Club and the "Corner House" are conspicuous. Bedford Farm, the residence of Sir George Farrar, is a good example of a modern house built in the Afrikander style.

Peak, the rounded masses of the Lion's Head and Signal Hill, the white and brown houses and buildings of the town, and even the bright waters of the Bay seem all to fall aside and leave nothing between it and him.

Table Mountain is, as it were, the head of a mallet, the handle of which is formed by the Table range that runs southward through the Cape Peninsula for some 30 miles to the actual Cape of Good Hope. The peninsula forms with the western coast of the mainland, from which it projects, Table Bay to the north and False Bay to the south. Capetown, looking northward to the equator across the Bay, with the curved wall of Table Mountain like a great refractor at its back, is unpleasantly hot in the summer; and all, or nearly all, of its well-to-do citizens make their homes in the suburbs, travelling each day out and back again to the town by rail or tramway to do their business. These suburbs are built for the most part, on the cool, well-wooded, eastern slopes of the Table range, looking over the sandy flats that join the peninsula to the mainland; but two, Sea Point and Green Point, lie west of the town and front the Atlantic. In climate and natural surroundings these eastern suburbs are scarcely less well endowed than the towns of the Mediterranean coastline of France, and many private gardens at Rondesbosch, Newlands, and Wynberg are famous for the beauty and variety of their flowering shrubs and plants. As the seat of the Union Legislature and of the Archbishopric of Capetown, the city contains the Houses of Parliament and the Cathedral of St. George; nor is it otherwise deficient either in public buildings or in the modern conveniences of town life.

Capetown, moreover, has the grace of antiquity—a quality possessed by no other town in South Africa, and by none in Australia or New Zealand, and shared only with the older cities of Canada. The judicious traveller will find on the shores of Table Bay the "Castle," constructed in 1666–74, when "javelins, bastions, curtains, horn works, and demilunes" were still the fashion; the *Stadthaus*, in Green Market Square, built in the time of Governor Tulbagh; and the tower of the "Town Church," furnished in 1727 with a clock sent out from Holland; and in these buildings, and in the eighteenth century houses of the older streets, he will catch glimpses of the quaint, constrained, isolated life led for more than a century by the little European community planted by the Dutch East India Company in 1652. While, further afield, in the fair valley of Constantia and dotted up and down the Peninsula, some of the best examples of the Afrikander homestead of the seventeenth century—Groot Constantia, Alphen, and Tokay—are to be seen among their avenues of oaks, their trim vineyards, and their glory of peach blossoms. Here, too, at Rondesbosch, is Groote Schuur, once the home of Rhodes, now by his gift the official residence of the Prime Minister of the Union. From the gardens of Groote Schuur, now enriched by Watts' fine sculpture of "Physical Energy," which forms most suitably the chief member of the Rhodes Memorial, the eye ranges through the screen of pine woods across grey-green and silver stretches to the blue peaks of the Hottentots' Holland and the vast interior of Africa. With this vision before her, Capetown sent her greeting to England:

> Hail! Snatched and bartered oft from hand to hand,
> I dream my dream, by rock and heath and pine,
> Of Empire to the northward. Ay, one land
> From Lion's Head to Line.

Pretoria

Among the towns of South Africa, Pretoria is the favourite of fortune. It owes its position as the seat of government to its convenient nearness to the Rand, the commercial and industrial centre of the Union. The town, which was laid out in 1855 on the northern bank of the little Aapies River, lies more than 1,000 ft. below the level of Johannesburg, and it has a temperature some 5° higher; the respective average means for the year being 64° and 59°. The suburbs, however, which are built on the slopes of the encircling hills, are appreciably cooler than the town itself. Prior to the establishment of the Crown Colony Administration, the respective claims of Johannesburg and Pretoria to be made the seat of government were debated for some time; but Mr. Chamberlain at length decided in favour of the latter; and the town, then endowed for the first time with municipal institutions, developed rapidly in the interval between the war and the Union.

During this period, Pretoria was the seat of the Transvaal Legislature and Judicature, and of all the administrative departments of the Government, with the exception of those of Mines and Native Affairs, which were established at Johannesburg. It was the head-quarters of the General Officer commanding the Imperial forces in South Africa, and it was here that the new Government House, required for the High Commissioner and Governor of the Transvaal and Orange River Colony, was built. Moreover, the establishment of the Central South African Railways' shops here in 1902 brought an appreciable increase of European population to the town.

At the date of the Union, the rateable value of the fixed property within the municipality was (in round numbers) £7,500,000, augmented by some £1,500,000 of Government property, in respect of which a grant-in-aid was paid in lieu of rates. The annual revenue (1909–10) was £211,000, the expenditure £187,000, and the debt of £1,000,000 4 per cent. Inscribed Stock, redeemable within thirty years (1939), was covered by municipal assets of the value of £1,150,000. The Council have power to levy a rate of 3d. in the £, but up to the present no rate higher than 2d. has been imposed.

To-day Pretoria is being furnished with the more ample equipment demanded by its new dignity and importance. With this end in view, both the Municipality and the Union Government have provided funds for a large capital expenditure. In 1910 the former obtained powers to raise a further million sterling upon the same terms as to interest and redemption of capital as the first. The works and undertakings for which the new loan was required include an increased water supply, sewerage, road construction, the canalisation of the Aapies River, tramways, abattoirs, and markets; and are, for the most part, of a directly reproductive character. The Government, on their side, have made provision for the erection of the following buildings at an approximate total cost of £1,500,000:

Union Government Buildings.
Post Office.
Museum and Government Library.
National College of Agriculture.
University College.
Railway Station.

In contemplating this list, the mind is carried back to the sudden change in the fortunes of Pretoria brought about twenty years ago by the establishment of the gold industry of the Rand. In 1889 the Volksraad met in a little thatch-roofed building. Three years later this meagre council chamber had been replaced by the new Raadzaal, erected at a cost of £200,000, and comparable in its spaciousness and architectural merit to any of the then existing public buildings of South Africa.

The Union Buildings, for which it will be remembered that a provision of £1,130,000 was made in the Estimates for 1911–12, are designed upon a scale which will make them almost as great an advance upon the Raadzaal as was the Raadzaal upon the original thatch-roofed home of the Volksraad. This stately edifice, as planned by

Mr. Herbert Baker to give expression to the Union of South Africa, is to be placed on the side of Meintjes Kop, the highest and terminating point of the spur that runs westward towards the town from the high ground of Bryntirion and the red roofs of the Governor-General's official home. The actual site recalls vaguely the entrance to the Acropolis of Athens. The slope of the hill is interrupted by a stretch of level ground, which, at a point near to the town, is itself broken by the passage of a ravine. On reaching the level ground, the ravine first widens, forming a hollow, in which the waters of the stream were once held up by a dam, and then closes again to climb upwards to the crown of the hill. On each side of the ravine, and on this shelf of level ground, Mr. Baker places a monumental block of public offices, and then unites them by a semicircular colonnade which follows the backward sweep of the hollow. The floor of the amphitheatre thus formed is to be laid out in gardens, and decorated with terraces, fountains, and statues.

The material of construction is the white stone of the country, and the dominant architectural forms are those of the Italian Renaissance. Each block of offices surrounds an open courtyard—an arrangement which secures coolness and an economic disposal of all available space—and their junction with the colonnade is marked on each side by a graceful cupola. The two lateral masses, though wedded to the hillside by the colonnade, present a lofty frontage to the town. Each block rises from a base of solid masonry, and the lines of its façade, broken by boldly projecting and richly columned members at either end, are emphasised by massive cornices and deep-set windows.

Meintjes Kop is only a mile from Church Square, the centre of Pretoria, and thus placed the Union buildings will be seen from every quarter of the town; while the prospect from the site itself is varied and far-reaching.

Beneath the spectator lies the town like an unfolded map ; westward, his eye follows the range of the Magaliesberg until—30 miles away—it sinks into the horizon, and southward it travels over curving uplands to the ridge behind which lies Johannesburg and the Rand.

Durban

To have won free access from the sea to this fine natural harbour by sheer determination, backed by engineering skill, is an achievement which would do honour to a far more numerous community than the little Colony of Natal. But it is unnecessary to repeat here what has been written in an earlier chapter[1] of Durban the port, and the admirable equipment of its harbour and docks. Durban, the town, is, perhaps, the pleasantest of all the large towns of the Union, and in its physical aspects and its social atmosphere it is suggestive rather of India than South Africa. The town itself is spread on the level ground around the Bay, and is laid out with wide streets, and well furnished with handsome buildings. It has the customary public offices and tramway service, parks and recreation grounds, a theatre, Turkish and swimming baths, and botanical gardens with a fine display of tropical growths. The residential quarter lies on the Berea, a ridge of high ground raised some three or four hundred feet above sea-level. Here, among the luxuriant foliage and brilliant hues of tropical plants, the chief citizens of Durban have made themselves delightful homes ; and to the near beauty and fragrance of their avenues and gardens is added a landscape of rare quality. For, from the Berea, we see spread out before us the town, the harbour with its shipping, the quiet waters of the Bay, the stretch of silver sands that mark the Point, the bold, tree-clad

[1] Part IV, Chap. II, p. 256.

headland of the Bluff, and beyond the blue line of the Indian Ocean. Durban is hot, the mean temperature for the year being 72° ; but the clubs and hotels are cool, and the needs of the visitor are well supplied by linen-clad Indian and native servants.

CHAPTER III

LAW, ORDER, AND DEFENCE

JUSTICE is administered in the Union on the dual basis of Roman-Dutch and English law. As the Cape was a Dutch Colony, the law of the land up to the time of the permanent British occupation, in 1806, was the Roman-Dutch common law of the Netherlands, modified, however, by local enactments. These latter comprised both the ordinances and placaats of the Governor and Council of the colony, and—since the Cape was a part of the administrative system of the Dutch East Indies—the "Statutes of India," that is to say, so much of the enactments of the Governor-General and Council of India, seated at Batavia, as was applicable to the Dutch East India system as a whole. In the British, as in the Roman Empire, it has been the practice to recognise the religion, customs, and law of the incorporated peoples; and, in accordance with this practice, the Roman-Dutch law was maintained in the Cape Colony.

None the less, the British occupation brought a change. The local Dutch statute law was abandoned perforce as obsolescent, and replaced almost entirely by local enactments based upon the existing circumstances of the colony or founded upon English statutes, and the Roman-Dutch common law, broadly speaking, came to be administered concurrently with English common law. Nor was it surprising that, with judges and advocates alike versed in the decisions and practice of the English Courts, English principles were more and more closely woven into the fabric of the Colonial law. And apart from the influence of the "case-law," thus built up through the Colonial Reports, circumstances—or rather its greater capacity to satisfy the conditions of modern life—gave the regulation of the field of commercial intercourse almost exclusively to English law. This gradual adoption of English legal principles and statute law would naturally have been a regular accompaniment of the expansion of the European race in the sub-continent; but the process was interrupted by the change of policy, which, in the middle of the last century, led, in Sir George Grey's words, to the "dismemberment" of European South Africa. From the establishment of the two Dutch Republics, and the erection of Natal into a separate British colony, has arisen what may be termed the "unequal penetration" of the Roman-Dutch law by English law. Not only is the statute law different in each of the four provinces, but the case law is divergent: that is to say, even where the same statutes and principles of law have been recognised, a different interpretation has been placed upon them by the judges of the respective colonies.

Thus, two processes are necessary to make one law for the Union as a whole. The Legislature must consolidate the statute law, and the Court of Appeal must reconcile the case law, of the four provinces. The Union Parliament has made already an energetic beginning of the first process; since eleven out of the forty-five enactments of the first session (1910–11) were, directly or indirectly, consolidating laws. That the Court of Appeal will not fail to carry out the second process would appear from the words used by the Chief Justice in a recent case.[1] The Natal judges had held that a landlord's lien (hypothec) held good against a tenant's goods, even when removed from the premises to the hands of an innocent third party, provided there was a "quick pursuit." No other South African court, however, had accepted the Natal doctrine, and the question for the Court of Appeal (said Lord De Villiers) was whether it

[1] *Webster* v. *Ellison;* heard early in 1911.

should be maintained " so far, at all events, as the province of Natal was concerned." His answer to this question is an emphatic assertion of the unifying mission of the Court of Appeal.

There is no reason, either of convenience or justice (he said) for applying the maxim *communis error facit jus* to a case like the present. Under the South Africa Act, the laws of the different provinces are preserved, but this Court as the Court of Appeal for the Union, cannot support any doctrine introduced into any particular province merely because it has found general acceptance there for a certain number of years. If the doctrine is at variance with the common law of South Africa, and if its disuse in the province in which it has been applied would not unsettle past transactions, it should not be upheld.[1]

That the task, thus begun, should prove to be lengthy as well as arduous, is only to be expected. Apart from the statutes and the case law, the existing differences of legal system and procedure, which include such matters as the powers and conditions of service of magistrates, and variations in the remuneration assigned to jurors and witnesses, have to be removed.

Writing from this point of view, and having in mind the financial and administrative economies to be secured, the Secretary for Justice remarks:[2]

The first years of Union will require an immense amount of harmonising and unifying, the continuation of the four divergent provincial systems prevailing at the date of the Union being neither expedient nor economical. In a number of directions, however, these systems are fixed by statute, and it will require a Union Act to sweep them away. Economy of administration can, therefore, necessarily only be gradual.

THE UNION DEPARTMENT OF JUSTICE

An excellent account of the work of the Ministry of Justice is given in the Report from which the above passage has been taken:

The Department of Justice, controlled by the Minister of Justice, embraces for administrative purposes the Superior

[1] 4 Buch. A.C., p. 343.
[2] Annual Report of the Department of Justice for 1910.

Courts, the Magistrates' Courts, Masters of the Supreme Court, the Police Sub-Department, the Prisons Sub-Department (including reformatories and industrial schools), the Sub-Departments of Patents, Trade-marks, Copyrights, Designs, and Companies, also Special and other Justices of the Peace, and Liquor Licensing Courts. It administers an expenditure of £2,543,493 of the public moneys per annum and controls 12,125 officers. At the date of the Union it had the following head office staffs in the various provinces: 51 in the Transvaal, 35 in the Cape, 18 in Natal, and 11 in the Orange Free State. The Superior Courts in the Provinces consist of the following: 9 judges in the Cape, 8 in Natal, 7 in the Transvaal, 3 in the Free State, and an Appeal Court, or Appellate Division of the Supreme Court at Bloemfontein, consisting of the Chief Justice of South Africa and two permanent or ordinary judges, and two additional judges of appeal. The Superior Courts staffs in the Provinces consist of 44 officers in the Cape, 37 in Natal, 39 in the Transvaal, 11 in the Free State, and 3 attached to the Appeal Court. Under Superior Courts the two Union executioners are also classed. With regard to magistrates, the Department controls 91 magistrates, 23 assistant magistrates, and 269 magistrates' clerks in the Cape; 26 magistrates, 17 assistant magistrates, and 259 magistrates' clerks in the Transvaal; 43 magistrates, 5 assistant magistrates, and 237 magistrates' clerks in Natal; and 25 magistrates, 12 assistant magistrates, and 122 magistrates' clerks in the Free State; or for the Union a total of 185 magistrates, 57 assistant magistrates, and 887 magisterial officers. With the advent of Union, four Attorney-Generals were appointed in the four Provinces, in whom, in terms of the South Africa Act of Union, the prosecution of crimes and offences is vested. The criminal staffs of the former Attorneys-General have been attached to them. The Secretary to the Law Department of Natal was appointed Attorney-General in Natal, the Secretary to the Law Department (Orange Free State) was appointed Acting Under-Secretary for that Province, and the Secretaries of the Cape and Transvaal Provinces were respectively appointed Acting Secretaries for Justice for the Union, Legal and Administrative.

Justices of the Peace

In addition to the 185 magistrates and 57 assistant-magistrates mentioned in the above statement, who are, of course, salaried officials, and whose duties are frequently administrative as well as judicial, there are a large number of justices of the peace. They are of two classes: Special or resident, and justices without judicial functions.

The justices of the first class are paid for their services, and they have jurisdiction in criminal cases. At the time of the Union their powers varied in the four provinces, ranging from power to inflict a maximum fine of £25 and a maximum term of imprisonment of two months, to power to inflict a maximum fine of £2 and a maximum term of imprisonment of one month. Of this class of justices, in 1910, there were 30 in the Transvaal, 17 in the Cape, 16 in the Free State, and 3 in Natal; or 66 in all.

The justices of the second class are much more numerous, but their duties are confined practically to attesting sworn statements, and, when appointed, sitting as members of Liquor Licensing Courts. In 1910 they numbered 4,132 in all, and were distributed among the several provinces fairly evenly on a basis of the respective populations.

The Police

The Police are a sub-department of the Ministry of Justice; and in 1911–12 numbered 8,743 men, maintained at an annual cost of £1,324,510. Under the Police and Defence Acts of 1912, however, the Force will be reconstituted, and in particular the mounted element will form the basis of the Permanent Defence Force of the Union. The Police Force, as provided for in the 1911–12 Estimates, is as follows:

The headquarters staff consists of the Chief Commissioner of Police for the Union (who is in the present instance the Commissioner of Police for the Transvaal Province), a Secretary, a Chief Paymaster and Accountant, and a Controller of Supplies and Transport.

The Police of the Cape Province are divided into: (1) The Cape Mounted Police; (2) the Urban Police, serving the Metropolitan district of the Cape Peninsula; (3) the

Rural Police ; and (4) the Diamond Detective Department at Kimberley. Of these bodies, the first is, with the exception of some 30 men, a distinctively mounted force ; it has 391 European privates, 132 native privates, and a total strength of 734 men, with the Commissioner of Police for the province at its head. In the Urban Police are included 46 sergeants, 319 constables, 32 detectives, and a finger-print expert. Of the Rural Police—about 1,000 strong—328 are district mounted troopers and 157 coloured constables. The Diamond Detective Department comprises the Chief of the Detective Department, who is also Commissioner of Urban Police for the Kimberley District, a female searcher, 16 sergeants, and 88 constables, etc.

The Police of Natal include a Chief Commissioner for the province, inspectors, detectives, 68 non-commissioned officers, a force of 594 troopers, 100 Indian and 104 native constables. There are two special bodies : The Railway Police and the Water Police. The cost of maintaining the former is shared between the Government and the Railway Commissioners, and it consists of a superintendent, 11 European and 34 native constables. The Water Police, whose services are required for Durban harbour, comprise 22 European and 64 native constables, etc.

The Chief Commissioner of Police has at his disposal in the Transvaal Province, 3 Deputy Commissioners, 15 inspectors, 23 sub-inspectors, 34 superintendents, 161 sergeants, 93 corporals, 1,776 constables, 72 detectives, 30 shoeing smiths, 13 female warders, etc. A large proportion of the police in this, and in the Free State Province, are drawn from the ranks of the South African Constabulary—the mounted and semi-military force which, under the Crown Colony administration, had attained a high state of efficiency. There is also a large native establishment, which is required to maintain order both in the native districts and among the numerous bodies of native labourers employed in the industrial centres. It comprises 47 head-constables, 107 corporals, 912 constables, and 169 labourers, drivers, etc.

The police force in the Free State Province is composed of the Commissioner of Police for the province, a sub-commissioner, 22 superintendent head-constables, 3 head-constables, 17 sergeants, 35 corporals, 397 constables, and 17 detectives; together with 178 native constables, 10 native detectives, and 50 native drivers and labourers, etc.

The Ministry of Defence

The bill embodying the proposals for the creation of a Citizen Army, prepared by General Smuts, who is Minister of Defence as well as Minister of the Interior, passed its third reading in the House of Assembly on May 6th, 1912. Before discussing these proposals, however, it will be useful to notice the nature of the defence forces which will be superseded by the new Defence Act.

Again taking the Estimates as the source of information, we find that the sum voted for the defence of the Union in the year 1911–12 was £440,699. The Forces for which provision was made were: The District Head-quarters Staffs, the Cape Mounted Riflemen, the Militia and Volunteers, and the Cadets. And, exclusive of this provision, various sums were voted for "special services." Of these forces, those comprised under the first two heads represented, with the exception of the Cape Garrison Artillery, the permanent and professional element. The strength of the District Head-quarters Staffs was 266, and the total strength of the Cape Mounted Rifles was 736—roughly, 1,000 in all. The C.M.R. was composed of a colonel commanding, a major, 5 captains, 18 lieutenants, 550 warrant and non-commissioned officers and riflemen, 70 native troopers, 63 native detectives, 20 native artillery drivers, a paymaster and quartermaster, an instructor in gunnery and officer commanding the artillery troop, a surgeon-captain, a bandmaster, and four native hospital attendants. The annual cost of these two forces was respectively £64,981 and £126,093.

The Militia and Volunteer Corps, which the Union took over from the four colonies, included a large percentage of men who had fought in the war, and, in particular, the survivors of the fine bodies of Irregular troops raised from among the dispersed British population of the Rand. The provision for these voluntary forces—£111,370—included: Pay and allowances, £5,640; capitation grants, £29,500; horse and special allowances, £37,440; ranges, £5,140; camps of exercise and bivouacs, £13,620. And a further provision of £5,250 for grants to rifle associations, out of which £1,000 was allocated to the Transvaal Bisley, was made under the head of "Special Services."

For the Cadets, who numbered some 11,000 in 1911, £19,645 was provided. This total was made up of: Capitation grants, £13,000; ammunition, £2,350, and camps of exercise, £2,175. The Cadet system, it may be added, has been developed with great success in Natal, and to a less degree in the Transvaal.

Under the head of "Special Services," £85,000 was provided as a contribution to the British Navy, £18,160 for the fixed defences of Table Bay, and a non-recurrent £10,000 to defray the expenses of the Union contingent sent to attend the Coronation of their Majesties, King George V and Queen Mary.

The Citizen Army

It is hardly necessary to remark that the European inhabitants of the Union, some 1,250,000 in all, comprise an exceptionally large proportion of men experienced in the actual business of war. The splendid devotion with which the entire male population of the former Republics, with the exception of very young boys and very old men, took the field is not likely to have been forgotten; but what, perhaps, is less fully realised is the fact that of the small British population of South Africa, no less than 46,858 fought for the Empire.[1] The Union, therefore, has, perhaps, better material for the creation of a national army, proportionate to its population, than any other civilised country.

The scheme embodied in the Defence Bill is said to have been modelled upon the Swiss system, but, however this may be, it is certain that the Imperial authorities—both central and local—have been taken into consultation; and in formulating his proposals, General Smuts has had the active assistance of Field-Marshal Lord Methuen, the term of whose service as Commander-in-Chief of the Imperial Forces of South Africa was prolonged for the purpose, and of Rear-Admiral Bush, the Commander of the South African Naval Station. The basis of the scheme is, of course, the recognition of the obligation of every able-bodied citizen to bear arms in defence of his country, and of the corresponding duty of the State to provide its citizens with the training, without which this obligation cannot be effectively discharged. The measure itself, however, provides for only a partial application of this principle. In the first place, the defence of the Union by sea is to be entrusted for all practical purposes, and, as heretofore, to the Imperial Navy; and, in the second, not the whole of the available manhood of the country, but only so much of it as, in the opinion of the Government, is sufficient to meet the

[1] The British Irregulars *raised* in South Africa were even more numerous, being, according to the Report of the War Commission, between 50,000 and 60,000. See *Lord Milner's Work in South Africa*, 1897–1902, Chap. VII, p. 324. (By the writer, Murray, 1906.)

requirements of the moment for land defence, is to be placed under effective military training. As the result of these limitations, the Union will be provided for the present with a small Permanent Force and Citizen Force, together capable of dealing successfully with any disturbances, concerted or otherwise, likely to arise among the native population, but not calculated to repel a foreign invader advancing by land, or one who may have effected a landing after defeating, or eluding, the Imperial Navy. On the other hand, General Smuts' proposals, although they contemplate only a limited application of the root principle of the scheme, have two undoubted merits. They make a judicious and immediate use of the existing military and semi-military material and organisations; and they provide, though they do not set in motion, the machinery for the more complete utilisation of the European fighting strength of the country. Moreover, if financial considerations are to be brought into the account, General Smuts' proposals must be credited with the further merit of economy, since it is estimated that the annual provision required to give effect to them will not exceed the existing provisions for defence and for the mounted element in the police forces by more than some £500,000.[1]

The Defence Forces of the Union, as established under the South Africa Defence Act, 1912, are:

I. The Permanent Force.
II. The Coast Garrison Force.

[1] In moving the second reading of the Defence Bill on February 2nd, 1912, General Smuts told the House of Assembly that the total annual cost of the scheme would be £1,172,000, of which £852,000 was provided already for existing services. But at the third reading (May 6th) he stated that the payment of the members of the Active Citizen Force, and the establishment of an Aviation School (amendments made in Select Committee), would raise the net additional expenditure to an approximate £500,000 per annum.

III. The Citizen Force.
IV. The Royal Naval Volunteer Reserve.
V. The Veteran Reserve and any other Special Reserve established under the Act.

Military Service and Peace Training

The following obligations in respect of military service and peace training are laid upon the citizens of the Union by the Defence Act :

Nature of Obligation.	Citizens Liable.
1. To render in time of war personal service in defence of the Union in any part of South Africa, whether within or outside the Union.	All between 17 and 60 years of age (both inclusive).
2. To complete four years' peace training for military service in the Defence Forces of the Union.	All between 21 and 25 years (both inclusive).
3. To serve as a member of a Rifle Association from their 21st year for four consecutive years.	All liable to the four years' peace training, who in their 21st year have not been entered for that training.
4. To be trained as a Cadet in appointed urban, and other populous areas where proper facilities for such training can conveniently be arranged.	All boys between 13 and 17 years (both inclusive) ; subject however to the principle of parental option.

The Act further provides :

(*a*) In respect of obligation 3 (rifle training), that every citizen liable to War Service (Obligation 1) may engage voluntarily to serve as a member of a Rifle Association.

(*b*) If a South African Division of the Royal Naval Volunteer Reserve is established, that every citizen entered for service in such Division will be deemed to fulfil Obligation 1, and every citizen trained in the Division will be deemed to fulfil Obligation 2.[1]

THE PERMANENT FORCE

The Permanent Force, for which the Act makes provision, consists of:

(*a*) Persons engaged for continuous service in the organisation and training of the Defence Forces; and

(*b*) Mounted troops charged in time of peace with the maintenance of order within the Union.

The mounted troops are to comprise five or more regiments of South African Mounted Rifles, and any other regiments or corps for which Parliament may provide. The Cape Mounted Riflemen will form the first regiment of the new force, and the Cape Mounted Police, the Cape District Mounted Police, and the Natal, Transvaal, and Orange River Colony Police will provide the material of the remaining regiments. Any portion of these mounted police forces that is not required for the Permanent Force will be transferred, if necessary, to the South African Police (established under the Police Act of 1912). Each of the regiments of the mounted troops is to be so organised that a permanent battery of artillery forms a component part of it, and that as many members as possible of the regiment may be trained efficiently in field artillery duties, in addition to their duties as mounted riflemen. In time of peace, each regiment will be assigned to a particular district of the Union, and the South African Mounted Riflemen, as a whole, will be employed in maintaining order; and for the discharge of this duty each member will have the

[1] Draft Bill [A.B. 7–'12] printed by the Union Government. These obligations are, of course, subject to the customary exemptions (ministers of religion, Members of Parliament, etc., and medically certified unfitness). In this account of the Defence Act the writer has endeavoured to incorporate the amendments made in Select Committee, as reported by cable in *The Times*, into the original Draft.

customary powers and obligations of a police officer or constable.

The Coast Garrison Force

The purpose of this force is to provide for the efficient protection of the defended ports of the Union. It is to consist of the South African Garrison Artillery, the South African Coast Defence Corps, and any other corps for which Parliament may make provision. The first division of the Garrison Artillery will be formed by the existing Volunteer Corps of the Cape Garrison Artillery ; and the Coast Defence Corps is to be composed of citizens trained specially to engineer, signalling and telegraphy, or harbour control duties. Provision is made in the Act for (1) the payment of members of this force who have attained proficiency ; and (2) the training and acting of the force as a whole, if and when necessary, in conjunction with H.M. Regular Forces stationed in the Union.

The Citizen Force

The Citizen Force is to consist of all persons liable to military service not being members of the Permanent Force, the Coast Garrison Force, the Veteran Reserve, or the Royal Naval Volunteer Reserve. It comprises three divisions:

I. The Active Citizen Force.
II. The Citizen Force Reserve.
III. The National Reserve.

The members of this Force, with the exception of the Active Citizen Force during training, are not entitled to be paid for their services during peace ; but they may receive such pay and allowances, etc., as may be prescribed.

The Active Citizen Force

The Active Citizen Force is to consist of men between the ages of 21 and 25 (who have been entered for, or are undergoing, the four years' peace training. The four years of training in all cases must be consecutive, but permission to complete one or more additional years may be obtained. The force is to be organised by units of the various combatant arms (Artillery, Engineer, Mounted Rifles, or Infantry), and by departmental services (Medical, Transport and Supply, Veterinary or Ordinance), as may be established. And, as far as practicable, the organisation of every unit of a combatant arm or departmental service is to be arranged territorially, in order that the several units may be composed of citizens residing in the same districts, and that a group of units in the same or adjoining districts may constitute a field force properly equipped to take the field.

The length of the annual training in camp is 30 days for the first year, and 21 days for the three following years. Subject to this the nature and number of the drills, rifle or gun practices, camps and manœuvres, to be performed by each member of the Active Citizen Force (or the Coast Garrison Force) are to be prescribed by the military authorities of the Union.

In moving the second reading of the Act, General Smuts stated that the total strength of the Active Citizen Force would be fixed at the outset at 24,000, with a yearly levy of 6,000 recruits to balance the passing out of the same number of men into the Citizen Force Reserve.

For administrative purposes, the Act provides for the establishment of a Council of Defence under the Minister of Defence, a Head-quarter Staff of the Defence Forces, and for the division of the Union into military districts and sub-districts, in each of which Instructional and Administrative Staffs are to be appointed as required. The Council of Defence consists of four members appointed by the Governor-General in Council and the Minister of Defence. It is to meet within one month of the passing of the Act, and henceforward once at least in every six months. It must be consulted and advise collectively upon all questions arising out of the administration of the Act, and upon the appointment and dismissal of officers.

The steps by which the Active Citizen Force is first to be brought into being, and then maintained at its required strength, are these. All existing militia and volunteer corps are to be embodied and trained as part of the Active Citizen Force, provided (*a*) a due proportion of the members of the respective corps engage to serve and undergo training for not less than twelve months; (*b*) a due proportion of other citizens enter voluntarily for the four years' Peace Training. If these conditions are fulfilled, the identity of these existing corps, with their honourable traditions of past service, will be preserved. Any corps in respect of which these conditions are not fulfilled will be disbanded as prescribed in the Act; but provision is made for allowing any members of existing Militia or Volunteer corps, or of Reserve or Cadet organisations, to be transferred at their own request to the Active Citizen Force, or an equivalent Defence Force. During the month of January next following the coming into operation of the Act, and during the same month in subsequent years, all citizens of the age of 21 will be required to register themselves. On the succeeding April 15th in each year the number of citizens liable to be called upon for Peace Training in each military district or sub-district, in order to keep up the full strength of the Active Citizen Force, will be proclaimed in the Government *Gazette*. If the required number is not obtained by voluntary offers of service, then the deficiency will be made good by a compulsory levy from the roll of citizens as thus registered; and if the number on the roll is greater than the deficiency to be made good, citizens to the requisite number will be chosen by ballot. It is anticipated, however, that the 6,000 recruits annually required will be obtained by the voluntary enlistment of

Cadets and other citizens, and that the need to put in force the provisions for compulsory enlistment for Peace Training will not arise.

The Citizen Force Reserve

The Citizen Force Reserve is to be constituted of two classes: *A* and *B*. Of these, the former consists of all citizens not older than 45 who have undergone the four years' Peace Training, and the latter, of all citizens not older than 45, who are serving, or have completed their service, as members of a Rifle Association. Members of Class *A*, unless engaged for the Field Reserve of the South African Mounted Riflemen, are to be reservists to the unit of the Active Citizen Force, in which they received their Peace Training. Members of Class *B*, unless engaged for the Police Reserve of the South African Mounted Riflemen, are to be reservists to the unit of that arm or service for which their avocations and circumstances respectively render them most suitable.

All members of this Force are liable, when called upon, to present themselves once annually for inspection.

The National Reserve

The National Reserve is to be constituted of all citizens, who, not being members of any other portion of the Defence Forces of the Union, are not younger than 17 nor older than 60, and are, therefore, liable to military service in time of war. In the event of this Force being called out, each member is required to serve as he may be assigned, whether in a unit of the combatant arm or in a departmental service, or otherwise.

The Royal Naval Volunteer Reserve

Provision is made for raising a body of Volunteers in pursuance of the Colonial Naval Defence Acts, 1865 and 1909 (Great Britain and Ireland), to be entered on the terms of being bound to general service in the Royal

Navy in case of emergency. If, and when, raised, this body will be called the South African Division of the Royal Naval Volunteer Reserve, and will form a part of the Royal Naval Reserve constituted under the Naval Forces Act, 1903 (Great Britain and Ireland). The Division will be maintained by the Union Government, by whom it may be placed at the disposal of His Majesty.

The Act also makes provision for the formation of a Reserve of Officers for the Defence Forces, a Veteran Reserve, and for the establishment of Field and Police Reserves for the South African Mounted Riflemen. Of these two latter, the first is to be composed of citizens (1) who have completed their period of engagement as active members of the Permanent Force; or (2) who are enrolled as members of Class *A* of the Citizen Force Reserve, or who, being suitable, engage themselves for the Field Reserve. The Police Reserve is to be formed out of members of Class *B* of the Citizen Force Reserve, who are not less than 30 years of age and otherwise suitable. All members of the Field Reserves will be liable in time of war to be called upon to serve in the Permanent Force.

The Rifle Associations

Every citizen liable to Peace Training, who being in his twenty-first year has not been entered for such training, must be enrolled as a member of a Rifle Association on July 1st, and undergo annually a prescribed course of instruction and exercise in the care and use of the rifle, and in the rudiments of drill until June 30th in his twenty-fifth year. Every other citizen liable to personal service in time of war, not being a member of the Permanent, Coast Garrison, or Active Citizen Forces, may engage voluntarily to serve as a member of a Rifle Association. The members of the Rifle Associations are to be allowed to elect their own officers, in the manner of the Burgher Commando; and each member will be entitled to receive a rifle, which remains, however, the property of the State, and an annual free issue of suitable ammunition.

The Cadet Training

In appointed Cadet-training areas, subject to the exercise of the principle of parental option, all boys between the ages of 13 and 16 are to be registered in the year in which the Act comes into operation, and in subsequent years all boys who attain their thirteenth year. The boys thus registered (with the exception of the medically unfit) will receive normally a four years' course of training in the Cadet corps; but a certificate of efficiency, which will exempt the holder from recruit training on joining the Active Citizen Force, may be acquired after three years. All necessary arms, etc., will be supplied to the Cadets at the public expense.

Military Instruction

Military instruction and training are to be given to members of the Defence Forces in English or Dutch, the language chosen being that with which the persons under instruction are the more familiar; but all ranks are to be taught to give and receive the executive words of command in both English and Dutch. The material for the Head-quarter Staff and the Administrative and Instructional Staffs of the several military districts and sub-districts will be drawn, no doubt, largely at first from the existing head-quarters staffs; but provision is made for securing a supply of officers in the future by the establishment of a South African Military College. This institution, although conducted at first upon the simplest lines, will be capable of expansion as circumstances may require. In moving the second reading of the Bill, General Smuts announced that a School of Instruction would be opened at Bloemfontein on July 1st, 1912, the head of which would be Brigadier-General Aston, Lord Methuen's late Chief of

30—(2139)

Staff; with a staff consisting of one British Regular officer, one officer from the Cape Mounted Rifles, and one officer from the Staats Artillerie of the late South African Republic. Provision is made also for the establishment of a School of Aviation, which is to be followed by the enrolment of a South African Aviation Corps.

Service in Time of War

The Permanent Force is liable at all times to be employed within or without the Union in time of war; and thus the Union Government will be enabled to place a useful military force at the disposal of any other member of the Empire, if the need for such assistance should arise. The Coast Garrison, Active Citizen, and Citizen Reserve Forces may be mobilised in time of war for service anywhere in South Africa. After the Active Citizen and Reserve Forces in any district have been called out, the then members of the National Reserve become liable to be called out for personal service in defence of the Union in the order following:

Class	I.	Citizens of	17 to 30	years.
,,	II.	,,	31 ,, 45	,,
,,	III.	,,	46 ,, 60	,,

The whole, or any part, of these Defence Forces may be called out for the prevention or suppression of internal disorder; and the Act gives the Government special powers in relation to defence (such as to commandeer supplies, to expropriate land, etc., etc.), and creates special military offences for which appropriate penalties are prescribed.

It may be noticed, in conclusion, that the native and coloured populations are absolutely excluded from taking any part in the defence of the Union. Thus the numerically predominant element of its inhabitants will neither receive any military training, nor be placed under any obligation to render service in time of war.

CHAPTER IV

EDUCATION

UNDER the Constitution Act the work of education is divided for the first five years of the Union between the Union Government and the Provincial Administrations. In the actual words of the clause in question (Sec. 85, iii), " Education, other than higher education, for a period of five years and thereafter until Parliament otherwise provides," is one of the classes of subjects with which the Provincial Councils and Committees are competent to deal.

PRIMARY AND SECONDARY EDUCATION

At the date of the Union fully developed systems of primary and secondary education were in operation in each of the four colonies. In the case of the Cape and Natal, these systems, of course, dated back to a period long anterior to the war (1889–1902), while those of the Transvaal and Free State had been established at the earliest possible moment after the war by the Crown Colony Administration. Remembering that these systems are designed almost exclusively to meet the requirements of the European inhabitants, and that the provision made for the instruction of native and coloured persons will be reserved for consideration under the head of " Native Education," a few sentences will serve to indicate the general character of the primary and secondary education of the Union.

To-day, then—since the educational deficiencies of the new colonies were rapidly removed under the Crown Colony Administration—there are to be found throughout the Union efficient primary schools in all towns and almost in all villages ; and arrangements are made under which even children living in isolated homesteads can be provided with "farm" schools. In all large, and in many lesser towns, High and other Secondary Schools have been established, and Government supervision is exercised over the numerous private institutions of this class. A glance at the respective returns of the schools of the Transvaal and Natal, in the year of the Union, will show how widely and how evenly the education net has been spread; for while in the case of Natal the schools are the product of a system which has grown up with the colony, the Transvaal schools are the fruit of the system introduced under Lord Milner only seven years before. In 1909, Natal, with a European population of less than 100,000 persons, had 52 Government schools, including secondary schools at Maritzburg and Durban, and there were in all 269 schools for European children under Government inspection. At the same date in the Transvaal, where the European inhabitants numbered some 400,000, there were directly under the Education Department 6 high schools, 150 town and village schools, 546 country schools, and 33 Government-aided schools, with a total enrolment of 49,198 scholars; while 1,243 pupils were receiving secondary education in the 21 superior schools, and a further 708 were being taught similarly in the 138 schools under the control of Boards where education above Standard VI was provided. (The figures refer, of course, exclusively to schools for white children.)

In each province there is an Education Department, with a Director of Education and a staff of inspectors, and the departments are assisted locally by School Boards or Committees and District Education Boards. While the details of management, and the proportions of the cost borne by localities, vary as yet in the several provinces, the main lines upon which the Government schools are administered are, with one exception, practically identical throughout the Union. This exception is the introduction effected in 1908 by General Hertzog, of what is known as " Compulsory Bi-lingualism " into the Free State system. General Hertzog's action has given rise to much political discussion ; and an endeavour is being made by the Union Government as a whole, and by the Minister of Education in particular, to formulate a common procedure in the respective use of Dutch and English as media of instruction, which will be acceptable to the education authorities in all four provinces.

The Bi-lingual Question

As the harmonious working of the principle of equal rights for the English and Dutch languages, laid down in the Constitution Act, is a matter which touches every aspect of the life of the Union, an account of what has been done to secure the desired result in the field of Education will be of interest. The Transvaal Education Act of 1907 embodied a compromise on the language question made between the Government and the (mainly British) Opposition of this colony ; and, as such, it represents the view of the Unionist (and mainly British) party in the Union Parliament of to-day. The Free State Education Law of 1908, as already noticed, gave effect to the opposite view. A comparison of the relevant provisions of these two laws will reveal, therefore, the issue in its clearest and most practical form ; and such a comparison is presented in the authoritative statements given below.

Memorandum by Dr. Viljoen, Director of Education in the Orange Free State Province, on the operation of the Education Act of 1908 :

Medium of Instruction

(*a*) Up to Standard IV.

. . . All pupils, up to and including Standard IV, are entitled to receive all their instruction in every subject through the medium of the language best spoken and understood by them. The second language, which is not the medium of instruction, may be gradually introduced in these classes in so far as is compatible with the age, health, and intelligence of the pupils. . . .

(*b*) Beyond Standard IV.

As regards the medium of instruction above Standard IV, up to and including the Standard of Matriculation, the law requires that as far as possible English and Dutch shall be used equally as mediums of instruction, that is to say, an equal number of principal subjects shall be taught through each of the two languages.

Memorandum by Mr. J. E. Adamson, Director of Education in the Transvaal Province, on the working of the language clauses of the Transvaal Education Act, 1907:

It may be observed, first, that there are really five propositions embodied in the relevant sections of the Act. These are:

1. The first steps in instruction must be taken through the native language.
2. English, when it is not the native language, must be introduced gradually as a medium and must be the medium above the Third Standard, save in respect of two subjects which may be taught through the medium of Dutch. (The Bible lesson is given in Dutch to Dutch children and in English to English children.)
3. Proficiency in English is to be a condition of probation from Standard to Standard.
4. Proper provision for the teaching of Dutch must be made in every school, and every child is to be taught Dutch, unless the parent asks for exemption.
5. The Board is to decide when any appeal or doubt arises; but there is the right of appeal to the Minister.

In view of this divergence, Mr. F. S. Malan, the Union Minister of Education, on November 2nd, 1910, addressed a letter to the Administrators of the four Provinces, in which he pointed out that at present different systems and different policies prevailed under the Education Acts in the several provinces, adding that "these controversies in regard to questions of language at public schools point to the desirability of dealing on broad uniform lines with those questions in all Provinces of the Union." Mr. Malan then suggested that the four Directors of Education should meet in conference at Capetown on

November 15th, 1910. This Conference was held on November 15th to 22nd; and the recommendations of the Directors of Education, together with the correspondence leading up to the Conference, were published and laid before Parliament.[1]

In the Parliamentary Session of 1911 a Conference of Ministerialist and Unionist representatives was held on the Language Question, and a compromise between the Transvaal and Free State systems, as formulated in the Majority Report, was adopted. While, however, General Botha was in England—he had left South Africa to attend the Imperial Conference and the Coronation of King George V—General Hertzog, the (Union) Minister of Justice and the author of the Free State Law of 1908, took occasion to advocate the claims of "Compulsory Bi-lingualism," especially among the more conservative of the Government's supporters. None the less, on November 21st, 1911, General Botha announced to the Congress of the South African Party, assembled at Bloemfontein, that the compromise on the Language Question would be put into effect. At the next sittings of the Provincial Councils, he then said, the respective education laws would be amended "so as to allow the language question to be settled on the basis of the Ministerialist and Unionist compromise embodied in the Majority Report of the recent Conference, or with the addition to it of the amendments, made with the concurrence of the Opposition, in the Transvaal Act."[2]

[1] Reports and Correspondence, on the Provisions in the Education Acts of the Four Provinces for the teaching of the English and Dutch languages and their use as media of instruction.

[2] As reported by Reuter. It was at this Congress that the new name "South African Party" (in lieu of "Nationalist") was adopted by General Botha's followers. The necessary steps for the dissolution of the Africander Bond, the Orangia Unie, and Het Volk, the three Nationalist organisations in the Cape, Free State, and the Transvaal respectively, were taken subsequently.

THE UNION OF SOUTH AFRICA

Higher Education

The Union Ministry of Education was an entirely new department, and the Education Departments of the four provinces carried on the whole work of education up to the appointment of an (acting) Under-Secretary for Education, which took effect on August 16th, 1910.

"The first difficulty which presented itself on the institution of this Department," writes[1] the (acting) Under-Secretary, "was the interpretation of the term Higher Education, as denoting the sphere of the Department's operations in accordance with the South Africa Act."

On this point a Conference of the Directors of Education of the four Provinces, held at Capetown on June 17th and 18th, 1910, passed the following resolutions:

(*a*) That this Conference is of opinion that for the purposes of the South Africa Act, Higher Education should include education beyond the standard of Matriculation, or a standard considered by the Minister to be equivalent thereto which is carried on in an institution established under a special statute, and any extension or continuation courses carried on in connection with such an institution which the Minister may approve, and courses for the training of teachers followed in institutions to be afterwards named.

(*b*) That, with a view to ultimate uniformity in the syllabuses and the standards of examination for Teachers' Certificates, as well as in the method of training, it is desirable that the four directors of Education should take the whole subject and related matters into consideration, and meet in September, 1910, to formulate the general lines of some scheme likely to be practicable for all the Provinces.

These resolutions were adopted by the Union Minister of Education as an "initial working principle." For the present, therefore, the training schools for teachers in the several provinces have been left under the jurisdiction of the respective Provincial Education Departments.

[1] Mr. George M. Hofmeyr, in his Report for the year ending December 31st, 1910 (addressed to the Minister of Education, the Hon. F. S. Malan), from which the above and following passages and tables are taken.

"It was further decided," the Under-Secretary continues, "that Agricultural Higher Education should remain under the Department of Agriculture, the Agricultural Colleges being brought into closer connection with, and their Principals being charged with the superintendence of, the Experimental Stations."

The Department is, therefore, for the present concerned with the following institutions, incorporated by Acts of Parliament in the years undermentioned:

(1) The University of the Cape of Good Hope, 1873.
(2) South African College, 1878.
(3) Victoria College, 1881.
(4) Diocesan College (discontinued from January 1st, 1911).
(5) Rhodes' University College, 1904.
(6) Huguenot College for Women, 1909.
(7) The Grey University College, 1910.
(8) The Transvaal University College, 1910.
(9) The South African School of Mines and Technology, 1910.
(10) The Natal University College, 1909.

Besides controlling these institutions (see Table on p. 464), the Minister of Education takes the initiative in miscellaneous matters affecting education in the four provinces, where joint action is needed; and in such cases the Union Department becomes the channel of correspondence. He also takes action in respect of any matters of general educational interest which may be brought to his notice.

Aims and Policy of the Department

In view of the abnormally high cost of higher education, and generally with a desire to improve the efficiency of the several institutions, it is the expressed purpose of the Department "to aim steadfastly at uniformity, at equality, and a reduction of the expenditure generally, even though it may be found necessary for this purpose to propose amendments of legislative enactments." And the Under-Secretary continues: "In the case of some institutions, determined efforts should be made to increase their revenue. The process, I submit, will require to

The Following Table shows the Number of Students Attending the Colleges of the Union during the Last Term of 1910

College.	Intermediate.	Arts Course.				Mining Course.	Forestry.	Medical.	Assaying.	Survey.	Law.	Engineering, etc.	Dentistry.	Education.	Miscellaneous.	Total.
		B.A.			M.A.											
		Literature.	Science.	Mixed.												
South African College	85	30	32	6	3	24	2	6	12	4	18	38	1	7	22	290
Victoria College	137	34	42	34	2	2	—	—	—	2	—	—	—	20	19	292
Diocesan College[1]	16	4	1	—	—	—	—	—	—	—	—	—	—	—	3	24
Rhodes' University College	58	19	19	7	—	4	—	—	—	1	9	—	—	—	8	125
Huguenot College	34	12	4	3	—	—	—	—	—	—	—	—	—	—	—	53
Grey University College	31	9	9	7	2	3	—	—	—	—	2	—	—	6	4	73
Transvaal University College	49	5	13	4	1	—	—	—	—	—	21	—	—	—	—	93
Natal University College	19	2	2	4	—	5	—	—	—	—	25	—	—	—	—	57
South African School of Mines and Technology ..	—	—	—	—	—	51	—	—	—	—	23	37	—	—	2	113
Totals	429	115	122	65	8	89	2	6	12	7	98	75	1	33	58	
											Grand		Total ..		..	1,120

[1] Incorporated into South African College after January 1st, 1911

be a gradual one, and will involve the consideration of many important details."

The Estimates for 1911–12, therefore, provided for no increases and no recurrent expenditure, except such as had been proved to be essential. And in this connection it may be noted that the amount voted under the sub-head " Higher Education " in this year was £109,520, and the total vote for the Ministry of Education was £327,094.

Steps have also been taken (the Under-Secretary reports) to introduce uniformity into the extent of the control exercised by the Department over the Councils, or other governing bodies, of the various colleges and institutions.

The approximate cost per student to the State is thus exhibited :

Colleges.	No. of Students.	Approximate Total Cost to State per annum.	Approximate Cost per Student per annum.	
			Including Law Students.	Excluding Law Students and their additional Cost.
		£	£	£
South African College	310	15,567 (a)	50	53
Victoria College	340	11,344	33	33
Rhodes University College	125	7,169	57	60
Huguenot College	53	2,580	49	49
Grey University College ..	72	8,640 (b)	120	121
Transvaal University Coll..	104	10,332	99	120
Natal University College ..	57	5,785 (c)	101	170
South African School of Mines, etc.	110	15,668	142	176
Total	1,171	Tl. 77,085	Avg. 66	Avg. 70

(a) £1,500 deducted for cost of new equipment of laboratories; (b) £900 deducted, being fees payable into Treasury; (c) £600 deducted, being fees payable into Treasury.

In this table the total number of students is that for the third term of 1910, and the total cost is that provided for in the Estimates for 1911–12.

In addition to the sums shown above, grants to teachers

in training at these colleges and other bursaries were given by the State of the total value of £6,780.

With further reference to these tables, the Under-Secretary gives the following additional information concerning the youngest and the oldest of the institutions mentioned. The Natal University College came into existence only in 1909. Like other University Colleges of the Union, it is affiliated to the University of the Cape of Good Hope. The affairs of the College are managed by a Council, consisting of sixteen members representative of all interests concerned in the prosperity of the institution, together with a Registrar. The College issues its own calendar, which gives information as to the fees, courses of lectures, etc.

The Diocesan College, Rondesbosch, which in the past forty years has done excellent work, ceased to exist at the end of 1910. Under an agreement made between the Councils of the Diocesan and South African Colleges, and approved by the Union Government, this college was amalgamated with the South African College from January 1st, 1911. As a condition of this agreement, three professors of the Diocesan College joined the staff of the South African College, and retiring allowances were granted to three other members of the Diocesan College staff, whose services were no longer required.

The University

The existing University, the University of the Cape of Good Hope, is governed by a Council, the members of which are partly nominated by Government and partly elected by Convocation; and it was incorporated by Act of Parliament of the Cape Colony in 1873, the year following that in which Responsible Government was granted to this colony. It is—as was the University of London until recent years—an examining and degree-conferring body, and not a teaching university. In this capacity it has served the whole of South Africa; and from time to time the other colonies or states have made annual contributions towards its expenditure, and obtained a measure of representation on its Council and Committees. For many years past, however, the need for a teaching university of the same character as those established in Australia and in Canada has been recognised; and the necessary steps for the creation of this institution are now being taken by the Union Government. On this subject the Under-Secretary for Education writes in his Report for 1910:

> A Bill will at an early date be introduced into Parliament providing the first measures towards the establishment of a University of South Africa to meet adequately the demands of the times.

The funds for the creation of the new University have been provided already, in a large measure, by the munificence of the late Mr. Alfred Beit. While he was still alive, Mr. Beit gave the beautiful Frankenwald Estate, lying between Johannesburg and Pretoria, to the Transvaal (then under the Crown Colony Administration) for the purposes of education; and by his will he left a legacy of £200,000 for the foundation of a teaching university on this estate. The Transvaal (Responsible) Government, however, was not in a position to accept this legacy under the conditions by which it was accompanied; and it is understood that, when the necessary legislation has been passed releasing the trustees of the will from their obligations to the Transvaal Government (or its legal successor), Mr. Otto Beit will be ready to apply this £200,000—making, together with the proceeds of the sale of Frankenwald, some £500,000 in all—to the founding of a teaching university within the Union.

In offering to sanction the appropriation of this large sum of £500,000 to the purposes of the new South African University, the Beit Trustees required naturally that the terms and conditions of its foundation should be such as could command their approval. The University and Higher Education Bill, as originally drafted, failed to satisfy this requirement, and Mr. Malan informed the Union Parliament on February 6th, 1912, that the introduction of the measure would be postponed, pending further negotiations with the Beit Trustees. The proposals to which objection was taken were understood to be the restriction of the Arts and Science faculties to post-graduates, the omission of the essential requirement of residence on the part of the students, and the introduction of the bi-lingual principle into the teaching system of the University. It is also reported that the Trustees have suggested that the proposals to be embodied in the new Bill should be drawn up by a commission of educational experts, to be appointed by the Union Government for this purpose.

Assuming that an agreement is reached between the Union Government and the Beit Trustees on these and other matters, it remains to be seen where the University of South Africa is to be placed. The opinion is held very generally that the slopes of Table Mountain, hard by Groote Schuur, once Rhodes' house, and now the official residence of the Prime Minister of the Union of South Africa, would afford an ideal site for a great seat of learning. And the proposal has the further recommendation that the South African College, only a few miles off at Capetown, both in respect of its teaching staff and its large attendance of students, would form an excellent nucleus out of which the University could be developed.

In this connection it may be noticed that the Government, in agreeing recently to the institution of Chairs of Anatomy and Physiology at the South African College, has reserved to itself the right to transfer these Chairs, if desired, to the seat of the new University.

Professional Training

The facilities at present afforded for professional training, exclusive of the ordinary courses for the Arts and Law degrees conferred by the (existing) University, are as follows :

Medicine. A fully-equipped South African School of Medicine is in process of formation at the South African College, Capetown. The object of the school is to furnish to aspirant physicians and surgeons the equivalent of the first two years of a medical course in the Universities of Great Britain and the Continent of Europe.

Teaching. At the Grey University College, Bloemfontein, and at the Victoria College, Stellenbosch, Chairs of Education have been established.

Mining, Engineering, etc. It would be strange if the Rand, where scientific knowledge, as applied to the processes of chemical and mechanical extraction of gold from the ore, is to be seen in all its latest and most original developments, had no School of Mines. In point of fact, the South African School of Mines and Technology, at Johannesburg—the successor of the Transvaal Technical Institute—promises to be one of the most efficient teaching institutions of its kind within the Empire. Incorporated by Act of Parliament in 1910, it provides a four years' course in mining, whereas the other colleges of the Union offer teaching only in the first two years' courses of the full four years' mining curriculum prescribed by the University. It also provides courses of over four years in mechanical and electrical engineering, civil engineering, metallurgy, and chemical engineering. The question whether the School should itself conduct examinations and grant certificates, both to the students of other institutions (such as the South African College, where there is a considerable number of mining students), as well as to its own members, is under consideration (*i.e.*, in 1911).

The qualification for admission to this Institution is the Matriculation of the Cape University, or its equivalent; and during the year 1910 there were 90 students in attendance. Of these, 51 were first year, 21 second year, 9 third year, 7 fourth year, and 2 occasional, students.

The distribution of the students among the various subjects of study is shown by the following table:

Subject.	No. of Students.
Mining	51
Mechanical and Electrical Engineering ..	26
Civil Engineering	3
Metallurgy	5
Chemical Engineering	3
Occasional	2

In addition to the foregoing courses, the School provides instruction in law, and carries on Evening and "Reef" Classes. At the close of 1910 no less than 1,700 men were receiving instruction by means of these secondary efforts of the Institution. In this and in other ways, writes the Under-Secretary for Education, the South African School of Mines has "sought to serve the technical and scientific interests of the community. It is now turning out students capable of eventually filling high positions in the mining industry, besides qualifying hundreds for less important posts. The School has, since its inception, received liberal support from the industry and the public."

Conferences were held in the latter part of 1910, at Capetown, with the object of fixing the standard of merit in mining and engineering, and in the theory and practice of surveying, which was to entitle students to obtain, by examination, certificates of competence in these branches of professional knowledge. In the first case, the members of the Conference consisted of representatives of the University Council, the South African College, and the South African School of Mines; in the second case, the delegates of the University Council conferred with the Surveyors-General of the four provinces of the Union. As the result of these Conferences, it is hoped that it will be possible, prior to the establishment of the new South African University, to enable students to provide themselves with evidences of capacity in the several professions in question, which will satisfy fully the reasonable requirements of employers in all four provinces of the Union.

Agricultural Education

The provision for agricultural education, made in the vote for the Ministry of Education for the year 1911–12, amounts to £95,629. The institutions to be maintained, and the sums respectively allocated to them, are as under:

	£
Transvaal College of Agriculture, Pretoria	500
Elsenberg Agricultural School	10,500
Robertson Experiment Section	1,540
Grootfontein Agricultural School	19,110
Cedera Agricultural School	9,431
Winkelspruit Experiment Station	1,691
Weenen Experiment Station	988
Potchefstroom Agricultural School and Farm ..	18,608
Standerton Stud Farm	5,468
Ermolo Stud Sheep Farm	3,628
Marico Experiment Orchard	774
Groot Vlei Experiment Station	2,575
Tweespruit Agricultural School and Farm	8,821
Besterspruit Experiment Station and Donkey Stud Farm	1,960
Frankenwald Estate	1,020

And to these amounts must be added the further sums of £4,015 and £5,000, appropriated respectively to agricultural scholarships and the purchase of pedigree stock.

Many of these institutions are scarcely of an educational character, and both they themselves and the fact of their appearance on the Education Vote have been the subject of comment in an earlier chapter.[1]

[1] Part IV, Chap. IV, p. 367.

The nature of the teaching staff and general equipment of the agricultural schools will be indicated sufficiently in the following particulars:

At Grootfontein Agricultural School (Cape Province) there are eleven lecturers and instructors, beside the Principal of the School and the Manager of the Farm; and provision is made for agricultural bursaries, etc.

At Cedera Agricultural School (Natal Province) there is a director, a house-master, thirteen lecturers and instructors, three clerical assistants, a farm manager, a matron, and a medical officer.

At Potchefstroom Agricultural School and Farm (Transvaal Province) there is a manager, a warden and lecturer, five lecturers, a poultry expert and his assistant, two assistant horticulturists, and a dairy assistant, etc.

At Tweespruit Agricultural School and Farm (Free State Province) the staff consists of a principal, a manager, six lecturers, an orchardist, etc.

The departmental receipts in respect of these institutions for the year 1911–12 were estimated to be:

	£
School Fees	7,750
Receipts from Experiment Farms, Sales of Stock, Stud Fees, etc., etc.	18,959
	£26,709

It must be remembered also that a National Institution for the Teaching of Agriculture—The Transvaal College of Agriculture—is in course of erection at the present time at Pretoria, the administrative capital of the Union.

The Rhodes Scholarships

As all the world knows, Rhodes provided in his will[1] for the foundation of a number of scholarships carrying

[1] Rhodes died on March 26th, 1902. The will (in its final form) was dated July 1st, 1899.

students from the oversea dominions, the United States, and Germany to Oxford. In the distribution of these benefactions, the (then) British Colonies in South Africa, and Rhodesia in particular, were treated naturally with especial generosity; and under the system established by the Rhodes Trustees no less than eight scholars are elected annually in South Africa—four in the Cape Province, one in Natal, and three in Rhodesia.

The purpose of the scholarships and the method of selecting candidates are both explained by Rhodes himself in the will.

Of the Colonial and American Scholarships, he says:

> Whereas I consider that the education of young Colonists at one of the Universities in the United Kingdom is of great advantage to them for giving breadth to their views for their instruction in life and manners and for instilling into their minds the advantage to the Colonies as well as to the United Kingdom of the retention of the Unity of the Empire And whereas . . . I attach very great importance to the University having a residential system such as is in force at the Universities of Oxford and Cambridge for without it those students are at the most critical period of their lives left without any supervision. . . . And whereas my own University—the University of Oxford—has such a system. . . . And whereas I also desire to encourage and foster an appreciation of the advantages which I implicitly believe will result from the Union of the English-speaking peoples throughout the world and to encourage in the students from the United States of North America who will benefit from the American scholarships to be established for the reason above given at the University of Oxford under this my Will an attachment to the country from which they have sprung but without I hope withdrawing them or their sympathies from the land of their adoption or birth Now therefore I direct my Trustees . . . to establish for male students the Scholarships hereinafter directed to be established each of which shall be of the yearly value of £300 and be tenable at any College in the University of Oxford for three consecutive academical years.

And of the German Scholarships, in a codicil executed in South Africa, he says:

> I note the German Emperor has made instruction in English compulsory in German schools. I leave five yearly scholarships

at Oxford of £250 per annum to students of German birth the scholars to be nominated by the German Emperor for the time being. Each Scholarship to continue for three years so that each year after the first three there will be fifteen scholars. The object is that an understanding between the three Great Powers will render war impossible and educational relations make the strongest tie.

The method of selection is set out in the following paragraph:

My desire being that the students who shall be elected to the Scholarships shall not be merely bookworms I direct that in the election of a student to a Scholarship regard shall be had to (i) his literary and scholastic attainments (ii) his fondness for and success in manly outdoor sports such as cricket football and the like (iii) his qualities of manhood truth courage devotion to duty sympathy for and protection of the weak kindliness unselfishness and fellowship and (iv) his exhibition during schooldays of moral force of character and of instincts to lead and to take an interest in his schoolmates for those latter attributes will be likely in after life to guide him to esteem the performance of public duties as his highest aim As mere suggestions for the guidance of those who will have the choice of students for the Scholarships I record that (i) my ideal qualified student would combine these four qualifications in the proportion of three-tenths for the first two-tenths for the second three-tenths for the third and two-tenths for the fourth qualification so that according to my ideas if the maximum number of marks for any Scholarship were 200 they would be apportioned as follows—60 to each of the first and third qualifications and 40 to each of the second and fourth qualifications (ii) the marks for the several qualifications would be awarded independently as follows (that is to say) the marks for the first qualification by examination for the second and third qualifications respectively by ballot by the fellow students of the candidates and for the fourth qualification by the headmaster of the candidate's school and (iii) the results of the awards (that is to say the marks obtained by each candidate for each qualification) would be sent as soon as possible for consideration to the Trustees or to some person or persons appointed to receive the same and the person or persons so appointed would ascertain by arranging the marks in blocks of 20 marks each of all candidates the best ideal qualified students.

No student shall be qualified or disqualified for election to a scholarship on account of his race or religious opinions.

The original distribution of the Colonial Scholarships

was set out in the following table, which formed part of the will:

Total No. Appropriated.	To be tenable by Students of or from	No. of Scholarships to be filled up in each year.
9	Rhodesia	3 and no more
3	The South African College School, in the Colony of the Cape of Good Hope	1 ,, ,,
3	The Stellenbosch College School, in the same Colony	1 ,, ,,
3	The Diocesan College School of Rondesbosch, in the same Colony ..	1 ,, ,,
3	St. Andrew's College School, Grahamstown, in the same Colony	1 ,, ,,
3	The Colony of Natal	1 ,, ,,
3	The Colony of New South Wales ..	1 ,, ,,
3	The Colony of Victoria	1 ,, ,,
3	The Colony of South Australia ..	1 ,, ,,
3	The Colony of Queensland	1 ,, ,,
3	The Colony of Western Australia ..	1 ,, ,,
3	The Colony of Tasmania	1 ,, ,,
3	The Colony of New Zealand	1 ,, ,,
3	The Province of Ontario, in the Dominion of Canada	1 ,, ,,
3	The Province of Quebec, in the Dominion of Canada	1 ,, ,,
3	The Colony or Island of Newfoundland and its Dependencies	1 ,, ,,
3	The Colony or Islands of the Bermudas	1 ,, ,,
3	The Colony or Island of Jamaica ..	1 ,, ,,

There were thus 60 scholarships in all, of which 24 were appropriated to South Africa, 18 to Australia, 3 to New Zealand, 6 to Canada, 3 to the Bermudas, and 3 to Jamaica; and the total number to be filled up each year was 20. In view of the obvious insufficiency of the number of scholarships appropriated to Canada, the Rhodes Trustees have assigned scholarships (on the same basis as those allocated to Ontario and Quebec) to the provinces of Nova Scotia, New Brunswick, Prince Edward Island, British Columbia, Manitoba, and the North-west Provinces (Alberta and Saskatchewan alternatively). The present total of the Colonial Scholarships is, therefore, seventy-eight.

As Rhodes allocated two (instead of three) scholarships to each of the forty-eight States or Territories of the United States, while making them, nevertheless, tenable for three years like the Colonial Scholarships, in the case of the American Scholarships there is no election in every third year.[1] Thus the respective complements of Colonial, American, and German Rhodes' scholars in residence at Oxford in any one term are 78, 96, and 15. Rhodes' foundation, therefore, has widened the current of Oxford undergraduate life by a constant stream of some 200 young men of promise, gathered from the Dominions, the United States, and Germany.

The trustees and executors appointed by Rhodes in his will were: The Earl of Rosebery, Earl Grey, Viscount Milner, Mr. Alfred Beit, Dr. (now Sir) L. S. Jameson, Sir Lewis Mitchell, and Mr. B. F. Hawksley. With the exception of Mr. Alfred Beit, who has died recently, they are the existing Rhodes Trustees.

The special conditions and regulations governing the election of the South African Rhodes scholars will be found in an Appendix.[2] It remains, however, to identify the four schools in the Cape Province to which scholarships were assigned in the will. The "South African College School" is better known as the South African College; the "Stellenbosch College School" is the Victoria College, Stellenbosch; the "Diocesan College School at Rondesbosch," better known as the Diocesan

[1] Elections of Rhodes' scholars were held in 1910 and 1911: none, therefore, will be held in the present year (1912).

[2] See p. 483.

College, has been absorbed into the South African College; and the "St. Andrew's College School at Grahamstown" has been converted into the Rhodes University College, Grahamstown.

Native Education

In all provinces of the Union, grants are made for the purposes of native education. According to the general practice in South Africa, the Education Departments do not, with slight exceptions, undertake directly the work of instructing the native and coloured population, but they utilise the services of religious bodies in this respect, subsidising their mission schools and other educational institutions. The classes of institutions thus recognised as entitled to Government assistance are: (1) Elementary schools, (2) trade and industrial schools, and (3) training colleges for native teachers; and all institutions receiving grants-in-aid are placed under the supervision of the Department, and expected to maintain a moderate standard of efficiency in respect of teaching staff and general equipment.

This system, naturally, has sufficed to bring only a fraction of the native population of the Union under regular instruction. At the date of the Union, the Cape Province, with a non-European population of nearly 2,000,000, had a native (and coloured) enrolment of rather more than 100,000 scholars. In Natal, on June 30th, 1909, there were 178 native, 19 coloured, and 35 Indian schools; and at these schools, out of a total non-European population of over 1,000,000, there were respectively 12,484 native, 902 coloured, and 3,245 Indian scholars in attendance. In the Transvaal, where the non-European population numbers 1,250,000, £10,833 13s. was spent on native education during the year ended May 31st, 1910; and at this date there were 230 Government-aided and 179 unaided native schools, with an average attendance of 9,942 and 5,803 pupils respectively. As will be seen from the subjoined table, this return marks the attainment of a considerable advance since the war.

Table showing progress of native education in the Transvaal:

Year.	Number of Schools.		Total.	Amount of Govt. Grant.
	Aided by Govt.	Unaided.		£ s. d.
1903.. ..	7	155	164 (*sic*)	——
1904.. ..	6	157	161 (*sic*)	——
1905.. ..	142	134	276	5,850 0 0
1906.. ..	197	177	374	6,592 13 0
1907–8 ..	221	125	346	8,554 17 10
1908–9 ..	243	141	384	10,976 17 3
1909–10 ..	230	179	409	10,883 13 0

In the Free State a small grant for native education is given directly to the religious bodies, by whom it is appropriated to the various teaching institutions; and in this province there were at the date of the Union some 10,000 native children under instruction out of a non-European population of 350,000.

The greatest progress in native education is to be found naturally in the Cape Province, the oldest and most settled portion of South Africa, where for more than a century such work has been carried on by missionary enterprise. Among the products of this enterprise, the Lovedale Missionary College is, perhaps, the most remarkable; and it may be cited as an institution with so long and successful a career that its records afford reliable data for measuring the extent to which a judicious system of education may be expected to raise the economic capacity of the native population of South Africa. The institution embraces (1) a collegiate department for training native clergy and teachers; (2) workshops in which young men are taught bookbinding, printing, and the trades of blacksmiths, carpenters, and the like, and young women sewing and laundry work and the like; and (3) elementary schools for boys and girls.

As the result of the work done by Lovedale and similar institutions, educated natives have been employed for many years past as carriers of letters, telegrams, and parcels; have occupied responsible positions as clerks, interpreters, schoolmasters, and sewing-mistresses by the hundred; and, in still larger numbers, have practised successfully the trades of carpenters, blacksmiths, bricklayers, tinsmiths, wagon-makers, shoemakers, printers, sailmakers, saddlers, and the like. Here, in the Cape Province, where to-day 5 per cent. of the non-European population are under instruction, newspapers are printed in the language of the natives; and in this province alone, as we have seen, natives and coloured people are admitted to the Parliamentary franchise, and a system of local government has been established among the purely native populations of its dependent territories.

These broad facts may be supplemented usefully in one or two particulars by the Returns of the "Statistical Register" of the Cape Province. Evidence of the standard of education which prevails respectively among the European and non-European population is afforded in the following figures: On a basis of the census of 1904, there were in this province (out of a total population of 2,409,804) 621,037 persons who could both read and write, of whom 434,827 were Europeans and 186,210 "other than Europeans"; 45,897 persons could read only, of whom 10,338 were Europeans and 35,559 "other than Europeans"; 1,735,491 persons could neither read nor write, of whom 133,569 were Europeans and 1,601,922 "other than Europeans"; and 7,379 persons remained unspecified. In the return showing the various "Occupations" of the people of the province, we find that, in 1904, some 6,000 natives and coloured people were engaged in "Professional," and some 20,000 in "Commercial," occupations; the corresponding figures for Europeans being 32,202 and 46,750 respectively. A third return—and one of special interest—is that which gives the number of registered Parliamentary voters in the 51 electoral divisions of the province, as proclaimed by the Governor-General in Council on June 23rd, 1910, *i.e.*, the voters by whom the first Union Members of Parliament for the province were elected. According to this return, the total electorate of the province numbered 142,367; and of this total 121,346 were European, and 21,021 "other than European," voters. The coloured vote was composed of: Kafir and Fingo, 6,633; Hottentot, 715; Malay, 911; Indian, 764; Chinese, 20; other mixed and coloured races, 11,978.

Higher Education for Natives

The claim of the native population to possess an institution of higher education has recently been brought before the Minister of Education. The proposal is that recommended by the Native Affairs Commission of 1903–5; namely, that an inter-State (or Union) College should be established, which would serve the whole native population of the Union, and, as such, would be equipped upon an adequate scale. The purpose of such an institution would be to enable any really exceptional men of the Bantu races to obtain University degrees, and in other respects to qualify for professional careers and possibly for public life.

In this connection it must be remembered that in South Africa, as in India, there is a certain danger attaching to the premature extension of educational privileges to the native races. In recent years a practical example of this danger has been afforded in the rise of Ethiopianism; and a brief account of this movement will provide evidence of the nature and reality of the danger in question.

The Ethiopian or Church Separatist Movement

Such an account is to be found in the Report of the Native Affairs Commission of 1903–5. The movement, says the Report, has had its origin in "a desire on the part of a section of the Christianised natives to be freed from control by European Churches. Its ranks are recruited from every denomination carrying on extensive operations in South Africa, and there is in each case little or no doctrinal divergence from the tenets of the parent Church, though it is alleged, and the Commission fears with truth, that relaxed strictness in the moral standard maintained frequently follows."

And the substance of the conclusions at which the Commission arrived is thus stated:

> The Ethiopian Movement, now represented by the African Methodist Episcopal Church, the Ethiopian Order in the Church of England, and numerous semi-organised schismatic fragments detached from every denomination . . . is the outcome of a desire on the part of the Natives for ecclesiastical self-support and self-control, first taking tangible form in the secession of discontented and restless spirits from religious bodies under the supervision of European Missionaries without any previous external incitation thereto. Further, that upon the affiliation of certain of these seceders and their followings to the African Methodist Episcopal Church lamentable want of discrimination was displayed by the first emissaries to South Africa in the ordination to the ministry of unsuitable men.
>
> The Commission is not disposed to condemn the aspiration after religious independence, unassociated with mischievous political propaganda, but at the same time does not fail to recognise that in the case of a subject race such an aspiration misdirected, on the one hand, by the leadership of ignorant and misguided men, and repressed by misunderstanding or harshness on the other, might be fraught with the seeds of social mistrust and discontent. . . .

The Commission would not advise any measure of legislative repression, unless unforeseen developments render it necessary, considering that effort should rather be directed towards securing efficient constitutional control and organisation in order that the influences at work may be wisely directed, and any individual cases in which pastors abuse the trust reposed in them, may be amenable to authoritative discipline. To this end the Commission would deprecate the recognition of detached secessional fragments acknowledging no efficient central authority.

The Native Press

Of the Native Press the same authoritative Report writes, that " the mere fact of its existence is a proof of rapid educational advancement." On the whole, it has " proved itself to be fairly accurate in tracing the course of passing events, and useful in extending the range of native information. . . . It cannot be said that nothing but good has accrued from it, but an infant Press could not be expected to be wholly free from mistakes and indiscretions. Although the organs claiming to express native opinion may not have been infallible, freedom of thought and speech within lawful limits is not likely to be assailed." And it adds that " though a minority in the Commission favoured the enactment of a law to punish persons responsible for publications in the native language creating distrust and animosity between the races, or likely to produce conflict between them," the motion was negatived after careful consideration.[1]

[1] " Report of the South African Native Affairs Commission, 1903–05." Cd. 2399. This Report (and the evidence taken by the Commission) contains valuable and authoritative information upon all questions connected with the native races of South Africa. It is one of the many contributions made by the Milner *régime* to the political and economic progress of South Africa.

APPENDIX TO CHAPTER IV

PART V

THE RHODES SCHOLARSHIPS IN SOUTH AFRICA

Extracts from the Memorandum issued by the Trustees of the Will of the late Mr. Cecil Rhodes for the information of educational authorities and intending candidates for Scholarships in South Africa :

The Scholarships are of the value of £300 a year ; and are tenable for three years.

The election of Scholars in South Africa must each year be completed at a date *not later than* the 1st of January, unless by special permission of the Trustees. The Scholars will begin residence at Oxford in October of the year for which they are elected.

CONDITIONS OF ELIGIBILITY

All candidates shall be British subjects, and unmarried. Subsequent to the election for the year 1912, all candidates from Rhodesia and Natal must have passed their nineteenth but not have passed their twenty-fourth birthday on October 1st of the year for which they are elected ; and such candidates (except any educated in England, from whom Responsions will be required) must have passed at least the Intermediate Examination of the Cape of Good Hope. The four University College Schools of the Cape Province, which elect under special regulations approved by the Trustees, have been asked to conform to the same lower limit of age and University standing in the selection of their scholars.

All candidates, unless specially excused, must either

(1) have passed the Oxford Responsions Examination or its equivalent ; or (2) have passed the Intermediate Examination of the University of the Cape of Good Hope, and have satisfied the Examiners at that examination in Latin and Mathematics ; or (3) have qualified for the Standing of a Junior or Senior Colonial Student at Oxford under the Colonial Universities Statute.

The following are the Decrees which determine the standing at Oxford of Students from the University of the Cape of Good Hope :

(1) FOR SENIOR STANDING.—" That any Member of the University of the Cape of Good Hope who shall have passed at the University *either* the Bachelor of Arts Honours Examination, *or*, after the first day of January, 1905, Part I of the Master of Arts Examination, shall be deemed to have taken Honours as required by the provisions of Statt. Tit. 11. Sect. VIII cl. 5 " (and, therefore, be eligible for admission to the status and privileges of a Senior Colonial Student).

(2) FOR JUNIOR STANDING.—"That any member of the University of the Cape of Good Hope who *either* (*a*) shall have passed the Bachelor of Arts Examination at that University; *or* (*b*) having passed the Intermediate Examination at that University shall have subsequently pursued a course of study extending over at least one year at one of the Colleges named in the Schedule appended to this Decree, shall be eligible for admission to the status and privileges of a Junior Colonial Student."

SCHEDULE

The South African College, Capetown.
The Victoria College, Stellenbosch.
The Diocesan College, Rondesbosch.
The Rhodes University College, Grahamstown.

If necessary, arrangements will be made for the conduct of an examination equivalent to Responsions at Capetown, Grahamstown, and convenient centres in Natal and Rhodesia. Oxford Examiners will prepare the papers and report on the examination in the usual way, and the University of Oxford agrees to accept in lieu of Responsions the Certificate of its Examiners that candidates have passed this Examination in Latin, Greek, and Mathematics. While it is necessary to pass in these three subjects in order to be exempt from Responsions, a candidate will be accepted as eligible to a Rhodes Scholarship who may have passed in Latin and Mathematics, but not in Greek. Such candidate will be held by the University of Oxford to have passed Responsions only when he shall have further passed in Greek, either at Responsions (in Oxford) or at an examination conducted, on behalf of the Rhodes Trustees, by the Oxford University Delegacy of Local Examinations.

The following Special Conditions of Eligibility apply within the Colony and Provinces named:

(*a*) THE PROVINCE OF THE CAPE OF GOOD HOPE.—Candidates must have been educated at one of the Schools to which the Scholarships are assigned. These Schools, however, may elect as Scholars students who have pursued their studies, after leaving School, at one of the higher institutions of the Province. Further information may be obtained from the School authorities.

(*b*) NATAL.—Candidates are required (i) to have been educated in the Province of Natal for six continuous years immediately previous to the date of election, or (ii) to have their legal domicile in Natal for six years, though acquiring their education, or any part of it, in any other Colony or Province of South Africa. The Committee of Selection is free to make allowance, at its discretion, for temporary absences from the Province or from South Africa during the six years referred to.

(*c*) RHODESIA.—In view of existing educational conditions, leave is occasionally given at present by the Trustees for candidates to compete for the Scholarships assigned to Rhodesia who are being educated in other parts of Africa or in England, provided that their parents reside in or are intimately connected with the Colony. In these instances, the candidate is allowed to take Responsions or its equivalent, either in England or in the Colony or Province where he is receiving his education. Application for leave to compete under these conditions must be made to the Trustees, either directly or through the Director of Education for Rhodesia. Other things being equal, preference will be given to candidates educated in Rhodesia.

SELECTION OF SCHOLARS

In accordance with the wish of Mr. Rhodes, the Trustees desire that "in the election of a student to a Scholarship regard shall be had to (i) his literary and scholastic attainments; (ii) his fondness for and success in manly outdoor sports, such as cricket, football, and the like; (iii) his qualities of manhood, truth, courage, devotion to duty, sympathy for and protection of the weak, kindliness, unselfishness, and fellowship; and (iv) his exhibition during schooldays of moral force of character, and of instincts to lead and to take an interest in his schoolmates." Mr. Rhodes suggested that (ii) and (iii) should be decided in any School or College by the votes of fellow-students, and (iv) by the Head of the School or College. Where circumstances render it impracticable to carry out the letter of these suggestions as to the method of selection, the Trustees hope that every effort will be made to give effect to their spirit, but desire it to be understood that the final decision must rest with the Committee of Selection, or the School making an appointment.

Each candidate for a Scholarship is required to furnish with his application:

(*a*) A certificate showing that he is within the prescribed limit of age.

(*b*) A full statement of his School and College career, including his educational qualifications, his record in athletics, and such testimonials from his masters at School and his professors at College, in reference to the qualities indicated by Mr. Rhodes, as seem best adapted to guide the judgment of those making the appointment.

If a careful comparison of these records and personal interviews do not furnish sufficient grounds for making a decision, those making the appointment are free to apply to the candidates, or to any selected number of them, such further intellectual or other tests as they may consider necessary.

THE UNION OF SOUTH AFRICA

The following regulations as to the selection of Scholars apply within the special Colony or Province named:

(1) The Province of the Cape of Good Hope.—Under the provisions of the Will of Mr. Rhodes the four Schools to which Scholarships are assigned in the Province of the Cape of Good Hope have the right of electing the Scholar. This is done under special regulations approved by the Trustees for each School. Further information may be obtained from the School authorities.

(2) Natal.—In Natal the final choice of the Scholars is entrusted to the following Committee of Selection:

His Excellency the Governor in his private capacity (Chairman).
The Honourable the Chief Justice.
The Superintendent of Education.

Application should be made to the Superintendent of Education.

(3) Rhodesia.—In Rhodesia the Director of Education receives the application of candidates, together with their credentials and testimonials, and submits them with his recommendations to the Trustees for approval.

The Chairman of the Committee of Selection, or the Head Master of the School making appointment, should at once notify to the Trustees and to Mr. F. J. Wylie, 9 South Parks Road, Oxford, the name of the elected Scholar, and should forward to the latter all the records, credentials, and testimonials relating to the Scholar on which the election was made. These papers should be transmitted immediately, as they are used in consulting the College authorities in regard to the admission of Scholars. It has been the experience of past years that Scholars have frequently been unable to gain admission to any of the Colleges of their preference owing to remissness in forwarding to Mr. Wylie the necessary information.

Unless specially exempted, Scholars will be expected (1) to reside in College for at least two years, and (2) to take any degree for which they may have qualified.

The Scholarship will be paid in quarterly instalments, the first on beginning residence at Oxford, and thereafter terminally, provided that the College to which any Scholar may belong be satisfied with his work and conduct. Marriage vacates the Scholarship. Should a Scholarship lapse through the failure of a student to give satisfaction to his College, from marriage, from resignation, or from any other cause, it will not be filled up until the year in which it would naturally expire. This provision is made in order not to interfere with the *rota* of succeeding Scholars.

It may be added that the office of the Rhodes Trust is Seymour House, Waterloo Place, London, S.W.; and that letters should be addressed to the Secretary.

CHAPTER V

IMMIGRATION, LAND SETTLEMENT, AND COST OF LIVING

THE preceding pages will have shown that South Africa, under its present political and economic conditions, does not present the free field for British emigrants which is offered by Canada, and, in a lesser degree, by Australia and New Zealand. The presence of the native races and the high cost of living to Europeans, taken together, practically exclude the British manual or unskilled labourer; and the number of British skilled artisans and mechanics, clerks, shop-assistants, governesses, women typewriters, telephonists, domestic servants and the like, which the (mainly British) industrial community can absorb, is very limited. On the other hand, the foregoing accounts of the mineral wealth and agricultural resources of South Africa will have made it no less plain, that to the man who combines the possession of capital, large or small, with youth, energy, and capacity, it is a land of infinite possibilities. Nor must it be forgotten that if South Africa has proved in the past "the grave of reputations," it has been, none the less, a nursery of great careers in geographical discovery, politics, war, and industry. And, here, where the names of Livingstone and Rhodes illustrate Napoleon's paradox, even to-day the youngest recruit may carry a field-marshal's bâton in his knapsack.

While, however, at the time of writing, the amount of employment open to British and other emigrants is very small, Ministers in the Union Government have recently made declarations indicating an intention to accelerate the operation of the agencies already at work for modifying the economic and social conditions which hitherto have most effectively restricted immigration. In particular, it would appear to be the desire of the Government to create a class of white labourers such as is to be found in Canada and Australia ; and it is proposed to inaugurate a scheme of land-settlement which will make it possible for immigrants with small capital to take up agricultural holdings. Whether these special efforts will or will not prove successful, time alone can show. But a change in the conditions hitherto governing the economic position of the European community in South Africa must be counted among the possibilities of the immediate future.

In these circumstances it will be convenient, first, to set down the little that can be said usefully in respect of the opportunities offered to new arrivals by South Africa under its existing conditions, and then to pass on to the discussion of the White Labour and Land Settlement proposals of the Government, and the changes to which they point.

As we have seen, then, manual labour is for the time being entirely ruled out, and the wage-earners who are required, whether they work with hands or head, are principally of the classes enumerated above. The normal demand for such emigrants, collectively, is small, and the extent to which one or other particular class may be wanted will vary, of course, from time to time. When, for example, the Camp Schools were established during the progress of the war, the local supply of teachers proved altogether inadequate, and it was necessary to obtain large numbers of elementary teachers from the United Kingdom and the Oversea Dominions. Or, again, a rapid extension of the telegraph or telephone system might make it necessary to supplement the local supply of operators, whether men or women, by candidates drawn from Europe. The latest and most reliable information in respect of all normal and exceptional openings for wage-earners is to be obtained from the Circulars issued at frequent intervals by the Emigrants' Information Office;[1] and from the same source all necessary particulars relating to "Assisted Passages," "Prohibited Immigrants," "Arrangements for Landing," "Shipping Companies," "Railways," "Clothing," and many other practical matters can be learnt whether by the intending emigrant or the traveller. Women emigrants are recommended to apply also to the South African Colonisation Society,[2] where they will hear of the special arrangements made to secure their comfort and safety during the voyage and upon arrival in South Africa, as well as obtain information in respect of vacancies for (female) domestic servants and other forms of employment open to women.

What Agriculture Offers

Setting aside the occasional appointments, public and private, open only to persons possessing special professional qualifications, the main field of activity offered by South Africa to the man of good education and moderate capital is agriculture in some form or other, coupled with the acquisition and development of land. The wide range of the products, animal and vegetable, for each of which South Africa provides a suitable soil and climate, has been brought to the notice of the reader on more than

[1] Emigrants' Information Office: 34 Broadway, Westminster, S.W. Office hours: 10 a.m. to 5 p.m.; Saturdays, 10 a.m. to 1.30 p.m. Letters to the Emigrants' Information Office need not be stamped. The circulars are free; but a small charge (from 1d. to 6d.) is made for "handbooks," and this must be prepaid.

[2] At 23 Army and Navy Mansions (No. 2), 115 Victoria Street, Westminster, S.W. This organisation is, like the Emigrants' Information Office, under the control of the Colonial Office.

one occasion; and the characteristics both of these various industries and of the localities in which they respectively flourish, have been indicated in an earlier chapter.[1] Nor is it necessary to recall how large and generous a measure of assistance is given by the State to agriculture, or to point out that a country which registers more hours of bright sunshine than Monte Carlo or California, should be alike pleasant and profitable to a man whose business lies almost wholly out of doors. On these heads, therefore, it will be sufficient merely to summarise facts elsewhere stated at length, by observing that the choice of occupation and investment, open to the "settler" of this class, ranges from tobacco and cotton-growing, or the breeding of ostriches and Angoras, to the familiar "corn and cattle" of an English farm, and that some of these special industries carry with them the prospect of ample profits. One further point, however, must be mentioned. In respect of labour, the farmer, or planter, has the advantage of the mine-owner. The cultivation of the soil and the tending of cattle form, apart from war, the general and natural occupation of the African natives. Mining, and especially the underground work of the mines, they dislike, and high wages are necessary to induce them to engage in it; but native labour can be obtained for agriculture more easily, and is paid at lesser rates. The agricultural employer, therefore, may expect to benefit fully from the great industrial asset possessed by South Africa in its large native population.

These are conspicuous advantages; and, at first sight, it seems strange that comparatively few Englishmen should have been found to take up the pursuit of agriculture in South Africa. The explanation lies in the circumstance that the initial expenditure required of the settler is considerably larger here than in other Dominions.

[1] Part IV, Chap. IV, p. 332.

Local experience enjoins that the area of the holding should be relatively large. While the average size of a farm in England is 66, and in the United States 143, acres, in the Transvaal it is 5,000 acres. And in South Africa it is generally held that 1,000 acres is the smallest area with which a man can hope to start farming successfully, and that, over and above the capital necessary for the purchase or rental of the land, he should have at least £1,000 for stock, buildings, and general equipment —or just five times the £200 considered necessary for the settler in Canada with a holding of from 160 to 640 acres. It will be understood, of course, that this estimate of the capital and acreage required refers to average men and average land, and does not pretend to cover exceptional cases. In respect of (Southern) Rhodesia, the advice on this head given to possible settlers is:

> Self-reliance, combined with a recognition of the necessity for individual effort, will be found as valuable a qualification in Rhodesia as in any other country. . . . It is impossible to say what is the least amount of capital a man can start with, with good prospects of success, as so much depends on the man himself. It is generally considered, however, that *by the time he is able to take up his farm* he should have at least £500 at his disposal if single, and naturally, the more capital a man has the better his chance in Rhodesia, as elsewhere. A married man obviously requires more capital.[1]

The total area of the unalienated Crown Lands within the Union is stated in the recent Closer Settlement Report to be 46,442,141 acres; but, of course, only a very small proportion of this great public estate would be ready for immediate occupation by settlers. The terms and conditions upon which public land can be acquired in the Union and Rhodesia do not differ greatly from those in force in the other oversea dominions; and in return for covenants to occupy personally, and erect buildings, cultivate, etc., they afford the customary facilities of the

[1] Handbook issued in September, 1911, by the British South Africa Company, 2 London Wall Buildings, E.C.

deferred payment of purchase money, rental with power to purchase, and advances for permanent improvements at low rates of interest. The latest as well as the most authoritative information on such points, however, should be obtained; and for this purpose application should be made to the Government Departments concerned.

In the case of the Union, requests for information in respect of vacant Crown Lands ready for occupation, the prices of the different classes of such lands, and the situation and prices of the private lands available for purchase, should be addressed to the Ministry of Lands; while information in respect of the varying conditions of the country as a whole, and the soil, climate, crops, etc., of particular districts, should be sought from the Ministry of Agriculture.[1] If the wholly British Colony of Rhodesia be preferred to the Union, applications for land should be made in writing to the Director of Land Settlement, at Salisbury or Buluwayo; and, " while the purchase of land cannot be completed in England, general information in regard to land available can be obtained at the Company's Emigration Offices at 138 Strand, London, W.C., and 140 Buchanan Street, Glasgow."

In cases where an approved applicant has decided to inspect land in Rhodesia, moreover, " arrangements are made (provided adequate notice of the date of his arrival in the country is given to the Estates Office, Salisbury or Buluwayo) to show him farms in districts which are recommended by the Company for settlement, and to assist him to procure a farm suited to his requirements. Free transport to districts which are recommended for settlement is provided for this purpose whenever possible. During the months from December to March travelling

[1] In both cases applications should be addressed to the Secretary, the Ministry of Lands (or Agriculture), Pretoria. General information, it is presumed, may be obtained in England from the High Commissioner for the Union of South Africa, 32 Victoria Street, London, S.W.

is difficult owing to the rains ; this time of year should not, therefore, be selected for inspecting land, as transport cannot be guaranteed."[1]

Cost of Living

The purchasing power of the sovereign varies considerably both as between the United Kingdom and the several oversea dominions, and as between the dominions themselves ; and it is lowest in South Africa. But the degree in which the cost of living to the European is higher in South Africa than it is elsewhere is not to be measured entirely by this fact and all that it implies. In part, the higher cost of living is the result of a higher standard of living. The reluctance of the white man to perform manual labour has some foundation in the circumstance that continuous physical effort is more exhausting in South Africa than in northern or central Europe, but it arises mainly from the feeling that to do such work is to put himself on the same social plane as the native and coloured person. It is Kafir's work, and so he speaks and thinks of it. But the same social ordinance which forbids the white man to become an unskilled labourer, also impels the married artisan to employ a Kafir "boy" to do the rough work of his house ; and the oversea Englishman in South Africa, as elsewhere, demands, and generally obtains, a "fuller life" than the home Englishman of the corresponding class. He has more opportunity to spend money upon things which minister to his comfort and his amusement, and, as a rule, he makes full use of it.

If, however, we limit the high cost of living to the effect of the low purchasing power of the sovereign, or, in other words, to the general expensiveness of things purchased in shops and of housing, whether the house be

[1] Handbook (as before). Excellent handbooks (cultural and general) for the information of prospective settlers in Rhodesia are to be obtained at the Company's Emigration Offices.

bought or rented, its origin can be traced to certain well-defined economic causes. They are:

(1) The undeveloped condition of agriculture, which makes it necessary for South Africa to import food-stuffs in large quantities; (2) the largeness of the proportion of the necessaries and conveniences of civilised life (other than food-stuffs), which, in the absence of local manufactures, have to be imported from Europe and America; and (3) the long distance of the inland centres of population from the ports, and the consequent high cost of the carriage of imported goods in a country destitute of waterways.

It will be noticed that these are causes which do not operate equally in all areas of the Union or among all classes of its European inhabitants. Obviously the farmer, whether English or Dutch, who grows most of the food necessary to support himself, his family, and his employés, is not troubled by the expensiveness of provisions in the towns. It is the Englishman, especially the English townsman, who, while engaging in all the characteristic pursuits of South Africa, wants to lose none of the advantages of England—from the "latest thing in ties" to the last new play as presented by an English company on tour—that is most affected by the high cost of living. For the information of English employés and wage-earners in general, the "South African Circular" (Oct. 1st, 1911) of the Emigrants' Information Office writes on this head:

> House rent and servants' wages in towns are dearer than in the United Kingdom, but board and lodging for single persons are reasonably cheap. The price of provisions varies a good deal; in easily accessible places it is not much more than at home, with the exception of vegetables; but in places away from railways, provisions, if not raised locally, are dear, owing to cost of transport, and fresh vegetables are very difficult to obtain. The cost of living generally is higher than in Europe. In Johannesburg, the average expenditure of an artisan and his wife, and three children under twelve years of age, for food, clothing, rent, etc.,

is estimated at £25 a month, exclusive of medical attendance, tobacco, or liquor. As the average artisan earns £26 a month, and the average clerk £20 to £24, either of them would have great difficulty in supporting a wife and young family.

And, again, the inland provinces (the Transvaal and the Free State) are naturally more affected by the cost of transport than the coastal provinces (the Cape and Natal); and the least expensive districts in the latter are those in the immediate neighbourhood of the ports. A convenient measure of the difference in the cost of living as between the coastal and the inland provinces of the Union, and as between the normal and abnormal districts in the former, is afforded by the scale of extra allowances on this account recommended by the Civil Service Reorganisation Commission. The passage, which is contained in the First Report of the Commission, is as follows:

Local non-pensionable allowances to meet the extra costs of living in the more expensive portions of the Union should be given as under:

	Transvaal.	O.F.S., Griqualand West, Bechuanaland, N.W. districts of Cape and N. of Zululand.
In cases of salaries up to £399	£60	£30
,, ,, from £400 to £700 ..	£90	£45
,, ,, from £701 to £1,000 ..	£120	£60
,, ,, above £1,000	£150	£75

In addition to the local allowance, a special allowance of £50 per annum (but not to exceed £120 in all) should be given to officers in receipt of salaries not above £700 per annum who were married prior to the date of Union and transferred from the Coast Provinces to the Transvaal. The few cases of such officers removed from the Orange Free State to the Transvaal, or from the Coastal Provinces to the Orange Free State, should be dealt with on their merits.

Remedial Measures

The evil of the high cost of living to Europeans, and, in particular, its injurious influence in checking the growth of the British industrial population in the Transvaal, were matters which claimed Lord Milner's closest attention ; and during the period of his administration of the new colonies, not only were measures taken immediately to lessen its effects, but agencies were set in motion calculated, directly or indirectly, to remove the causes from which it arose. It was to these measures and agencies, taken collectively, that Lord Milner looked for the attainment of his economic policy—a policy which aimed at making South Africa no less accessible to the British emigrant than other oversea dominions. With the one material exception of the weakening of the unskilled labour supply of the Union by the withdrawal of the Chinese coolies from the gold industry, Lord Milner's economic agencies have been maintained since he left South Africa (1905). As compared, then, with the years immediately preceding and following the war, the cost of living in general, and the cost of living in the inland provinces in particular, has been reduced appreciably ; and, in the absence of any unforeseen development, it will be reduced still further, as the remedial agencies thus initiated become increasingly effective.

The nature of these agencies will appear most clearly if they are considered in relation to the three economic causes of the high cost of living, which have been distinguished above :

(1) The inadequacy of the local food supplies. The agencies tending to remedy this defect in the economic system of the Union are (*a*) a generous measure of State-aid to agriculture—administered upon enlightened lines—and (*b*) the extension of the railway system, with the increase of railway facilities for the cheap and rapid transport of provisions from the producer to the consumer.

(2) The exceptional dependence of South Africa upon imports other than food-stuffs ; and (3) the high cost of their carriage. Pending the establishment of local

manufacturing industries, Governmental action has been, and is being, employed to cheapen the cost of transport of imported goods both by sea and land. The endeavour to reduce freights has continued from the date when, early in 1903, Lord Milner discussed the "Shipping Ring" Question with Mr. Chamberlain, to the time of writing, when negotiations are in progress between the Union Government and the representatives of the Conference lines. And so determined are the present Union Government to reduce the cost of sea carriage, that they have announced their intention of building a fleet of ocean-going steamships in the event of their failing to obtain reasonable concessions from the shipping companies concerned. As regards land transport, South Africa, as we know, having no Mississippi or St. Lawrence—practically no navigable rivers at all—is almost entirely dependent upon railways. The National Convention, therefore, with a full consciousness of this salient economic fact, has pledged the Union Government in the South Africa Act to abolish railway taxation (or, in other words, to carry goods and passengers on the State railways at cost price) within the four years terminating on May 30th, 1914. The degree in which the Union may be expected to benefit ultimately from this measure has been discussed already at some length, and it is unnecessary, therefore, to refer again in detail to this subject.[1] It may, however, be noted under this head that an important development of manufacturing industry has been made quite recently by the formation of the Union Steel Corporation (of South Africa), Ltd., in the Transvaal. The immediate purpose of this undertaking is to manufacture commercial forms of steel and iron for the local market from the large quantities of scrap discarded by the mining industry and the railways, and hitherto exported to England and other iron manufacturing countries. Its ultimate object is to

[1] See Part IV, Chap. II: p. 263.

produce and manufacture iron and steel from the ample deposits of iron ore with which the Transvaal and other parts of South Africa are enriched. Recognising the economic importance of the enterprise, the Union Government have determined to support it; and for the first sixteen years the Corporation will benefit by the following forms of State assistance. The Government will (1) raise the railway rates on exported scrap; (2) carry scrap consigned to the Corporation at reduced rates; (3) sell to the Corporation at £1 per ton all the scrap from the railway shops and plants in the Transvaal and in a large district south of this province, and, if required, scrap from other portions of the railway system at correspondingly low prices; and (4) purchase from the Corporation such iron and steel goods as are required by the railways and can be produced at the Corporation's works. The grant of these advantages is accompanied by the condition that the Corporation, after its third year of operations, will carry out certain experiments in the smelting, working, and treatment of the native iron ores. It is obvious that the importance of the enterprise lies in the fulfilment of this condition; since, if the results so obtained are successful, South Africa will begin to supply herself from her own deposits with the bulk of the iron and steel goods that she requires and now imports.

White Labour

Apart from the unsuccessful attempts to utilise white unskilled labour in the gold industry made in the period 1901–4, such labour has been employed from time to time since the war by Government, and by municipalities and other public authorities. The regular employment of white unskilled labour by the State is, however, practically confined to the railway service; and an interesting account of what has been done in the past, and a sympathetic statement of the policy of the present Union Government on this difficult and important question, are to be found in the pages of the Railway Report:[1]

> During the general depression in the year 1909 (the General Manager writes) the Cape Government provided work for several hundred white men of the labouring class. A large number are still retained in the service as "European labourers," but their employment originally was intended merely to alleviate distress.
>
> In the Transvaal and Orange Free State Provinces what is known as the "White Labour" policy, having for its object the employment of white men on classes of work previously performed almost exclusively by natives, has actually been in operation for some years past.
>
> As far back as 1904 indigent white men were engaged by the Central South African Railway Administration . . . but no decided action . . . was taken until 1907, when, with the advent of Responsible Government in the Transvaal, the Railway Committee approved of the policy of employing white men, in lieu of natives, on *all* new construction works and also on maintenance work at stations and depôts on open lines. At the end of December, 1910, the number of white labourers employed in the Transvaal and Orange Free State was 1,775 on open lines and 1,391 on construction work. . . . The Railway Administration, therefore, now employs over 3,000 white men on work previously performed by natives.
>
> . . . On economical grounds it would, therefore [*i.e.*, because of the largely increased cost of such labour], be difficult to defend the change . . . but, as a matter of fact, the employment of white labour has one result—apart from its effect on the gradual decrease of indigency—that will ultimately affect the future working of the railways, and that is, the creation, in course of time, of a class of white labourer equal to the European unskilled labourer. This in itself justifies the policy of employing white men, which should gradually be extended to all new construction work, to all maintenance work on open lines throughout the Union, and to certain other kinds of labour. With the provision of mechanical appliances for handling traffic and for other services, white labour can, and will I hope, be made use of.

At the same time, the General Manager reveals in his subsequent remarks that the application of the white labour policy even to the railway system has proved in

[1] Report of the General Manager of Railways and Harbours for the year ended December 31st, 1910. Pretoria, 1911.

many ways unsatisfactory. He notices with disapproval the tendency among the white labourers "to agitate continuously for increased pay"; and recognises that, at most, the extended use of such labour must be "gradual," and that it will for long remain "casual" in character, since white men regard such work as merely "a temporary relief to tide them over" until they can secure some permanent and more congenial employment.

This statement of the intention of the Union Government to foster the employment of white unskilled labour on the railways has been followed by definite pronouncements in favour of a white labour policy on the part of certain Ministers, and by the passage through Parliament of the Land Settlement Act, 1912—the measure in which the Government's scheme of land settlement is embodied. Among these utterances, that of General Smuts, the Minister of the Interior, is the most striking.

The Government's Scheme of Land Settlement

Speaking at Pretoria, on December 7th, 1911, this distinguished member of General Botha's ministry, is reported to have said:

> The time is coming when we shall have to get a decent class of white labour into the country. The Government ought to establish the necessary machinery for getting the labour in the first place here in South Africa. If we cannot get labour here to work at a reasonable living wage, then we can get it from other countries where it is plentiful. Nobody can tell me that what has been done in Australia, New Zealand, and Canada is impossible here in South Africa.[1]

Earlier, in the same speech, General Smuts advocated in no less emphatic language the active promotion by the State of white immigration:

> My deliberate, considered opinion (he said) is that a good sound policy of white immigration into South Africa is one of the finest acts of statesmanship which can be achieved. If we

[1] *The Times*, December 9th, 1911. Turned, however, from the *oratio obliqua* of *The Times* to *oratio recta*.

look at the last Census [1] we cannot get away from the fact that, if we are not careful, there is going to be a turn of the tide in South Africa. The population figures in South Africa for the last 200 years have always shown an increase in white as compared with black. The last figures point to the alarming fact that the coloured population of South Africa has increased in a greater proportion than the white. There is no finer country in the world, and no finer home for a great white race. If we wish to maintain and strengthen the lead we have at present, the people of South Africa will have to turn their attention deliberately to this great issue of immigration. We do not want the riff-raff of Europe here and the scum of the cities, but the same sound stock that has founded this race in South Africa. We want these people to supply new energy and new blood. [2]

The main outline of the scheme of land settlement, as disclosed in subsequent speeches and in the debates in Parliament, appear to be these: A sum of £5,000,000, of which £1,000,000 is to be expended annually for the next five years, is to be appropriated under the terms of the Act "to make further provision for the allotment for settlement purposes of Crown Land, including land acquired for such purposes, and for the improvement and disposal of such land, or for other purposes in connection therewith." In moving the second reading of the Bill in the House of Assembly, Mr. Fischer, the Minister of Lands, declared that the Government recognised the need of augmenting the supply of white labour both from within South Africa and from without. Four classes of settlers were comprised in the scheme. First, the poor white class—poor South Africans without capital—who are to be settled on the land by a system of labour colonies; second, suitable immigrants from oversea with a capital of £200 to £250, who are to be provided with the additional capital necessary for stocking their farms, etc., by advances from the Land Banks; third, local applicants and oversea immigrants with a moderate command of capital; and, fourth, applicants with large capital.

It is obvious that the value of a measure of this sort

[1] See p. 388. [2] *Ibid.*

depends entirely upon the manner in which its provisions are administered.

In the course of the second reading of the Bill, General Botha, when charged with being opposed to immigration, replied that, on the contrary, he had always encouraged it. But if General Botha was correctly reported, the form of encouragement which he proposed to adopt was one that seems scarcely likely to secure the end in view. " The Government," he said, " would first assist the poor whites in South Africa, and when they had assisted these and no further assistance was required, and they had some money over, then they could see about immigration."[1] If the Land Settlement Act is to be administered in the spirit of this remark—if, that is to say, all that is to be offered to the British immigrant is the land and money left over after the needs of the " poor whites " of the Union have been satisfied—some other instrument will be needed to give effect to General Smuts' " good, sound policy of white immigration into South Africa."

It must be noticed, however, that before the Bill came to its third reading a slight concession was made in favour of possible British applicants for land. In Select Committee, a clause enabling the Minister of Lands to receive such applications through the High Commissioner's Office in London, if he so desired, was moved by Mr. Fischer himself; and this clause, after some debate in which it was at first opposed by the Government and strongly advocated by the Opposition, was ultimately inserted in the Act.

[1] Reuter telegram in *The Times* of February 13th, 1912.

CHAPTER VI

SOCIAL AND INTELLECTUAL DEVELOPMENT

SCIENCE, Literature, and Art enter into the life of the educated South African, whether of British or Dutch descent, even as they do into the lives of their kinsmen in Europe. In some respects, South Africa—that is, of course, European South Africa—is at an advantage as compared with the other oversea dominions. The country and its aboriginal inhabitants present a wide and stimulating field of ethnological and archæological research; the clash of conflicting racial ideals has stirred her adopted sons early to literary expression, and the development of her domestic architecture, at all events, has been guided by a tradition derived from seventeenth and eighteenth century models of her own.

Prior to the Union, a varying measure of recognition and financial aid was given to religious bodies by the several Colonial governments, and some assistance out of public funds was rendered to scientific associations, libraries, museums, and art galleries. In South Africa, as in the other dominions, public opinion sets in the direction of making religious organisations entirely independent of the State; but, with this exception, it may be expected that under the Union the encouragement extended by the State to agencies and institutions calculated to promote the higher intellectual development of the community will be both more ample and more effective.

RELIGION

The towns and villages of the Union are well supplied with places of worship, varying in character from the

Gothic cathedrals[1] at the seats of the Anglican bishoprics to the primitive "Dopper" church of the Back-veld. The Dutch, naturally, are members of the Dutch Reformed Church and the Z.A. Gereformeerde Kirk, and the membership of these two bodies constitutes a majority of the European Christian community. A large and influential element of the British population belong to the Church of England; and the Roman Catholic Church and the various Nonconformist communions each possess an appreciable number of European members, although their main numerical strength lies in their coloured and native adherents.

The two following tables, which have reference to the Cape Province, will serve to indicate the relative membership of the principal religious bodies in South Africa.

Principal Religions in the Cape Province, according to the Census of April, 1904:

[Total population, 2,409,804; European population, 579,741.]

	All races.	White.
Congregationalist	112,202	4,986
Dutch Reformed Church	399,487	296,792
Methodist, Wesleyan	277,285	35,860
Presbyterian	88,653	26,357
Church of England	281,433	126,552
Roman Catholic	37,069	28,480
Greek Church	1,049	1,028
Jews	19,537	19,509
Mohammedans	22,623	48
Buddhist	489	3
Hindu	2,033	11
Confucian, or Church of China	773	—
No Religion (so stated)	1,015,760	532

[From the Statistical Register of the Cape Province, 1910.]

[1] Most of these buildings are unfinished.

Altogether, some thirty distinct religions and sects are recorded.

Number of ministers, etc., of the principal religious denominations in operation, in 1907, in the Cape Province :

	No. of Ministers.	Ministers' Stipends and Fees.	Contributions exclusive of support of Ministers.	Average Attendance at principal Sunday Service.
		£	£	
Baptist	15	2,977	2,798	2,612
Congregational ..	30	6,493	3,372	12,288
Dutch Reformed Ch...	116	34,971	57,575	37,254
Church of England ..	153	30,915	25,813	31,014
Methodist, Wesleyan .	195	No information.	No information.	121,055
Presbyterian	34	7,077	5,219	11,080
Roman Catholic ..	55	3,007	769	5,823
Total for all twenty two denominations	708	101,321	112,301	247,610

[From the same.]

In the Estimates for 1911–12 certain grants to religious bodies made by the Union Government are included in the Vote for the Ministry of Finance. Under the sub-head "Allowances to Chaplains (Cape of Good Hope)," the sum of £1,928 is appropriated ; and grants to various churches in the Free State amount to £7,961, the Dutch Reformed Church receiving £5,928 of this latter sum. With reference to the first of these items, it may be explained that during the first half century of British occupation the Cape Government made regular grants in support of the ministers of the various denominations ; and in 1853, when representative institutions were introduced, the annual sum provided on this account amounted to £16,060. The Constitution Ordinance of that year required that an amount not less than this sum should be appropriated out of revenue annually to the service of "Religious Worship." In 1875, however, after the establishment of Responsible Government, the "Voluntary Principle" was adopted by the Cape Parliament, and an Act was passed providing for the gradual extinction of the grants. As the stipends of existing incumbents, and of such of their successors as were appointed within five years of the taking effect of the Act, were secured to them until death or resignation, the process has been lengthy; and the vote, although reduced to the figure shown above, still appears upon the annual estimates.

The most striking and characteristic effort of the churches in South Africa is, of course, the instruction of the native races in the doctrines of Christianity. Although some serious indiscretions were committed by individual missionaries, and by missionary organisations, in the early part of the nineteenth century, the introduction of Christianity, taken as a whole, has not only advanced the moral and material condition of the natives, but has increased appreciably their economic value to the European community. In particular, as we have had occasion to notice elsewhere,[1] practically the whole of the work of native education, as at present conducted, has been relegated to the various missionary organisations by the Union and other South African Governments.

Museums and Libraries

The advancement of science, literature, and art, no less than that of religion, has been promoted, in the main, by private enterprise. The collection of MSS. and rare books presented by Sir George Grey is a chief possession of the South African library at Capetown, and the public library at Johannesburg has been enriched similarly by the gift of the Seymour Memorial Library.

[1] "Education," Part V, Chap. IV, p. 477.

The Chartered Company and Rhodes provided funds for the scientific investigation of the Semitic remains in Mashonaland and Matabeleland; and Rhodes himself, besides being responsible directly or indirectly for the two or three monuments of real merit which the country at present possesses, exercised a powerful and salutary influence upon the recent development of architecture in South Africa. The Johannesburg Art Gallery, the one public collection of paintings of artistic value, has been brought into existence by the efforts of Lady Lionel Phillips. These instances by no means exhaust the list of private benefactors of this class, and take no account of the many generous gifts devoted to educational purposes.

A general idea of the chief scientific and literary institutions of the Union, and of the degree in which they are assisted by the State, may be gathered from the estimates for 1911–12. Among various miscellaneous provisions included in the Education vote is the grant-in-aid of £250 to the Royal Society of South Africa. Under sub-heads Q and Z of the vote for the Minister of the Interior, we find provision is made for grants-in-aid to museums, libraries and reading-rooms, art galleries, and art associations and herbariums. The museums include those of Capetown (the South African Museum), Grahamstown, Kimberley, King William's Town, Port Elizabeth, Maritzburg, Durban, and Bloemfontein; and the grants which they receive amount collectively to £9,170. The total sum granted to public libraries and reading-rooms is £15,400, and the chief recipients are the South African Library at Capetown (£1,050), and the libraries at Pretoria and Johannesburg (each £1,300). In the Cape Province the library grants run from £15 to £150, and in the Transvaal from £25 to £100; and in both of these provinces the public libraries are fairly numerous. In this connection it must be remembered that the

Union Government is spending £90,000 on building the new Museum and Government Library at Pretoria.

The grants to art institutions amount collectively to £1,070. The South African Fine Arts Association, and Gallery, at Capetown, receive respectively £100 and £500; the Grahamstown Fine Arts Association, £100; the Durban Art Gallery, £135; the Maritzburg Art Gallery, £135; and the Arts and Crafts Association, £100. Here, again, it must not be forgotten that both the (late) Transvaal Government and the Municipality of Johannesburg have given, and are giving, generous assistance to the Johannesburg Art Gallery, both in respect of the actual building and the subsequent maintenance of the institution.

The following return, which is taken from the Statistical Abstract of the Cape Province, will indicate the character of the public libraries in South Africa, and the degree in which they are utilised by the people.

Position of the 157 public libraries in the Cape Province at December 31st, 1909:

No. of Volumes added in 1909.	Volumes (excluding Periodicals) now on shelves.	Average Monthly Circulation.	Subscribers.	Average Visitors *per diem*.
30,492	610,676	57,160	10,151	4,360

Games and Sports

What Horace said of the Roman emigrant of his day: *Cœlum non animum mutant, qui trans marc currunt*—our colonists change their temperature, but not their temperament—may be applied with equal truth to the oversea British. In South Africa, as in the other dominions, the new arrival from the Homeland will find the social distinctions and pursuits with which he is familiar ; and in the matter of outdoor games and sports he will gain probably more than he has lost.[1] There will be some changes, of course, arising out of the changed conditions—climatic and general—of the new land. Little or no riding to hounds is to be enjoyed, and the shooting is of an entirely different order. He will bowl down a cricket pitch of matting, and discover greens on the golf courses formed of fine sand instead of grass. But notwithstanding such shortcomings, he will find that cricket, football, lawn-tennis, golf, shooting, fishing, riding, bicycling, motoring, yachting, and horse-racing can be pursued with the same keen enjoyment as in England. Nor must it be forgotten that under the new Union Defence Act the Englishman in South Africa will have excellent opportunities of learning the use of the rifle, and of becoming reasonably efficient in the duties of a citizen-soldier.

Literature

Apart from the historical and scientific works among which the many volumes of Mr. George McCall Theal's History stand pre-eminent, European South Africa can show an output in creative literature and *belles lettres* which, alike in quality and quantity, is remarkable for so young a country. Thomas Pringle's poetry, from which couplets and stanzas have been quoted in an early chapter,[2] has little genuine poetic quality, but its descriptive power gives it a certain value and interest. And although " Afar in the Desert " and his other South African poems have failed to retain the favour with which they were received by his contemporaries, Pringle none the less has performed an appreciable service to South

[1] An exception must be made in the case of the purely Dutch country districts.

[2] Part I, Chap. II, p. 20 *et seq.*

African letters by showing at the outset that the conflict of European civilisation with the primitive manners and emotions of the native African races was a subject rich in poetic motives. To him, therefore, is due in large measure the fact that to-day there exists in South Africa a group of Civil servants and Native Commissioners capable of writing verse of no inconsiderable interest and promise. But our attention is diverted from both the work of this group and from the many examples of South African prose-fiction of lesser but still considerable merit by two arresting works of creative literature—*The Story of an African Farm* and *Jock of the Bushveld.* The former of these, written now some thirty years ago, has made the name of "Olive Schreiner" (Mrs. Cronwright Schreiner) familiar throughout the English-speaking world; the latter has added a fresh distinction to the otherwise notable career of Sir Percy Fitzpatrick.

The presentation of the gospel of the emancipation of women, which first won for *The Story of an African Farm* the attention of the reading public, has lost its novelty to-day, but the literary quality of this first work of "Olive Schreiner"—a careful, almost photographic, exactness of representation, which recalls an interior of the Dutch School of painting—remains unaffected by the lapse of time. The book is, as every one knows, a picture of life in a Dutch homestead in the old colony, in which the salient features of Dutch Afrikander thought and character, and the typical moods of nature in the Karoo, are faithfully reproduced. At the time when it was written, Miss Olive Schreiner, the daughter of a Lutheran minister at Capetown, was exceedingly young, and she was actually living amid the surroundings which she has depicted. A perusal of this book, therefore, affords the reader a useful insight into the thought and manners of the rural Dutch Afrikanders. And in illustration of this aspect of "Olive Schreiner's" work, I may, perhaps, be allowed to recall some passages to which I drew attention in an earlier book.[1]

"To begin, let us observe how Nature is drawn in its most characteristic aspect in South Africa.

"First the long period of drought:

From end to end of the land the earth cried for water. Man and beast turned their eyes to the pitiless sky, that, like the roof of some brazen oven, arched overhead. On the farm, day after day, month after month, the water in the dams fell lower and lower; the sheep died in the fields; the cattle, scarcely able to crawl, tottered as they moved from spot to spot in search of food. Week after week, month after month, the sun looked down from the cloudless sky, till the Karoo bushes were leafless sticks, broken into the earth, and the earth itself was naked and bare; and only the milk bushes, like old hags, pointed their shrivelled fingers heavenwards, praying for the rain that never came.

"Then the torrential rain:

Outside the rain poured; a six months' drought had broken, and the thirsty plain was drenched with water. What it could not swallow ran off in mad rivulets to the great "sloot" that now foamed like an angry river across the flat. Even the little furrow between the farmhouse and the kraals was now a stream, knee-deep, which almost bore away the Kafir woman who crossed it. . . . The fowls had collected, a melancholy crowd, in and about the wagon-house, and the solitary gander who alone had survived the six months' want of water, walked hither and thither, printing his webbed foot-marks on the mud, to have them washed out the next moment by the pelting rain.

"And, afterwards, the 'princely day' which follows the breaking of the drought:

The long morning had melted slowly into a rich afternoon. Rains had covered the Karoo with a heavy coat of green that hid the red earth everywhere. In the very chinks of the stone walls dark green leaves hung out, and beauty and growth had crept even into the beds of the sandy furrows and lined them with weeds.

"Then there is the farmhouse on the Karoo and its inhabitants. In spite of the strangeness of the surroundings, how familiar the author has made it all;

[1] In *South Africa: A Study in Colonial Administration and Development*, 1895.

and how well we seem to know the chief personages of the story. Tant' Sannie, the Boer woman, with her grossness of person and language; Em and Lyndall, the two little English girls, whose dead father married Tant' Sannie to provide them with a protector; the kindly old German overseer and his son Waldo; the vagabond Englishman who supplants him; Greg, who takes half the farm off Em's hands when she has grown up; even the Hottentot maids and the Kafirs, the Boer visitors, and the strangers who come and go—all assume reality under the deft portraiture of "Olive Schreiner."

"Tant' Sannie does not occupy much space upon the canvas, but the figure is distinct and vigorous. As she sits in her elbow chair, sipping her coffee, with her feet comfortably resting upon the wooden stove, and her Hottentot maid in attendance, she seems to personify the Africander Dutch in every inch of her huge frame. Apart from her external appearance, she displays the typical mental attitude in its two characteristic traits—she is superstitious and averse to change. For this gross creature, in spite of her violent language and material philosophy, 'was a firm believer in the chinks in the world above, where not only ears, but eyes, might be applied to see how things went on in this world below. She never felt sure how far the spirit world might overlap this world of sense, and, as a rule, prudently abstained from doing anything which might offend unseen auditors.'

"Equally characteristic is her reproval of Em for using soda, instead of milk-bushes, to make soap.

Not that I believe in this new plan of putting soda in the pot. If the dear Father had meant soda to be put into soap, what would he have made milk-bushes for, and stuck them all over the veld as thick as lambs in the lambing season?

"And again—

My mother boiled soap with bushes, and I will boil soap with bushes. If the wrath of God is to fall upon this land (said Tant' Sannie, with the serenity of conscious virtue), it shall not be through me. Let them make their steam-wagons and their fire-carriages; let them go on as though the dear Lord didn't know what He was about when He gave horses and oxen legs—the destruction of the Lord will follow them. I don't know how such people read their Bibles. When do we hear of Moses and Noah riding in a railway?"

As a study of South African life, Sir Percy Fitzpatrick's work, alike in its scenery and characters, provides an excellent counterpart to that of "Olive Schreiner." It is a story of trekking and hunting, and transport riding in the days of gold discovery in the early eighties; and the scene is laid not in the flat and arid Karoo, but in the well-watered valleys and mountains of the north-east Transvaal and seawards beyond the great ranges.

Apart from Jock, the central figure of the canvas, and the wild life of the Bushveld and the Berg, there is excellent character drawing; but the white men are English, not Dutch, and the natives are natives as developed under English, not Dutch, influence. In Jim Makokel, who "catch 'em lion 'live,'"; in Sam (Sam no good; Sam leader Bible); and in the rest we have a vivid and truthful presentation of the working of the native mind. Among the white men, "old Charlie Roberts," the driver, "a really first-class man" with the transport oxen; and "Rocky the Hunter," "who had the knack of getting at the heart of things, and putting it all into the fewest words," leave, perhaps, the clearest images in the mind. And it is in respect of "old Charlie's" calling that Sir Percy writes some wise words, which state with simple directness the truth that physical as well as moral superiority is needed to make the relationship of the whites and blacks in South Africa of mutual benefit.

Patience, understanding, judgment, and decision: these are the qualities it calls for, and here, again, the white man justifies his claim to lead and rule; for, although they are as ten or twenty to one, there is not a native driver who can compare with the best of the white men.

Almost as intimate a knowledge is shown of the animals, wild and tame. Jock, the hunting dog, must be seen as he lives and moves in the pages of the book; but lesser types may be recalled. Among these are "the old Rooster whose name was Pezulu," and Snowball, the hunting horse, who was "no fool," and "had no unpractical prejudices: he objected to work—that was all." Old Zwaartland, too, the front ox, who "with his steady, sober air, perfect understanding of his work, and firm, clean, buck-like tread," still led the front span in the hour of dire extremity; and Zole, "the little, fat schoolboy, always out of breath, always good-tempered and quiet, as tame as a pet dog," are figures that linger in the memory. Nor must even this glimpse between the covers of Sir Percy's book fail to mark the close observation of wild animals and their ways, and the knowledge of the hunter's craft, that will delight and instruct the sportsman in equal measure.

Architecture

On more than one occasion the reader has been reminded that the Cape Peninsula, Stellenbosch, Fransch-hoek, and the Paarl—the districts in which the first genuine Dutch and French settlers made their homes—can show many modest but pleasing examples of seventeenth and eighteenth century buildings. The "Castle," on the shores of Table Bay, which replaced the original fort constructed by Van Riebeck, was planned by no less a person than Vauban, the famous military engineer. Its most noticeable architectural features are the entrance gateway, from which rises an octagonal clock-tower surmounted by a low cupola, and the portico in the court-yard which leads to what was once the residence of the Dutch Governor of the colony. This latter, with its serpentine balustrade and base united by graceful columns, is, perhaps, the best single "piece" to be found among the architectural designs of this class in the Cape Province. Apart from the "Castle," the only noteworthy public buildings of the period are the Stadthaus and the tower of the Dutch Church, both of which are of considerable interest. On the other hand, the older streets of Capetown hold numerous eighteenth century private houses, whose fronts are decorated with pilasters and pediments, and furnished with stoeps, heavy window frames, enriched doorways, and other characteristic features.[1]

THE AFRIKANDER HOUSE

It is, however, in the homesteads of the Cape Peninsula and the near mainland that interest centres, since these houses display certain departures from their contemporary Dutch and Flemish models—departures the joint product of Eastern influences and local conditions—which entitle them to be regarded as constituting collectively a separate architectural type, analogous to that to which in the United States the term "Colonial" is applied. Notable examples of such Afrikander houses in the Cape Peninsula are afforded by Tokay (now occupied by the Woods and Forestry Department), Groot Constantia (now the Government Wine Farm), Alphen, and (the reproduced) Groote Schuur. The features which differentiate the Afrikander house are the wide, central passage, or hall, running from front to back, and the Stoep, or raised platform of brick, upon which this passage opens at either end. The purpose of both these features is to enable the household to secure fresh air and coolness during the eight out of the twelve months of the year, in which the sun shines almost continuously. They were introduced, no doubt, from Batavia, the head-quarters of the Dutch East India Company, where the need for air

[1] One of the best examples of the Afrikander townhouse is the residence of the late Mde. Coepmans de Wet in Strand St.

and coolness was much greater, and where they remain all-important elements in the homes of the Dutch and English residents of to-day. From the central hall, which in the Afrikander house is commonly divided mid-way by a wooden screen for domestic convenience, there open the doors which give access to the living rooms thrown out on either side ; and thus by opening the entrance doors at either end of the hall a current of air is obtained which can be circulated easily throughout the whole of the house. The stoeps run the entire length of the eastern and western fronts, and here the members of the household can sit at ease in the open air, choosing the eastern or western stoep according to the position of the sun. The Afrikander stoep, however, differs in one respect from the Batavian portico. With the exception of the occasional instances in which it bears a screen of vines or creepers carried upon columns, such as may be seen at Tokay, it is left entirely open and uncovered, with nothing more than a low parapet and possibly seats of stuccoed brick at either end.

Apart from these differentiating features, the Afrikander house is frankly Dutch or Flemish, with here and there a wealth of plaster enrichment to be traced probably to the Huguenot influence. It is usually one-storeyed, and the low walls, built solidly of brick by slave labour, are plastered and whitewashed, so that their gleaming surfaces contrast sharply with the coal black thatch of the roofs. The gables show the characteristic Flemish curves, and in the Huguenot districts especially are frequently decorated with scrolls, geometrical patterns, rosettes, and other forms of plaster work. The windows are fitted with sashes, and show the thick bars and massive frames that dignify the windows of the English house of the late Stuart period ; and the doors, with their fanlights above, are set between pilasters and surmounted by an entablature or hood, variously enriched.

The courtyards formed at the back and front of the house by servants' quarters, stables, granaries, wine presses and stores, and the like outbuildings, are shaded by trees, and the whole group of buildings is approached by wide avenues of oaks.

South African Domestic Architecture

The originating impulse which has led to the growth of a style of domestic architecture in South Africa, based upon the Afrikander homestead, was given by Rhodes. In the first place, he acquired for his own home a typical house of this class in Groote Schuur, then the residence of a Dutch lady, and known as "The Grange," Rondesbosch. In the almost complete destruction of Groote Schuur by fire, and its subsequent restoration, Rhodes found the opportunity of securing for South Africa the services of Mr. Herbert Baker, the gifted architect, who in recent years has developed out of the Afrikander model a style of domestic architecture not unworthy of adoption by the new nation.

In addition to the reconstructed Groote Schuur, Bedford Farm, Sir George Farrar's country house, near Johannesburg, and Government House, near Pretoria, the official residence of the Governor-General of the Union, may be mentioned as recent and notable examples of Mr. Baker's work. There are, however, many "Baker" houses, large and small, to be found in the Transvaal and elsewhere.

The Rhodes Memorial

The Union Buildings, which, like the Rhodes' Memorial, have been designed by Mr. Baker, have been described elsewhere. It only remains, therefore, to remind the reader that, when they are completed, Pretoria will be possessed of an architectural work of a magnitude and dignity hitherto unattempted in South Africa.

The three public monuments of recognised merit which South Africa possesses to-day are the Alan Wilson Memorial on the Matopos, the Kimberley Siege Memorial, and the Rhodes Memorial on Table Mountain. Some reference has been made to each of these in the preceding pages, but a few words of more precise description may be added here in respect of this last work—the latest and most considerable of the three.

The Rhodes Memorial, designed by Mr. Baker, is placed on the slope of Table Mountain above the teak bench, which was Rhodes' favourite seat in the grounds of Groote Schuur. The site commands a threefold outlook—in front, the green and silver isthmus, with the vast, dim mainland of Africa beyond; on the left, the circle of Table Bay and the Atlantic; on the right, the wooded spurs and foothills of the Table Range, False Bay, and the Pacific. The value of this wide outlook was recognised by the first Dutch settlers, and they kept a look-out post, which survives[1] in the block-house on the Devil's Peak, at a higher point of this same angle of the mountain, to warn them of the approach of raiding Hottentots or of other unwelcome visitors arriving by land or sea. There, fashioned like its mountain home of granite, a winged colonnade, built in the Doric manner of Sicily, is set high above four flights of wide steps, which fall to a massive rounded bastion at their base. At the end of each plinth is a recumbent lion in bronze; and the approach to the central member of the colonnade is thus guarded by eight lions—four on either side. The downward sweep of the lowest flight of steps is broken by a boldly projecting mass of granite, from which leap out the eager forms—man and horse—of Watts' "Physical Energy." Within the colonnade stands a sculptured

[1] Unless it has been removed in recent years.

By permission of *W. Basil Worsfold*

BEDFORD FARM IN 1904 (THE RESIDENCE OF SIR GEORGE FARRAR)

head emblematic of the Spirit of Table Mountain, which, like the bronze lions, comes from the hand of the late J. M. Swan, while at the sides are spaces ready to be filled with bronze panels depicting actual scenes in the life of Rhodes.

Thus enshrined, it may well be that:

His great and brooding spirit still shall quicken and control,
Living he was the land, and dead his soul shall be her soul.

INDEX

INDEX

INDEX

34A—(2139)

INDEX

INDEX

INDEX

INDEX

THE END

Printed by Sir Isaac Pitman & Sons, Ltd., Bath.
(2130)

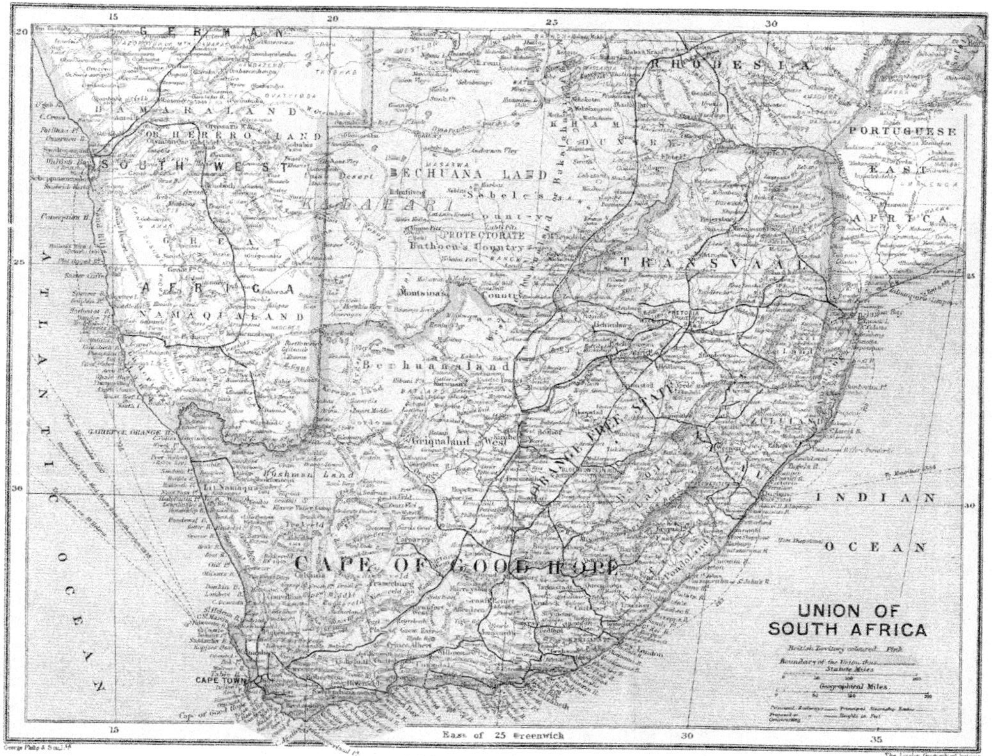

George Philip & Son Ltd
Sir Isaac Pitman & Sons, Ltd. London & Bath
The London Geographical Institute

PITMAN'S
CATALOGUE OF GENERAL LITERATURE

ART

GREAT PAINTERS OF THE 19th CENTURY AND THEIR PAINTINGS. By LÉONCE BÉNÉDITE, Keeper of the Musée National de Luxembourg. With over 400 illustrations and 13 coloured plates. In large demy 4to, cloth gilt, gilt top, 10s. 6d. net.

"It is a splendid survey of the progress of painting in Europe and America during the nineteenth century, and combines art criticism with biography in a scholarly and instructive manner. *Western Mail.*

THE HISTORY OF MUSIC: A Handbook and Guide. By WALDO SELDEN PRATT. With 130 illustrations and three maps. In demy 8vo, cloth gilt. 7s. 6d. net.

"A most convenient and valuable work of reference . . . the book may be said to cover the whole extensive field to which it is devoted, in a remarkably thorough and comprehensive fashion." —*Westminster Gazette.* "Indispensable in the music-lover's library."—*Pall Mall Gazette.* "A book which for terseness and inclusiveness has never been equalled in music literature."—*Sheffield Telegraph.*

COLOUR PRINTING AND COLOUR PRINTERS. By R. M. BURCH With a chapter on Modern Processes by W. GAMBLE. With 23 colour prints and 8 half-tone illustrations. In super royal 8vo, cloth gilt. 12s. 6d. net.

"In his excellent work recently published on this subject a work which is to be heartily commended for the thorough knowledge it displays of colour printing in all its phases as well as for the clear and pleasant style in which this knowledge is communicated—Mr. Burch has traced the history of the colour print from the first doubtful experiments of the Fifteenth Century down to the present day."—*Morning Post.*

WHO'S WHO IN THE THEATRE. A Biographical Record of the Contemporary Stage. By JOHN PARKER. In demy 8vo, cloth gilt, 750 pp., 6s. net; leather, 8s. 6d. net. In addition to the biographical section, the book contains many other interesting features, including a calendar of notable theatrical events; full details of the new play productions and important revivals of the year in London, the Provinces, New York, and Paris; Genealogical Tables of famous theatrical families, compiled by Mr. J. M. Bulloch, the Editor of *The Graphic*; a most exhaustive Dramatic and Musical Obituary; full particulars of the principal theatres in London, New York, Paris, and Berlin, with the seating plans of the leading London theatres, etc.

WHO'S WHO IN MUSIC. A Biographical Record of Contemporary Musicians. Compiled and Edited by H. SAXE WYNDHAM and GEOFFREY L'EPINE. Large crown 8vo, cloth, 6s. net; Leather, 8s. 6d. net.

The first edition of the work contains about 1,000 biographies of the leading British, American, Continental, and Colonial Musicians, including Concert and Opera Singers, Concert Agents, Critics, Managers, with detailed accounts of their careers. There are many other features of interest and value.

BIOGRAPHY

JOHN BUNYAN: His Life, Times and Work. By the Rev. JOHN BROWN, B.A., D.D. With portrait and illustrations by WHYMPER. Cheap edition. In demy 8vo, cloth gilt, 7s. 6d.

"The best life of John Bunyan."—*Literary World.*

(*See also* **Dainty Volume Library,** page 4.)

MRS. GASKELL. Haunts, Homes, and Stories. By MRS. ELLIS H. CHADWICK. New, revised and cheaper edition. In demy 8vo, cloth gilt, 37 illustrations. 5s. net.

"The volume is certain of an enduring place among those which deal with the literary history of this country, and it is certainly indispensable to any who wish to understand the woman of whose life it tells, or the value of her work and influence . . . indeed a sympathetic and faithful picture not only of Mrs. Gaskell, but also of the days in which she lived."—*Manchester Daily Despatch.*

THE LIFE OF DANTE. By the late E. H. PLUMPTRE, D.D., Dean of Wells. Edited by ARTHUR JOHN BUTLER. In fcap. 8vo, lambskin gilt, 2s. 6d. net. Also in cloth, 1s. 6d. net.

THE LIFE OF SAMUEL JOHNSON, LL.D. By JAMES BOSWELL, Newly edited with notes by ROGER INGPEN. With 568 illustrations, including 12 photogravure plates, fully indexed. In two vols., crown 4to, half morocco, 21s. net. (Also in two vols., handsome cloth gilt, 18s. net.)

"A singularly complete and attractive edition. The greatest judgment has been shown in selecting pictures which should illustrate Johnson's period, and bring before the reader's eye the actual features of the men and women among whom he moved. Altogether the New 'Boswell' is one which will be certain to secure a fresh band of admirers for a work which will ever remain one of the treasures of our literature."—*Westminster Gazette.*

BISHOP WALSHAM HOW. A Memoir. By his Son, FREDERICK DOUGLAS HOW. Cheap Edition. In crown 8vo, cloth gilt, 6s.

"Extremely well done . . altogether a book which cannot be read without profit and encouragement."—*Guardian.*

GEORGE MACDONALD. A Biographical and Critical Appreciation. By JOSEPH JOHNSON. In crown 8vo, cloth gilt 2s. 6d. net.

THE LIFE OF SIR ISAAC PITMAN (Inventor of Phonography). By ALFRED BAKER. "Centenary Edition." In demy 8vo, cloth gilt, with about 50 illustrations, including photogravure and steel plates. 2s. 6d. net.

"The book is very well done. It gives a life-like picture of a strenuous reformer, an original personality, an inventor to whom every newspaper, every public body, and every great business house owes an incalculable debt."—*Christian World.*

LIFE OF REGINALD POLE. By MARTIN HAILE. Second, Revised, and Cheaper edition. In demy 8vo, cloth gilt, with eight illustrations, 7s. 6d. net.

"An excellent book, based on a first-hand acquaintance with documents, some of which are here utilised for the first time. It gives a vivid and most faithful picture of the last Archbishop of Canterbury who acknowledged the See of Rome."—*Daily Chronicle.*

THE LETTERS OF PERCY BYSSHE SHELLEY. Containing about 480 letters. Collected and edited by ROGER INGPEN. With 42 illustrations and two photogravures. New and cheaper edition, with corrections and additional matter. In two volumes, large crown 8vo, cloth gilt, gilt top, 12s. 6d. net. Hand-made paper *edition de luxe*, half leather, large demy 8vo, 21s. net.

"Mr. Ingpen has done all that can be done to provide us with a perfect edition of one of the most interesting series of letters in English literature. The edition is worthy of the magnificent material with which it deals."—*Daily News.*

REMINISCENCES OF MY LIFE. By Sir CHARLES SANTLEY. In demy 8vo, cloth gilt, gilt top, with 15 illustrations, 16s. net.

"Not a trace of the weary veteran is discernible in this entertaining volume."—*The World.*

COLLECTIVE BIOGRAPHIES

GREAT ASTRONOMERS. By Sir ROBERT BALL. Illustrated. In demy 8vo, cloth gilt, gilt top, 3s. 6d. net

THE HEROIC IN MISSIONS. Pioneers in six fields By the Rev. A. R. BUCKLAND, M.A. In crown 8vo, cloth gilt, 1s. 6d.

MODERN PAINTERS AND THEIR PAINTINGS. By SARAH TYTLER. For the use of Schools and Learners in Art. In crown 8vo, quarter cloth gilt, 4s. 6d.

MUSICAL COMPOSERS AND THEIR WORKS. By the same Author. For the use of Schools and Students in Music. Revised. In crown 8vo, quarter cloth gilt, 4s. 6d.

THE OLD MASTERS AND THEIR PICTURES. By the same Author. For the use of Schools and Learners in Art. New and enlarged edition. In crown 8vo, quarter cloth gilt, 4s. 6d.

THE ORGAN AND ITS MASTERS. A short account of the most celebrated organists of former days, and of the present time, together with a brief sketch of the development of organ construction, organ music, and organ playing. By HENRY C. LAHEE. In large crown 8vo, cloth richly gilt, gilt top, with 14 full-page plate illustrations. 6s. net.

MODERN COMPOSERS OF EUROPE. Being an account of the most recent musical progress in the various European nations with some notes on their history, and critical and biographical sketches of the contemporary musical leaders in each country. By ARTHUR ELSON. In large crown 8vo, cloth gilt, gilt top, with 24 full-page plate illustrations. 6s. net.

PITMAN'S DAINTY VOLUME LIBRARY

Each in fcap. 8vo, limp lambskin gilt, gilt top, with Photogravure Frontispiece, 2s. 6d. per volume net.

DANTE. THE DIVINA COMMEDIA AND CANZONIERE. Translated by the late DEAN PLUMPTRE. With Notes, Studies, Estimates, and Life. In five volumes.

THE LIFE OF DANTE. By the same Author. In one volume.

THE TRAGEDIES OF ÆSCHYLOS. Translated by DEAN PLUMPTRE. In two volumes.

THE TRAGEDIES OF SOPHOCLES. Translated by DEAN PLUMPTRE. In two volumes

BOSWELL'S LIFE OF JOHNSON. (Abridged.) With an Introduction by G. K. CHESTERTON In two volumes.

THE POETRY OF ROBERT BROWNING. By STOPFORD A. BROOKE, M.A., LL.D. In two volumes.

TENNYSON: HIS ART AND RELATION TO MODERN LIFE. By STOPFORD A. BROOKE, M.A., LL.D. In two volumes.

JOHN BUNYAN: HIS LIFE, TIMES AND WORK. By JOHN BROWN, D.D. In two volumes.

CLOUGH, ARNOLD, ROSSETTI, AND MORRIS: A Study. By STOPFORD A. BROOKE, M.A., LL.D. In one volume, with four illustrations, 306 pp., 3s. 6d. net.

FICTION

BY WHAT AUTHORITY? By ROBERT HUGH BENSON. 6s.

THE LIGHT INVISIBLE. By ROBERT HUGH BENSON. 3s. 6d.

RICHARD RAYNAL, SOLITARY. By ROBERT HUGH BENSON, 3s. 6d.

THE KING'S ACHIEVEMENT. By ROBERT HUGH BENSON 6s.

THE QUEEN'S TRAGEDY. By ROBERT HUGH BENSON. 6s.

THE SENTIMENTALISTS. By ROBERT HUGH BENSON. 6s

A MIRROR OF SHALOTT. By ROBERT HUGH BENSON. 6s.

LORD OF THE WORLD. By ROBERT HUGH BENSON. 6s.

MY LADY OF AROS. A Tale of Mull and the Macleans. By JOHN BRANDANE. Coloured frontispiece. Cheaper Edition 2s. net.

MEN OF THE MOSS-HAGS. By S. R. CROCKETT. Illustrated. 6s.

A DAUGHTER OF THE SNOWS. By JACK LONDON. 6s.

ANNE OF GREEN GABLES. By L. M. MONTGOMERY. 6s.

ANNE OF AVONLEA. By the same Author. Coloured frontispiece. 6s.

KILMENY OF THE ORCHARD. By the same Author. With four coloured illustrations. 6s.

THE STORY GIRL. By the same Author. Coloured frontispiece. 6s.

THE GLORY OF THE CONQUERED. The Story of a Great Love. By SUSAN GLASPELL. 6s.

THE UNDER TRAIL. By ANNA ALICE CHAPIN. With illustrations. 6s.

THE PLEASURING OF SUSAN SMITH. By HELEN M. WINSLOW. With illustrations. 3s. 6d. net.

THE ISLAND OF BEAUTIFUL THINGS. A Romance of the South. By W. A. DROMGOOLE. With four coloured illustrations. 6s.

POLLYANNA. The "Glad" Book. By ELEANOR H. PORTER. Illustrated, 6s.

HISTORY

THE ENGLISH IN CHINA. Being an account of the Intercourse and Relations between England and China, from the year 1600 to the year 1843 and a summary of Later Developments. By J. BROMLEY EAMES, M.A., B.C.L. In demy 8vo, cloth gilt, gilt top, with maps and illustrations. 20s. net.

OUTLINES OF THE ECONOMIC HISTORY OF ENGLAND. A Study in Social Development. By H. O. MEREDITH, M.A., M.Com. In demy 8vo, cloth gilt, 5s. net.

THE BRITISH MUSEUM: ITS HISTORY AND TREASURES. A view of the origins of that great Institution, sketches of its Early Benefactors and Principal Officers, and a survey of the priceless objects preserved within its walls. By HENRY C. SHELLEY. Author of *Inns and Taverns of Old London*. With fifty illustrations. Size 6¼ in. by 9⅝ in., elaborate cloth gilt, gilt top, 12s. 6d. net.

PRINCE CHARLES EDWARD. His Life, Times, and Fight for the Crown. By J. CUTHBERT HADDEN. In demy 8vo, cloth gilt, with illustrations, 7s. 6d. net.

THE ROMANTIC STORY OF THE MAYFLOWER PILGRIMS AND ITS PLACE IN THE LIFE OF TO-DAY. By A. C. ADDISON. With numerous original illustrations. Size 6½ in. by 9⅝ in., cloth gilt, gilt top, 7s. 6d. net.

INNS AND TAVERNS OF OLD LONDON. Setting forth the historical and literary associations of those ancient hostelries, together with an account of the most notable coffee-houses, clubs, and pleasure gardens of the British metropolis. By HENRY C. SHELLEY. In large crown 8vo, cloth gilt, gilt top, with coloured frontispiece and 48 other illustrations. 7s. 6d. net.

OLD COUNTRY INNS. By HENRY P. MASKELL and EDWARD W. GREGORY. With 50 illustrations by the authors. In large crown 8vo, cloth. New Edition with Lists of Old Inns, "Trust" Inns, and Glossary of Signs. 3s. 6d. net.

"Messrs. Maskell and Gregory have written this history of theirs very well indeed. They classify the inns of England according to their origin, rating them as manorial, monastic, Church inns, and so on. They discourse in a pleasant gossipy strain on coaching inns, wayside inns, haunted inns, the inns of literature and art, historical and fanciful signs and curious signboards; of inn furniture, etc."—*Bookman.*

FLEET STREET IN SEVEN CENTURIES. Being a History of the growth of London beyond the Walls into the Western Liberty and of Fleet Street to our time. By WALTER GEORGE BELL. Author of *The Thames from Chelsea to the Nore.* With a Foreword by Sir Wm. P. Treloar, Bt. With 46 illustrations. Drawings by T. R. Way, Hanslip Fletcher, R. Anning Bell, T. E. Knightley; reproductions of old prints, original documents, maps and photographs. In demy 8vo, cloth, gilt top, 15s. net.

"Mr. Bell has chosen the picturesque and episodic method for history, yet it is, *par excellence*, an antiquary's book. Records, rolls, inventories, the proceedings of wardmotes have been laboriously ransacked; but the result is never tedious, and **we have seldom taken up a book of this kind which has had such a power of leading the reader on from page to page.**"—*Athenæum.*

THE FRANCO-PRUSSIAN WAR AND ITS HIDDEN CAUSES. By ÉMILE OLLIVIER, *of the Académie Française.* Translated from the French with an Introduction and Notes by G. BURNHAM IVES. In demy 8vo, cloth gilt, with eight illustrations, 558 pp., 8s. 6d. net.

A HOSPITAL IN THE MAKING. A History of the National Hospital for the Paralysed and Epileptic (Albany Memorial), 1859-1901. By B. BURFORD RAWLINGS, Author of *The Chronic Indigence of Hospitals*, etc. In crown 8vo, cloth, with illustrations, 5s. net.

A HUNDRED YEARS OF IRISH HISTORY. By R. BARRY O'BRIEN With Introductions by JOHN E. REDMOND, M.P. New Edition. In crown 8vo, cloth, 184 pp., 1s. 6d. net.

THE DISSOLUTION OF THE MONASTERIES. As illustrated by the Suppression of the Religious Houses of Staffordshire. By FRANCIS AIDAN HIBBERT, M.A., *of St. John's College, Cambridge; Headmaster of Denstone.* In crown 8vo, cloth gilt, 5s. net.

JOHN PYM. By C. E. WADE, M.A., *Barrister-at-law.* With frontispiece. Demy 8vo, cloth gilt, gilt top, 7s. 6d. net.

"Mr. Wade gives a fresh and effective picture of this statesman's career, keeping throughout in touch with his authorities; and his graphic narrative will fill the gap in the bookshelf made by the disappearance of Forster's Life written nearly 80 years ago, and long obsolete."—*The Times.*

MAKERS OF NATIONAL HISTORY. Edited by The Ven. W. H. HUTTON, B.D. Each volume in this series—the aim of which is to do fuller justice to men whose lives have not hitherto been adequately dealt with—is in crown 8vo, cloth gilt, with a frontispiece, 3s. 6d. net.

CARDINAL BEAUFORT. By the Rev. L. B. RADFORD, D.D.

"Studiously impartial . . . carefully written."—*Glasgow Herald.*

VISCOUNT CASTLEREAGH. By ARTHUR HASSALL, M.A.

"It is brilliantly written . . . exceptionally clear and vivid . . . a book which was needed."—*The Morning Leader.*

ARCHBISHOP PARKER. By W. M. KENNEDY, B.A.

"Exceedingly well conceived, clearly expressed, and compiled with great care." *The Guardian.*

GENERAL WOLFE. By EDWARD SALMON.

"A picture and an estimate of Wolfe which could not be more complete."—*Canada.*

FRANCIS ATTERBURY, Bishop of Rochester (1662-1732). By the Very Rev. H. C. BEECHING, M.A., Litt.D., Dean of Norwich.

"A most delightful as well as a most valuable book."—*Guardian.*

EDWARD THE FOURTH. By LAURENCE STRATFORD, B.A.

THOMAS BECKET, Archbishop of Canterbury. By The Ven. W. H. HUTTON, B.D., Canon of Peterborough, and Archdeacon of Northampton.

NATURAL HISTORY, ETC.

THE A B C OF POULTRY. By E. B. JOHNSTONE. In crown 8vo, cloth, cheap edition, 1s. net.

"A capital addition to the many books devoted to the outdoor life."—*World.*

CATS FOR PLEASURE AND PROFIT. By Miss FRANCES SIMPSON. Third Edition. In crown 8vo, with 25 beautifully reproduced photographs of famous prize-winning cats. 2s. net.

"The author explains that her object has been 'to help those who desire to combine pleasure with profit.' This aim is very successfully achieved."—*Pall Mall Gazette.*

REPTILES OF THE WORLD. Tortoises and Turtles, Crocodilians, Lizards and Snakes of the Eastern and Western Hemispheres. By Professor RAYMOND L. DITMARS. With frontispiece in colour, and nearly 200 illustrations from photographs taken by the author. In royal 8vo, cloth gilt, gilt top. 20s. net.

BRITISH FERNS. A pocket help for the Student and Collector (comprising all the native species and showing where found). By FRANCIS G. HEATH. Size 6¼ in. by 3¾ in., cloth, with 50 illustrations. 2s. net.

PEEPS INTO NATURE'S WAYS. By JOHN J. WARD. Being chapters on insect, plant and minute life. Illustrated from photographs and photo-micrographs taken by the Author. Cheaper Edition. In demy 8vo, cloth gilt, gilt top, 3s. 6d. net.

MISCELLANEOUS

THE DREDGING OF GOLD PLACERS. By J. E. HODGSON, F.R.G.S. With 17 illustrations. In demy 8vo, cloth gilt, gilt top, 5s. net. Principally intended for Company Directors, Property Managers, Prospectors, and the investing public.

ASTONISHING ANATOMY. An anatomical and medical skit profusely illustrated. Letterpress consisting of Imaginary Interviews, Free Medical Advice (after the style of the Weekly Press), etc., etc. By "Tingle." Crown 8vo, 1s. net.

ATHLETIC TRAINING FOR GIRLS. Compiled and Edited by C. E. THOMAS. In crown 8vo, cloth gilt, with many illustrations. Illustrated. 3s. 6d. net.

"A book that every schoolmistress, every mother, and every girl should be compelled to read, for it is full of the best advice and thoroughly practical hints."—*World.*

COMMON COMMODITIES OF COMMERCE. Each handbook is dealt with by an expert writer. Beginning with the life history of the plant, or other natural product, he follows its development until it becomes a commercial commodity, and so on through the various phases of its sale on the market, and its purchase by the consumer. Each is in crown 8vo, cloth, about 120 pp., with map, coloured frontispiece, chart and illustrations, 1s. 6d. net. **Tea,** from Grower to Consumer, by ALEXANDER IBBETSON. **Coffee**

from Grower to Consumer, by B. B. KEABLE. **Cotton.** From the Raw Material to the Finished Product. By R. J. PEAKE. **Oil;** Animal, Vegetable, Essential and Mineral. By C. AINSWORTH MITCHELL. **Sugar—Cane and Beet.** By GEO. MARTINEAU, C.B., **Rubber.** Production and utilisation of the raw material. By C. BEADLE and H. P. STEVENS, M.A., Ph.D. **Iron and Steel.** Their production and manufacture. By C. HOOD. **Silk.** Its production and manufacture. By LUTHER HOOPER. **Wool.** From the Raw Material to the Finished Product. By J. A. HUNTER. **Tobacco.** From Grower to Smoker. By A. E. TANNER. **Coal:** Its Origin, Method of Working, and Preparation for the Market. By FRANCIS H. WILSON, M.Inst., M.E., *Editor of Mining Engineering.*

Other Volumes in preparation

DICKENS IN YORKSHIRE. Being Notes of a Journey to the Delightful Village of Dotheboys, near Greta Bridge. By C. EYRE PASCOE. In foolscap 4to, with four illustrations in colour and 11 black and white illustrations. 1s. 6d. net.

THE FEDERAL SYSTEMS OF THE UNITED STATES AND THE BRITISH EMPIRE. Their Origin, Nature, and Development. By A. P. POLEY, B.A., *of the Inner Temple and Midland Circuit, Barrister-at-Law.* In demy 8vo, cloth gilt, 12s. 6d. net.

FRENCH PROSE WRITERS OF THE NINETEENTH CENTURY AND AFTER. With Biographical and Critical Notices in French, and Literary and Bibliographical Notes in English. By VICTOR LEULIETTE, B-ès-L., A.K.C. In crown 8vo, cloth gilt, 350 pp. 3s. net.

FOR HOME SERVICE AND OTHER STORIES. By LYDE HOWARD. With coloured frontispiece and black and white illustrations. In foolscap 4to, cloth, decorated, coloured top, and end papers, 2s. 6d. net.

HOME GYMNASTICS FOR OLD AND YOUNG. By T. J. HARTELIUS, M.D. Translated and adapted from the Swedish by C. LÖFVING. With 31 illustrations. Fifth Edition, revised. With a prefatory note by ARTHUR A. BEALE, M.B. In stiff boards, 1s. 6d.

HOW TO CHOOSE A HOUSE. How to Take and Keep it. By CHARLES EMANUEL, M.A., and E. M. JOSEPH, A.R.I.B.A. In crown 8vo, cloth, with illustrations. Cheap edition, 1s. net.

"This book seems to us to contain well nigh all the information that a person desiring to acquire a property could desire."—*Record.*

HYPNOTISM AND SUGGESTION. In Daily Life, Education, and Medical Practice. By BERNARD HOLLANDER, M.D. In crown 8vo, cloth gilt, 6s. net.

"We specially welcome the book before us. It is the work of a man of established reputation, who has devoted himself for years to the subject, and whose aim is to tell the English-speaking world what Hypnotism really is, what it can do, and to what conclusions it seems to point. It is written in a thoroughly scientific spirit. No fact is shirked, and no evidence is either suppressed or rated above its real value."—*Globe.*

LIGHTER MOMENTS. From the note-book of BISHOP WALSHAM HOW. Edited by his son, FREDERICK DOUGLAS HOW. In small crown 8vo, cloth gilt, gilt top, 2s. 6d.

OVERHEARD AT THE ZOO. By GLADYS DAVIDSON. With 2 coloured plates and 26 black and white illustrations. In foolscap 4to, cloth, 2s. 6d. net.

The author has catered for all children who love animals. Her aim has been to present the animals' own point of view, so far as it may be divined by sympathetic study.

OVERHEARD IN FAIRYLAND, or The Peter Pan Tales. By MADGE A. BIGHAM. With coloured illustrations by RUTH S. CLEMENTS. In large crown 8vo, cloth, 2s. 6d. net.

PUBLIC SCHOOL LIFE. Each in foolscap 8vo, cloth, with 32 full page plate illustrations. 2s. net.

WESTMINSTER. By W. TEIGNMOUTH SHORE.
ETON. By AN OLD ETONIAN.
HARROW. By ARCHIBALD FOX.
RUGBY. By H. H. HARDY.

THE ROYAL HIGH SCHOOL, EDINBURGH. By J. J. TROTTER. With 32 illustrations. In crown 8vo, cloth gilt, 3s. 6d. net.

THE REVERIES OF A BACHELOR: Or, A Book of the Heart. By the late IK MARVEL. With an Introduction by ARLO BATES. In foolscap 8vo, gilt top, limp lambskin, 2s. 6d. net. Also in cloth, 1s. 6d. net.

ROODSCREENS AND ROODLOFTS. By F. BLIGH BOND, F.R.I.B.A., and The Rev. DOM BEDE CAMM, O.S.B. With over 88 full-page collotype reproductions, and upwards of 300 other beautiful illustrations. In demy 4to, two vols., handsome cloth gilt, gilt top, 32s. net.

"A magnificent work."—*Evening Standard.*

THE BOOK OF THE CHILD. An Attempt to Set Down what is in the Mind of Children. By FREDERICK DOUGLAS HOW. In foolscap 8vo, leather, with dainty cover design, gilt corners, 3s. 6d. net; cloth, 2s. net.

MYSTICISM AND MAGIC IN TURKEY. An Account of the Religious Doctrines, Monastic Organisation, and Ecstatic Powers of the Dervish Orders. By LUCY M. J. GARNETT, Author of *Turkey of the Ottomans.* In crown 8vo, cloth gilt, gilt top, with illustrations. 6s. net.

THE PERSIAN PROBLEM. By H. J. WHIGHAM. With maps and illustrations. In demy 8vo, cloth gilt, 12s. 6d.

SCIENCE AND THE CRIMINAL. By C. AINSWORTH MITCHELL, B.A., F.I.C. In crown 8vo, cloth gilt, 250 pp., with 28 illustrations, 6s. net.

"The systems of personal identification are discussed, and the uses of photography, anthropometry, and finger prints are indicated. The selection of the cases and the manner in which the whole book is written show good judgment."—*Lancet.*

SYMBOLISM OF ANIMALS AND BIRDS Represented in English Church Architecture. By ARTHUR H. COLLINS, M.A. [Oxon.]. With 120 illustrations. In demy 8vo, cloth gilt, gilt top, 5s. net.

THE SUNLIT ROAD: Readings in Verse and Prose for Every Day in the Year. By the Rev. W. GARRETT HORDER. In demy 16mo, cloth gilt, gilt corners, 3s. net; leather gilt, gilt corners, 4s. net.

"A dainty and delightful little 'day book' for quiet moments. It is the most charming book of its kind we have seen for a very long time."—*Lady.*

PITMAN'S STUDIES IN ELOCUTION. A guide to the theory and practice of the art of public speaking and reciting, with over 100 selections for Reciters and Readers. By E. M. CORBOULD (Mrs. Mark Robinson). In crown 8vo, cloth gilt, gilt top, silk register, 2s. 6d. net.

"This treasury of prose and verse will appeal to all who cultivate the art of elocution or appreciate a choice store of literary gems."—*Educational News.*

THE BEGINNINGS OF THE TEACHING OF MODERN SUBJECTS IN ENGLAND. By FOSTER WATSON, M.A. (Professor of Education in the University College of Wales; Aberystwyth). In crown 8vo, cloth, 7s. 6d. net.

THE INNER LIFE OF GEORGE ELIOT. A Study of the Mental and Spiritual Development of the Novelist. By CHARLES GARDNER, M.A. In crown 8vo, cloth gilt, gilt top, 5s. net.

POETRY, CRITICISM, & LITERARY HISTORY

THE POETRY OF ROBERT BROWNING. By STOPFORD A. BROOKE. Original issue. In demy 8vo, cloth gilt, 10s. 6d.

"The most satisfactory and stimulating criticism of the poet yet published."—*Times.*

(*See also* Dainty Volume Library, page 4.)

TENNYSON: HIS ART AND RELATION TO MODERN LIFE. By the same Author. Original issue. In demy 8vo, cloth gilt, 7s. 6d.

"Will make a strong appeal to all lovers of our great Laureate."—*Quarterly Review.*

(*See also* Dainty Volume Library, page 4.)

A STUDY OF CLOUGH, ARNOLD, ROSSETTI, AND MORRIS. With an Introduction on the Course of Poetry from 1822 to 1852. By the same Author. In demy 8vo, cloth gilt, 6s. net.

(*See also* Dainty Volume Library, page 4.)

EXPERIMENTS IN PLAY WRITING. Six plays in Verse and Prose with an Introductory Essay. By JOHN LAWRENCE LAMBE. In demy 8vo, cloth gilt, 5s. net.

THE POEMS OF JAMES HOGG. The Ettrick Shepherd. Selected and edited, with an introduction, by WILLIAM WALLACE, LL.D. With photogravure portrait frontispiece. In crown 8vo, cloth gilt, gilt top, 5s.

WITH THE WILD GEESE. Songs of Irish Exile and Lament. By EMILY LAWLESS. With an Introduction by STOPFORD A. BROOKE. In square 8vo, cloth gilt, 4s. 6d. net.

THE WOOING OF A GODDESS. By B. BURFORD RAWLINGS. In foolscap 8vo, leather, 2s. 6d. net.

MODERN FRENCH LITERATURE. By B. W. WELLS, Ph.D. In crown 8vo, cloth gilt, 520 pp., 6s. net.

MODERN ITALIAN LITERATURE. By LACY COLLISON-MORLEY, Author of *Guiseppe Baretti and his Friends*. In crown 8vo, cloth gilt, 360 pp., 6s. net.

A SHORT HISTORY OF GREEK LITERATURE. From Homer to Julian. By WILMER CAVE WRIGHT, Ph.D., late of Girton College, Cambridge. In crown 8vo, cloth gilt, 544 pp., 6s. net.

GREEK INFLUENCE ON ENGLISH POETRY. By the late Professor JOHN CHURTON COLLINS. Edited with Introduction, by Professor M. MACMILLAN. In crown 8vo, cloth gilt, with portrait. 3s. 6d. net.

POLITICS, ETC.

THE CASE AGAINST WELSH DISENDOWMENT. By J. FOVARGUE BRADLEY. Third impression. In demy 8vo, 1s. net.

NONCONFORMISTS AND THE WELSH CHURCH BILL. By the same Author. 2nd impression. Crown 8vo, 1s. net.

THE RISE AND DECLINE OF WELSH NONCONFORMITY. An Impartial Investigation. By VIATOR CAMBRENSIS. Demy 8vo, 1s. net.

FAMOUS SPEECHES. First Series. From Cromwell to Gladstone. Selected and Edited with Introductory Notes by HERBERT PAUL. In demy 8vo, cloth, 470 pp., 7s. 6d. net.

"A book of selections such as this is delightful reading. Mr. Herbert Paul has chosen discreetly in the wide field from Cromwell to Gladstone, and has prefaced each orator with a judicious criticism."—*Spectator*.

FAMOUS SPEECHES. Second Series. From Lord Macaulay to Lord Rosebery. Selected and Edited with Introductory Notes by HERBERT PAUL. In demy 8vo, cloth, 398 pp., 7s. 6d. net.

THE TRUTH ABOUT HOME RULE. By PEMBROKE WICKS, LL.B., *Barrister-at-law*. With a Preface by the Rt. Hon. Sir EDWARD CARSON, K.C., M.P. In crown 8vo, cloth gilt, with illustrations, 3s. 6d. net.

SCIENCE

GREAT ASTRONOMERS. By Sir ROBERT BALL, D.Sc., LL.D., F.R.S. With numerous full-page and other illustrations. In demy 8vo, cloth gilt, gilt top, 3s. 6d. net.

"Sir Robert Ball's gifts as a narrator are very great. He is, of course, a master of his subject . . . . The most earth-bound mortal who opens this book must go on with it."—*Daily Chronicle.*

IN STARRY REALMS. By the same Author. The Wonders of the Heavens. With numerous full-page and other illustrations. In demy 8vo, cloth gilt, gilt top, 3s. 6d. net.

"The style of popular exposition adopted throughout is indeed admirable, the illustrations are excellent, the binding is tasteful, and the print good."—*Saturday Review.*

IN THE HIGH HEAVENS. By the same Author. A popular account of recent interesting astronomical events and phenomena, with numerous full-page and other illustrations. In demy 8vo, cloth gilt, gilt top, 3s. 6d. net.

"It has," says *The Scotsman*, "the freshest knowledge and the best scientific thought."

ASTRONOMY FOR EVERYBODY. By Professor SIMON NEWCOMBE, LL.D. With an Introduction by Sir ROBERT BALL. Illustrated. A popular exposition of the wonders of the Heavens. In demy 8vo, cloth gilt, gilt top, 3s. 6d. net.

BY LAND AND SKY. By the Rev. JOHN M. BACON, M.A., F.R.A.S. The Record of a Balloonist. With four illustrations. In demy 8vo, cloth gilt, gilt top, 3s. 6d. net.

SOCIOLOGY

SOCIALISM. By Professor ROBERT FLINT, LL.D. New, Revised and Cheaper Edition. In demy 8vo, cloth gilt, 6s. net.

"A new, revised and cheaper edition of Professor Flint's masterly study will be generally welcomed. References show that the additional notes are well up to date."—*Daily Mail.*

THE PEOPLE OF THE ABYSS. By JACK LONDON. A study of the social and economic conditions of life in the East End of London. By the author of *The Call of the Wild.* With 24 illustrations from actual photographs. In crown 8vo, cloth gilt, 6s.

". . . Mr. Jack London, who is already known to the British public as a fine descriptive writer, has done for the East End of London what he did for the Klondyke—has described it fully and faithfully, looking at it as intimately as dispassionately."—*Daily Chronicle.*

WHAT IS SOCIALISM? By "SCOTSBURN." An attempt to examine the principles and policy propounded by the advocates of Socialism. In demy 8vo, cloth gilt, 7s. 6d.

THE SOCIAL WORKER'S GUIDE. (See page 20.)

THEOLOGICAL

THE PRAYER BOOK DICTIONARY. An Indispensable Volume of Reference dealing with the origins, history, use, and teaching of the several authorised editions of the Book of Common Prayer within the Anglican Communion. Its scope embraces all accompanying ceremonies and supplementary rites, the ornaments of the Church and of all ministers, Church structures and fittings in their relation to worship, ecclesiastical persons and bodies, and the legislative judicial or administrative authorities now or heretofore empowered or exercising powers in regard to the above. Edited by GEORGE HARFORD, M.A., *Vicar of Mossley Hill, Hon. Canon of Liverpool;* and MORLEY STEVENSON, M.A., *Principal of Warrington Training College, Hon. Canon of Liverpool.* Assisted by J. W. TYRER, M.A., *Formerly Vicar of St. Luke the Evangelist, Walton.* Preface by THE LORD BISHOP OF LIVERPOOL.

Articles by nearly 150 Contributors, including :—The Bishop of Ossory; Lord Hugh Cecil; Dr. Hermitage Day; The late Dr. Dowden (Bishop of Edinburgh); Canon Driver; The Bishop of Ripon; The Provost of King's College, Cambridge; The Bishop of Lichfield; The Rev. T. A. Lacey; The Bishop of Moray and Ross; The Bishop of Aberdeen; Bishop Montgomery; The Bishop of Durham; The Bishop of Exeter; Canon Simpson; Chancellor P. V. Smith; Canon Staley; Dr. Eugene Stock; The Dean of Canterbury; Canon Bullock Webster; The Rev. James Baden Powell; Professor H. B. Swete; Dr. H. P. Allen; Professor Du Bose; Dr. Guy Warman; Dr. St. Clair Tisdall; Mr. Robert Bridges; Mr. Francis Burgess; Mr. Edwin H. Freshfield, F.S.A.; Mr. J. A. Fuller Maitland, M.A., F.S.A.; Sir T. Sydney Lea, Bart.; Sir Charles Nicholson, F R.I.B.A.; Mrs. Romanes; Professor J. E. Vernham. The work is complete in One Volume, crown 4to, half leather gilt, gilt top, 850 pp., 25s. net. *Write for 16 pp. Prospectus containing lists of Contributors and articles, specimen pages, etc.*

" A very successful attempt to meet a real want."—*Guardian.* " Thorough and scholarly."—*Church Times.* " The book will take its place at once amongst our indispensable works of reference . . . a great and scholarly achievement."—*The Churchman.* " We do not think that any Clergyman can afford to be without this highly scholarly volume."—*Church of Ireland Gazette.* " Its contents answer practically every question that we can ask about the book. It will make for itself a place on our reference shelves next to Hastings."—*Record.*

THE BOOK OF ISAIAH. NEWLY TRANSLATED WITH INTRODUCTIONS, CRITICAL NOTES AND EXPLANATIONS BY G. H. BOX, M.A. Together with a Prefatory Note by S. R. DRIVER, D.D Regius Professor of Hebrew in the University of Oxford, and Canon of Christ Church. In demy 8vo, cloth gilt, with two maps, 7s. 6d. net.

The Athenæum says it " deserves high commendation," and that " the advantage of having the prophecies placed before us in something like the original grouping of lines far outweighs the drawback of what might here and there be regarded as arbitrary or unnecessary alterations . . the book recommends itself by its scholarly character, its clearness of exposition, and the fearless, yet reverent spirit of investigation by which it is animated."

THE EZRA-APOCALYPSE. Being Chapters 3—14 of the Book commonly known as IV. Ezra (or II. Esdras). Translated from a critically revised text, with critical Introductions, Notes, and Explanations; with a General Introduction to the Apocalypse, and an Appendix containing the Latin text. By G. H. Box, M.A., Author of *The Book of Isaiah*, etc. Together with a Prefatory Note by W. Sanday, D.D., LL.D., Litt.D., *Lady Margaret Professor and Canon of Christ Church, Oxford.* In demy 8vo, cloth gilt, 488 pp., 10s. 6d. net.

"Already known to the student by his excellent edition of *Isaiah*, Mr. Box has now ventured successfully as we think, into a field which Dr. Charles had almost made his own; and Dr. Charles, we are sure, will not be backward in greeting him as a worthy *confrère*. Mr. Box's treatment of the various problems presented by the book is marked by the same clearness and thoroughness which characterised his *Isaiah* . . . Mr. Box has laid the readers of 2 Esdras under the highest obligations, and has produced a work, the only thorough English work on the subject, which does honour to English scholarship and will be indispensable to all students of this portion of the Apocrypha."—*Spectator.*

THE RELIGION AND WORSHIP OF THE SYNAGOGUE. An Introduction to the Study of Judaism from the New Testament Period. By W. O. E. Oesterley, D.D., and G. H. Box, M.A. In demy 8vo, cloth gilt, with eight illustrations. Second, Revised, and Cheaper Edition, 7s. 6d. net.

"It is not often that a large book can be written on a large subject in the field of religion, which is so entirely new and fresh as this important volume. . . . Its novelty and freshness lies in its point of view. It is a study of Judaism by Christian scholars of the Church of England, written for a Christian public, and it is a sympathetic, even a loving study."—*Church Times.*

"Its authors have written with good will and with quite exceptional knowledge."—*Jewish Chronicle.*

THE EVOLUTION OF THE MESSIANIC IDEA. A Study in Comparative Religion. By the Rev. W. O. E. Oesterley, D.D. In crown 8vo, cloth gilt, 3s. 6d. net.

"Dr. Oesterley's new work deserves the serious consideration of students. . . It is stimulating, earnest, frank, full of interesting information. . . . Likely to prove very useful to a wide circle of readers."—*Athenæum.*

THE FUTURE LIFE AND MODERN DIFFICULTIES. By F. Claude Kempson, M.B. In crown 8vo, cloth gilt, with diagrams, 3s. 6d. net.

"The author shows the simplest educated reader that there is nothing whatever in scientific discoveries to weaken our faith in Christianity."—*The Record.*

THE SAMSON-SAGA AND ITS PLACE IN COMPARATIVE RELIGION. By the Rev. A. SMYTHE PALMER, D.D. In crown 8vo, cloth gilt, with two illustrations, 5s. net.

THE KINGDOM WITHIN. Being Teaching for our Day Recorded Exclusively by St. Luke. By AGNES STANLEY LEATHES. In crown 8vo, cloth gilt, gilt top, 3s. 6d. net.

"Her studies are thoughtful and yet simple; they relate the primary teachings of Christ to the facts of modern life. There are useful chapters on spiritual healing, and on the value of the results of psychic research, a value often over estimated by the apologist for Christianity."—*Church Times.*

THE GOSPEL OF JOY. By the Rev. STOPFORD A. BROOKE, M.A., LL.D. In crown 8vo, cloth gilt, gilt top, 6s.

THE OLD TESTAMENT AND MODERN LIFE. By the same Author. In crown 8vo, cloth gilt, gilt top, 6s.

THE LIFE SUPERLATIVE. By the same Author. In crown 8vo, cloth gilt, gilt top, 6s.

THOUGHTS ON SOME OF THE MIRACLES OF JESUS. As MARKS OF THE WAY OF LIFE. By the Right Hon. and Most Rev. COSMO GORDON LANG, D.D., Lord Archbishop of York. In crown 8vo, cloth gilt, gilt top, 6s.

"A delightful book, full of helpfulness and cheer."—*Methodist Times.*

THOUGHTS ON SOME OF THE PARABLES OF JESUS. By the same Author. In crown 8vo, cloth gilt, gilt top, 6s.

"We can only express our wonder at the freshness of treatment which he has been able to bring to a familiar subject."—*The Times.*

FAMOUS SERMONS BY ENGLISH PREACHERS. From the VENERABLE BEDE to H. P. LIDDON. Edited with Historical and Biographical Notes by Canon DOUGLAS MACLEANE, M.A. In demy 8vo, cloth gilt, 6s. net.

"This is a delightful collection, and the reading public owe a debt of gratitude to Canon Macleane. Canon Macleane's Introductions to the Sermons are by no means the least valuable part of the work . . . it deserves, and will no doubt receive, a hearty welcome from all reading men interested in the history of our Church."—*Record.*

LAY SERMONS FROM "THE SPECTATOR." By M. C. E. With an introduction by J. ST. LOE STRACHEY. In crown 8vo, cloth gilt, gilt top silk register, 5s. net.

". . . The prime merit of these essays is their simplicity—a quality which should commend them to many who instinctively reject sermons as sermons are too often preached."—*Pall Mall Gazette.*

THE PRESENCE OF CHRIST. By the late Bishop THOROLD. In crown 8vo, cloth gilt, 3s. 6d.

THE TENDERNESS OF CHRIST. By the same Author. In crown 8vo, cloth gilt, 3s. 6d.

"Deals with questions of universal and abiding import. His style, too, has a rare charm."—*Pall Mall Gazette.*

THE GOSPEL OF CHRIST. By the same Author. In crown 8vo, cloth gilt, 3s. 6d.

"May well take its place amongst the classics of experimental religion."—*Record.*

ON LIFE'S THRESHOLD: TALKS TO YOUNG PEOPLE ON CHARACTER AND CONDUCT. By the same Author. Translated by EDNA ST. JOHN. In crown 8vo, cloth gilt, gilt top, 3s. 6d.

THE SIMPLE LIFE. By the same Author. Translated from the French by MARIE LOUISE HENDEE. With biographical sketch by GRACE KING. New Edition. In foolscap 8vo, cloth gilt, 1s. net.

THE COMMANDMENTS OF JESUS. By the Rev. R. F. HORTON. Popular edition. In crown 8vo, cloth gilt, 2s. 6d. net.

THE TEACHING OF JESUS. By the same Author. Popular edition. In crown 8vo, cloth gilt. 2s. 6d. net.

HELP FOR THE TEMPTED. By Professor AMOS R. WELLS. With an Introduction by the Rev. F. B. MEYER, B.A. In foolscap 8vo, cloth gilt, gilt top, 2s. 6d.; also in paper covers, price 1s. 6d.

THE GOSPEL AND THE AGE. By the late W. C. MAGEE, D.D., Archbishop of York. In large crown 8vo, cloth gilt, gilt top, 3s. 6d.

"Will arrest the attention of the world."—*Spectator.*

THE INDWELLING CHRIST. By the late HENRY ALLON, D.D. In large crown 8vo, cloth gilt, gilt top, 3s. 6d.

"Worthy to take their place among the masterpieces of the old divines."—*Daily Telegraph.*

CONSIDERATIONS FOR LENT. Readings for the Forty Days' Fast. By The Rev. VERNON STALEY, Hon. Canon of Inverness Cathedral. Author of *The Catholic Religion*, etc., etc. In foolscap 8vo, cloth, 1s. 6d. net. Leather gilt, gilt top, 2s. 6d. net.

The plan of the work is to give the reader food for reflection founded on Christian doctrine, in the best sense of the term, and to turn each day's reading, or portion, to bear upon character and practical religion.

CONSIDERATIONS FOR ADVENT. Devotional Readings for the Season. By the same Author. Cloth, 1s. 6d. net; leather, 2s. 6d. net.

A BOOK OF THE LOVE OF JESUS. By Mgr. R. H. BENSON. In foolscap 8vo, leather gilt, gilt top, 3s. 6d. net; cloth 2s. net.

"An anthology of some old Catholic devotions, slightly modernized, which will appeal to many by reason of its simplicity and beauty."—*To-day.*

A BOOK OF THE LOVE OF MARY. By F. M. Groves. Preface by His Eminence Cardinal Bourne. In foolscap 8vo., cloth, with frontispiece, 2s. net. Leather gilt, gilt top, photogravure frontispiece, 3s. net.

"We give a cordial and grateful welcome to this beautiful little book about Our Lady, and her churches, pictures, images, shrines, guilds, wells and salutations, and the poems, prayers and days that honour her."—*Catholic Times.*

A LITTLE HISTORY OF THE LOVE OF THE HOLY EUCHARIST. By the same Author. In foolscap 8vo, leather gilt, gilt top, 3s. 6d. net.

In a previous work Mrs. Groves traced the history of the devotion to Our Lady in these islands, showing the various forms it took and the traces it has left in the language and the social customs of the country. In the present work she renders a like service to the history of the devotion to the Blessed Sacrament of the Altar in England during the centuries between the planting of Christianity here and the reign of Mary Tudor.

IN OUR LADY'S PRAISE. An Anthology of Verse. Compiled by E. Hermitage Day, D.D., F.S.A. With Preface by the Right Hon. Viscount Halifax. In foolscap 8vo, cloth, with photogravure, 2s. net; leather gilt, gilt top, with photogravure frontispiece, 3s. net.

IN ANSWER TO PRAYER. Testimonies of Personal Experiences. By Bishop Boyd-Carpenter, the late Dean of Salisbury, Canon Knox Little, M.A., the late Rev. Dr. John Watson ("Ian Maclaren"), Rev. Dr. R. F. Horton, the late Rev. Hugh Price Hughes, and others. Cheaper edition. In crown 8vo, cloth gilt, gilt top, 2s.

THE LITERARY STUDY OF THE BIBLE. An account of the leading forms of literature in the Sacred Writings. By Professor R. G. Moulton, M.A., Ph.D. Cheaper Edition. In demy 8vo, cloth gilt, 6s. net.

"A valuable help to the study of the Sacred Writings. . . We heartily recommend this book."—*Daily Chronicle.*

THE PRACTICAL WISDOM OF THE BIBLE. Edited with an introduction by J. St. Loe Strachey (Editor of *The Spectator*). In demy 16mo, cloth gilt, gilt top, 2s. 6d. net; leather 3s. 6d. net.

"No one, after reading the elegant and carefully produced volume can doubt that Mr. Strachey has done a good work in a thoroughly good manner."—*Standard.*

THE ST. PAUL'S HANDBOOKS. Edited by E. Hermitage Day, D.D., F.S.A. Each in crown 8vo, cloth gilt, 2s. 6d. net.

This new series makes a strong appeal to the large number of busy Churchpeople who desire to obtain clear guidance for themselves upon those questions of faith and practice which emerge from time to time into the field of controversy. The volumes will be written by Priests and Laymen who have received the Faith from the Catholic Church in the English Provinces.

THE MINISTRY OF THE CHURCH. By E. Hermitage Day, D.D., F.S.A.

Other Volumes in preparation.

THE SOCIAL RESULTS OF EARLY CHRISTIANITY. By C. SCHMIDT. Translated by Mrs. THORPE. With Preliminary Essay by R. W DALE, LL.D. In crown 8vo, cloth gilt, 3s. 6d. net.

"An easy book to read, and the educated layman will find it full of vital interest, while the more exacting student will have the further satisfaction of being provided with full and precise references to the original authorities in which many startling assertions are made."—*Nottingham Daily Express.*

EDUCATION AND SOCIAL LIFE. By the Rev. J. WILSON HARPER, D.D. In crown 8vo, cloth, 4s. 6d. net.

MODERNISM. A RECORD AND REVIEW. By the Ven. A. LESLIE LILLEY, M.A., *Archdeacon of Ludlow.* In demy 8vo, cloth gilt, 6s. net.

"Mr. Lilley is admirably suited, both by knowledge and sympathy, to be the medium through which the modernist position may be made known to the English public."—*Church Times.*

BODY AND SOUL. An Enquiry into the effects of Religion upon health with a description of Christian works of healing from the New Testament to the present day. By PERCY DEARMER, D.D. Ninth impression. Cheaper Edition. In crown 8vo, cloth gilt, 2s. 6d. net.

"Here is the book for which we have so long waited. . . We may say at once that the work could hardly have been better done. It takes a comprehensive survey of the main question, and of matters related to it. It is arranged with an admirable clearness." —*Church Times.*

THE CHURCHMAN'S GUIDE. A Handbook for all persons, whether Clerical or Lay, who require a Book of Reference on questions of Church Law or Ecclesiology. Edited by ARTHUR REYNOLDS, M A. In crown 8vo, cloth, 368 pp., 3s. 6d. net.

"The work is extremely well done. Within the space of 333 pages, well and clearly printed in double columns, the editor has managed to include nearly a thousand articles and definitions. The articles on various legal points are lucid and authoritative; those on ecclesiology interesting and practical; those on historical points are commendably free from bias. In fact it is a trustworthy and convenient guide on the many matters on which the churchman constantly finds himself in need of information."—*Church Times.*

CHURCH ACCOUNTS. A Simple, Concise Method of Account Keeping, for use by the Clergy, Churchwardens, and other Officials. With Model Accounts. Compiled by the Rev. W. G. DOWSLEY, B.A. Size 15½ in. by 9½ in., half-leather, 106 pp., with interleaved blotting-paper, 6s. 6d. net.

"An exceedingly useful volume. . . . As to its thoroughness there can be no doubt; . . . for large and highly organised parishes it would be difficult to devise anything better."—*Guardian.*

THE SOCIAL WORKERS' GUIDE. A Handbook of Information and Counsel for all who are interested in Public Welfare. Edited by the Rev. J. B. HALDANE, M.A., Secretary of the Southwark Diocesan Social Service Committee, with assistance from Fifty Experts. In crown 8vo, cloth, 500 pp., with over 500 articles. 3s. 6d. net.

"A book of reference of more than average value. The need of such a book is patent, and we do not know of any other publication which attempts to supply it. The notes are arranged in alphabetical order, and, generally speaking, they are wonderfully exhaustive."—*Guardian.*

HOW TO TEACH AND CATECHISE. A Plea for the Employment of Educational Methods in the Religious Instruction of Children. By the Rev. J. A. RIVINGTON, M.A., formerly Second Master at St. Paul's Cathedral Choir School. With a Preface by the LORD BISHOP OF GLOUCESTER. Cheaper Edition. In crown 8vo, cloth gilt, 1s. 6d. net.

"This is an invaluable little book . . . it might well be put into the hands of every Sunday School teacher."—*Churchman.*

A POPULAR HISTORY OF THE CHURCH IN WALES From the Beginning to the Present Day. By the Rev. J. E. DE HIRSCH-DAVIES, B.A. In crown 8vo, cloth gilt, 356 pp., 5s. net.

"It shows wide reading no less than special study. It is written with the simplicity befitting a popular history, and its interest never flags. It makes the Welsh Church, in strength and weakness, depression or re-awakening, live before our eyes." *Church Times.*

THE LONDON CHURCH HANDBOOK. Being a Compendium of Information upon Church Affairs in the County of London [Dioceses of London and Southwark]. In crown 8vo, cloth, 412 pp., 2s. net.

THE SPRING OF THE DAY. SPIRITUAL ANALOGIES FROM THE THINGS OF NATURE. By the late HUGH MACMILLAN, D.D., LL.D. In crown 8vo, cloth gilt, 3s. 6d. net.

THE CLOCK OF NATURE. By the late HUGH MACMILLAN, D.D., LL.D. In crown 8vo, cloth gilt, 3s. 6d. net.

An attempt to bring out the wise lessons which the objects of Nature teach, and to illustrate the spiritual revelation of God in Christ by the revelation of God in Nature.

THE POETRY OF PLANTS. By the late HUGH MACMILLAN, D.D., LL.D. In crown 8vo, cloth gilt, 3s. 6d. net.

A collection of popular studies, showing the many points of beauty and interest about some of the commonest of our trees and wild flowers.

TRAVEL, TOPOGRAPHY, AND SPORT

THE ADVENTURER IN SPAIN. By S. R. CROCKETT. With 162 illustrations by GORDON BROWNE and from photographs taken by the Author. In large crown 8vo, cloth gilt, 6s.

WANDERINGS ON THE ITALIAN RIVIERA. The Record of a leisurely tour in Liguria. By FREDERIC LEES. With coloured plate, and 60 illustrations, map. In large crown 8vo, cloth gilt, gilt top, 7s. 6d. net.

"The Italian Riviera . . . is practically unknown to the majority of visitors, and Mr. Lees has done it and the public a service in writing this very readable and pleasant volume. All intellectual people will appreciate the description of local customs, art and architecture, literature and folk lore, which Mr. Lees has set himself to expound."—*World.*

THE IMMOVABLE EAST. Studies of the People and Customs of Palestine. By PHILIP J. BALDENSPERGER. With Biographical Introduction by FREDERIC LEES. With 24 full-page plate illustrations and map. In demy 8vo, cloth gilt, gilt top, 7s. 6d. net.

"Nothing so intimate has yet appeared upon the subject as this book. To those who know already something of the people and the life described, there is no book we should recommend more strongly to enlarge their knowledge."—*The Athenæum.*

Countries and Peoples Series

Each in imperial 16mo, cloth gilt, gilt top, with about 30 full-page plate illustrations, 6s. net.

ITALY OF THE ITALIANS. By HELEN ZIMMERN.

"The knowledge and judgment displayed in the volume are truly astounding, and the labour the author has expended on it has made it as indispensable as Baedeker to the traveller, as well as invaluable to the student of modern times."—*Daily Telegraph.*

FRANCE OF THE FRENCH. By E. HARRISON BARKER.

"A book of general information concerning the life and genius of the French people, with especial reference to contemporary France. Covers every phase of French intellectual life—architecture, players, science, and invention, etc."—*Times.*

SPAIN OF THE SPANISH. By Mrs. VILLIERS-WARDELL.

"Within little more than 250 pages she has collected a mass of ordered information which must be simply invaluable to any one who wants to know the facts of Spanish life at the present day. Nowhere else, so far as we are aware, can a more complete and yet compendious account of modern Spain be found."—*Pall Mall Gazette.*

SWITZERLAND OF THE SWISS. By FRANK WEBB.

"Mr. Webb's account of that unknown country is intimate, faithful, and interesting. It is an attempt to convey a real knowledge of a striking people—an admirably successful attempt."—*Morning Leader.*

GERMANY OF THE GERMANS. By ROBERT M. BERRY.

"Mr. Berry abundantly proves his ability to write of *Germany of the Germans* in an able and informing fashion. What he does is to state, so far as can be done within the scope of a single handy volume, particulars of all aspects of life as lived in Germany to-day." —*Daily Telegraph.*

TURKEY OF THE OTTOMANS. By LUCY M. J. GARNETT.

"There could hardly be a better handbook for the newspaper reader who wants to understand all the conditions of the 'danger zone.'"—*Spectator.*

BELGIUM OF THE BELGIANS. By DEMETRIUS C. BOULGER.

"A very complete handbook to the country."—*World.*

HOLLAND OF THE DUTCH. By the same author.

SERVIA OF THE SERVIANS. By CHEDO MIJATOVICH.

"It is a useful and informative work and it deserves to be widely read."—*Liverpool Daily Courier.*

JAPAN OF THE JAPANESE. By Professor J. H. LONGFORD. With map.

"A capital historical résumé and a mine of information regarding the country and its people."—*London and China Telegraph.*

Other Volumes in preparation.

The "All Red" Series

Each volume is in demy 8vo, cloth gilt, with 16 full-page plate illustrations, maps, etc., 7s. 6d. net.

THE COMMONWEALTH OF AUSTRALIA. By the Hon. BERNHARD RINGROSE WISE (formerly Attorney-General of New South Wales).

"The 'All Red' Series should become known as the Well-Read Series within a short space of time. Nobody is better qualified to write of Australia than the late Attorney-General of New South Wales, who knows the country intimately and writes of it with enthusiasm. It is one of the best accounts of the Island Continent that has yet been published. We desire to give a hearty welcome to this series."—*Globe*.

THE DOMINION OF NEW ZEALAND. By Sir ARTHUR P. DOUGLAS, Bt., formerly Under-Secretary for Defence, New Zealand, and previously a Lieutenant, R.N.

"Those who have failed to find romance in the history of the British Empire should read *The Dominion of New Zealand*. Sir Arthur Douglas contrives to present in the 444 pages of his book an admirable account of life in New Zealand and an impartial summary of her development up to the present time. It is a most alluring picture that one conjures up after reading it."—*Standard*.

THE DOMINION OF CANADA. By W. L. GRIFFITH, *Secretary to the Office of the High Commissioner for Canada*.

"The publishers could hardly have found an author better qualified than Mr. Griffith to represent the premier British Dominion . . . an excellent plain account of Canada, one of the best and most comprehensive yet published . . trustworthy."—*Athenæum*.

THE BRITISH WEST INDIES. Their History, Resources, and Progress. By ALGERNON E. ASPINALL, *Secretary to the West India Committee*.

". . . hence the value of such a book as Mr. Aspinall has compiled so skilfully. Its treatment of current topics is copious, up-to-date, and full of varied interest . . . every visitor to the West Indies will be well advised if he takes Mr. Aspinall's book as his guide."—*Times*

THE UNION OF SOUTH AFRICA. With chapters on Rhodesia and the Native Territories of the High Commission. By W. BASIL WORSFOLD, *Sometime Editor of the "Johannesburg Star."*

". . . The promoters of 'All Red Series' got the right man for the work. Mr. Worsfold's considerable experience of the making of the country from within, combined with his training as a journalist, have enabled him to cope with the task in a way that would have been impossible to a less skilled and well-informed annalist. Into 500 pages he has compressed the main outlines of the history and geography of that much-troubled dominion, the form of its new Constitution, its industrial developments, and social and political outlook. The volume is an encyclopædia of its subject."—*Yorkshire Post.*

THE EMPIRE OF INDIA. By SIR J. BAMPFYLDE FULLER, K.C.S.I., *Formerly Lieutenant-Governor of Eastern Bengal.*

"Sir Bampfylde Fuller was well qualified to write such a book as this which will serve admirably for an introduction to the study of Indian conditions and politics. Sir Bampfylde Fuller presents a complete picture of the Indian Empire—the country, its people, its government, and its future prospects."—*Times.*

"No western mind more practically versed in and sympathetic with the Indian spirit could be found than his, and his long administrative experience could not fail to lead him to compile a well balanced volume."—*Times of India.*

WINTER LIFE IN SWITZERLAND. Its Sports and Health Cures. By Mrs. M. L. and WINIFRED M. A. BROOKE. In crown 8vo, cloth, 290 pp., with coloured frontispiece and many full-page plates, maps, and other illustrations, 3s. 6d. net.

"This book is so full of description and useful information on all points as to be an indispensable possession to anyone intending a winter visit to Switzerland . . . this invaluable little book."—*Throne.*